Sue Pemberton
Julianne Hughes
Series Editor: Julian Gilbey

Cambridge International
AS & A Level Mathematics:

Pure Mathematics 2 & 3

Coursebook

CAMBRIDGE
UNIVERSITY PRESS

CAMBRIDGE
UNIVERSITY PRESS

University Printing House, Cambridge CB2 8BS, United Kingdom

One Liberty Plaza, 20th Floor, New York, NY 10006, USA

477 Williamstown Road, Port Melbourne, VIC 3207, Australia

314–321, 3rd Floor, Plot 3, Splendor Forum, Jasola District Centre, New Delhi – 110025, India

103 Penang Road, #05-06/07, Visioncrest Commercial, Singapore 238467

Cambridge University Press is part of the University of Cambridge.

It furthers the University's mission by disseminating knowledge in the pursuit of education, learning and research at the highest international levels of excellence.

www.cambridge.org
Information on this title: www.cambridge.org/9781108407199

© Cambridge University Press 2018

This publication is in copyright. Subject to statutory exception
and to the provisions of relevant collective licensing agreements,
no reproduction of any part may take place without the written
permission of Cambridge University Press.

First published 2018

20 19 18 17 16 15 14 13 12

Printed in Italy by Rotolito S.p.A.

A catalogue record for this publication is available from the British Library

ISBN 978-1-108-40719-9 Paperback
ISBN 978-1-108-56291-1 Paperback + Cambridge Online Mathematics, 2 years
ISBN 978-1-108-46222-8 Cambridge Online Mathematics, 2 years

Cambridge University Press has no responsibility for the persistence or accuracy of URLs for external or third-party internet websites referred to in this publication, and does not guarantee that any content on such websites is, or will remain, accurate or appropriate. Information regarding prices, travel timetables, and other factual information given in this work is correct at the time of first printing but Cambridge University Press does not guarantee the accuracy of such information thereafter.

® *IGCSE is a registered trademark*

Past exam paper questions throughout are reproduced by permission of Cambridge Assessment International Education. Cambridge Assessment International Education bears no responsibility for the example answers to questions taken from its past question papers which are contained in this publication.

The questions, example answers, marks awarded and/or comments that appear in this book were written by the author(s). In examination, the way marks would be awarded to answers like these may be different.

....

NOTICE TO TEACHERS IN THE UK
It is illegal to reproduce any part of this work in material form (including photocopying and electronic storage) except under the following circumstances:
(i) where you are abiding by a licence granted to your school or institution by the Copyright Licensing Agency;
(ii) where no such licence exists, or where you wish to exceed the terms of a licence, and you have gained the written permission of Cambridge University Press;
(iii) where you are allowed to reproduce without permission under the provisions of Chapter 3 of the Copyright, Designs and Patents Act 1988, which covers, for example, the reproduction of short passages within certain types of educational anthology and reproduction for the purposes of setting examination questions.

Contents

Series introduction   vi

How to use this book   viii

Acknowledgements   x

1 Algebra   1
1.1 The modulus function   2
1.2 Graphs of $y = |f(x)|$ where f(x) is linear   7
1.3 Solving modulus inequalities   8
1.4 Division of polynomials   11
1.5 The factor theorem   14
1.6 The remainder theorem   18
End-of-chapter review exercise 1   22

2 Logarithmic and exponential functions   25
2.1 Logarithms to base 10   26
2.2 Logarithms to base a 30
2.3 The laws of logarithms   33
2.4 Solving logarithmic equations   35
2.5 Solving exponential equations   38
2.6 Solving exponential inequalities   40
2.7 Natural logarithms   42
2.8 Transforming a relationship to linear form   44
End-of-chapter review exercise 2   50

3 Trigonometry   52
3.1 The cosecant, secant and cotangent ratios   53
3.2 Compound angle formulae   58
3.3 Double angle formulae   61
3.4 Further trigonometric identities   66
3.5 Expressing $a\sin\theta + b\cos\theta$ in the form $R\sin(\theta \pm \alpha)$ or $R\cos(\theta \pm \alpha)$ 68
End-of-chapter review exercise 3   74

Cross-topic review exercise 1   76

4 Differentiation — 80
4.1 The product rule — 81
4.2 The quotient rule — 84
4.3 Derivatives of exponential functions — 87
4.4 Derivatives of natural logarithmic functions — 91
4.5 Derivatives of trigonometric functions — 95
4.6 Implicit differentiation — 99
4.7 Parametric differentiation — 103
End-of-chapter review exercise 4 — 108

5 Integration — 111
5.1 Integration of exponential functions — 112
5.2 Integration of $\dfrac{1}{ax+b}$ — 115
5.3 Integration of $\sin(ax+b)$, $\cos(ax+b)$ and $\sec^2(ax+b)$ — 118
5.4 Further integration of trigonometric functions — 121
P2 5.5 The trapezium rule — 126
End-of-chapter review exercise 5 — 131

6 Numerical solutions of equations — 133
6.1 Finding a starting point — 135
6.2 Improving your solution — 140
6.3 Using iterative processes to solve problems involving other areas of mathematics — 149
End-of-chapter review exercise 6 — 155

Cross-topic review exercise 2 — 158

P3 7 Further algebra — 165
7.1 Improper algebraic fractions — 166
7.2 Partial fractions — 168
7.3 Binomial expansion of $(1+x)^n$ for values of n that are not positive integers — 174
7.4 Binomial expansion of $(a+x)^n$ for values of n that are not positive integers — 177
7.5 Partial fractions and binomial expansions — 179
End-of-chapter review exercise 7 — 182

P3 8 Further calculus — 184
8.1 Derivative of $\tan^{-1} x$ — 186
8.2 Integration of $\dfrac{1}{x^2+a^2}$ — 187
8.3 Integration of $\dfrac{k\,f'(x)}{f(x)}$ — 189

8.4	Integration by substitution	191
8.5	The use of partial fractions in integration	194
8.6	Integration by parts	197
8.7	Further integration	201
	End-of-chapter review exercise 8	203

(P3) Cross-topic review exercise 3 — 205

(P3) 9 Vectors — 209

9.1	Displacement or translation vectors	211
9.2	Position vectors	220
9.3	The scalar product	225
9.4	The vector equation of a line	231
9.5	Intersection of two lines	236
	End-of-chapter review exercise 9	240

(P3) 10 Differential equations — 243

10.1	The technique of separating the variables	245
10.2	Forming a differential equation from a problem	252
	End-of-chapter review exercise 10	262

(P3) 11 Complex numbers — 265

11.1	Imaginary numbers	267
11.2	Complex numbers	269
11.3	The complex plane	273
11.4	Solving equations	283
11.5	Loci	288
	End-of-chapter review exercise 11	297

(P3) Cross-topic review exercise 4 — 300

Pure Mathematics 2 Practice exam-style paper — 307

Pure Mathematics 3 Practice exam-style paper — 308

Answers — 310

Glossary — 353

Index — 355

Series introduction

Cambridge International AS & A Level Mathematics can be a life-changing course. On the one hand, it is a facilitating subject: there are many university courses that either require an A Level or equivalent qualification in mathematics or prefer applicants who have it. On the other hand, it will help you to learn to think more precisely and logically, while also encouraging creativity. Doing mathematics can be like doing art: just as an artist needs to master her tools (use of the paintbrush, for example) and understand theoretical ideas (perspective, colour wheels and so on), so does a mathematician (using tools such as algebra and calculus, which you will learn about in this course). But this is only the technical side: the joy in art comes through creativity, when the artist uses her tools to express ideas in novel ways. Mathematics is very similar: the tools are needed, but the deep joy in the subject comes through solving problems.

You might wonder what a mathematical 'problem' is. This is a very good question, and many people have offered different answers. You might like to write down your own thoughts on this question, and reflect on how they change as you progress through this course. One possible idea is that a mathematical problem is a mathematical question that you do not immediately know how to answer. (If you do know how to answer it immediately, then we might call it an 'exercise' instead.) Such a problem will take time to answer: you may have to try different approaches, using different tools or ideas, on your own or with others, until you finally discover a way into it. This may take minutes, hours, days or weeks to achieve, and your sense of achievement may well grow with the effort it has taken.

In addition to the mathematical tools that you will learn in this course, the problem-solving skills that you will develop will also help you throughout life, whatever you end up doing. It is very common to be faced with problems, be it in science, engineering, mathematics, accountancy, law or beyond, and having the confidence to systematically work your way through them will be very useful.

This series of Cambridge International AS & A Level Mathematics coursebooks, written for the Cambridge Assessment International Education syllabus for examination from 2020, will support you both to learn the mathematics required for these examinations and to develop your mathematical problem-solving skills. The new examinations may well include more unfamiliar questions than in the past, and having these skills will allow you to approach such questions with curiosity and confidence.

In addition to problem solving, there are two other key concepts that Cambridge Assessment International Education have introduced in this syllabus: namely communication and mathematical modelling. These appear in various forms throughout the coursebooks.

Communication in speech, writing and drawing lies at the heart of what it is to be human, and this is no less true in mathematics. While there is a temptation to think of mathematics as only existing in a dry, written form in textbooks, nothing could be further from the truth: mathematical communication comes in many forms, and discussing mathematical ideas with colleagues is a major part of every mathematician's working life. As you study this course, you will work on many problems. Exploring them or struggling with them together with a classmate will help you both to develop your understanding and thinking, as well as improving your (mathematical) communication skills. And being able to convince someone that your reasoning is correct, initially verbally and then in writing, forms the heart of the mathematical skill of 'proof'.

Mathematical modelling is where mathematics meets the 'real world'. There are many situations where people need to make predictions or to understand what is happening in the world, and mathematics frequently provides tools to assist with this. Mathematicians will look at the real world situation and attempt to capture the key aspects of it in the form of equations, thereby building a model of reality. They will use this model to make predictions, and where possible test these against reality. If necessary, they will then attempt to improve the model in order to make better predictions. Examples include weather prediction and climate change modelling, forensic science (to understand what happened at an accident or crime scene), modelling population change in the human, animal and plant kingdoms, modelling aircraft and ship behaviour, modelling financial markets and many others. In this course, we will be developing tools which are vital for modelling many of these situations.

To support you in your learning, these coursebooks have a variety of new features, for example:

- Explore activities: These activities are designed to offer problems for classroom use. They require thought and deliberation: some introduce a new idea, others will extend your thinking, while others can support consolidation. The activities are often best approached by working in small groups and then sharing your ideas with each other and the class, as they are not generally routine in nature. This is one of the ways in which you can develop problem-solving skills and confidence in handling unfamiliar questions.
- Questions labelled as P, M or PS: These are questions with a particular emphasis on 'Proof', 'Modelling' or 'Problem solving'. They are designed to support you in preparing for the new style of examination. They may or may not be harder than other questions in the exercise.
- The language of the explanatory sections makes much more use of the words 'we', 'us' and 'our' than in previous coursebooks. This language invites and encourages you to be an active participant rather than an observer, simply following instructions ('you do this, then you do that'). It is also the way that professional mathematicians usually write about mathematics. The new examinations may well present you with unfamiliar questions, and if you are used to being active in your mathematics, you will stand a better chance of being able to successfully handle such challenges.

At various points in the books, there are also web links to relevant Underground Mathematics resources, which can be found on the free **undergroundmathematics.org** website. Underground Mathematics has the aim of producing engaging, rich materials for all students of Cambridge International AS & A Level Mathematics and similar qualifications. These high-quality resources have the potential to simultaneously develop your mathematical thinking skills and your fluency in techniques, so we do encourage you to make good use of them.

We wish you every success as you embark on this course.

Julian Gilbey
London, 2018

Past exam paper questions throughout are reproduced by permission of Cambridge Assessment International Education. Cambridge Assessment International Education bears no responsibility for the example answers to questions taken from its past question papers which are contained in this publication.

The questions, example answers, marks awarded and/or comments that appear in this book were written by the author(s). In examination, the way marks would be awarded to answers like these may be different.

How to use this book

Throughout this book you will notice particular features that are designed to help your learning. This section provides a brief overview of these features.

This book covers both Pure Mathematics 2 and Pure Mathematics 3. One topic (5.5 The trapezium rule) is only covered in Pure Mathematics 2 and this section is marked with the icon **P2**. Chapters 7–11 are only covered in Pure Mathematics 3 and these are marked with the icon **P3**. The icons appear in the Contents list and in the relevant sections of the book.

In this chapter you will learn how to:
- formulate a simple statement involving a rate of change as a differential equation
- find, by integration, a general form of solution for a first order differential equation in which the variables are separable
- use an initial condition to find a particular solution
- interpret the solution of a differential equation in the context of a problem being modelled by the equation.

Learning objectives indicate the important concepts within each chapter and help you to navigate through the coursebook.

KEY POINT 8.4

The formula for integration by parts:
$$\int u \frac{dv}{dx} dx = uv - \int v \frac{du}{dx} dx$$

Key point boxes contain a summary of the most important methods, facts and formulae.

double angle formulae

Key terms are important terms in the topic that you are learning. They are highlighted in orange bold. The **glossary** contains clear definitions of these key terms.

EXPLORE 6.1

Anil is trying to find the roots of $f(x) = |\tan x| = 0$ between 0 and 2π radians.

He states:

$f(3.1) > 0$ and $f(3.2) > 0$. There is no change of sign so there is no root of $|\tan x| = 0$ between 3.1 and 3.2.

Mika states:

$y = |\tan x|$ and it meets the x-axis where $x = \pi$ radians so Anil has made a calculation error.

Discuss Anil and Mika's statements.

Explore boxes contain enrichment activities for extension work. These activities promote group-work and peer-to-peer discussion, and are intended to deepen your understanding of a concept. (Answers to the Explore questions are provided in the Teacher's Resource.)

PREREQUISITE KNOWLEDGE

Where it comes from	What you should be able to do	Check your skills
IGCSE® / O Level Mathematics	Perform long division on numbers and find the remainder where necessary.	1 Using long division, calculate: a $4998 \div 14$ b $10287 \div 27$ c $4283 \div 32$
IGCSE / O Level Mathematics	Sketch straight-line graphs.	2 Sketch the graph of $y = 2x - 5$.

Prerequisite knowledge exercises identify prior learning that you need to have covered before starting the chapter. Try the questions to identify any areas that you need to review before continuing with the chapter.

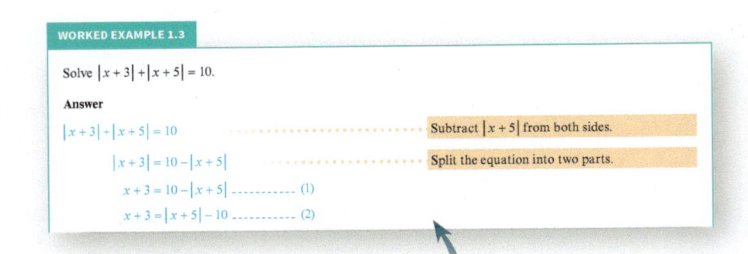

WORKED EXAMPLE 1.3

Solve $|x + 3| + |x + 5| = 10$.

Answer

$|x + 3| + |x + 5| = 10$ — Subtract $|x + 5|$ from both sides.

$|x + 3| = 10 - |x + 5|$ — Split the equation into two parts.

$x + 3 = 10 - |x + 5|$ .... (1)

$x + 3 = |x + 5| - 10$ .... (2)

Worked examples provide step-by-step approaches to answering questions. The left side shows a fully worked solution, while the right side contains a commentary explaining each step in the working.

TIP

Solving trigonometric inequalities is not in the Cambridge syllabus. You should, however, have the skills necessary to tackle these more challenging questions.

Tip boxes contain helpful guidance about calculating or checking your answers.

How to use this book

> **REWIND**
> You might need to look back at Chapter 3 before attempting question 11.

> **FAST FORWARD**
> You need to have studied the skills in Chapter 8 before attempting questions 13 and 14.

Rewind and **Fast forward** boxes direct you to related learning. **Rewind** boxes refer to earlier learning, in case you need to revise a topic. **Fast forward** boxes refer to topics that you will cover at a later stage, in case you would like to extend your study.

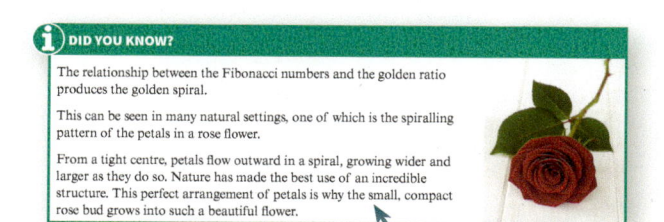

> **DID YOU KNOW?**
> The relationship between the Fibonacci numbers and the golden ratio produces the golden spiral.
> This can be seen in many natural settings, one of which is the spiralling pattern of the petals in a rose flower.
> From a tight centre, petals flow outward in a spiral, growing wider and larger as they do so. Nature has made the best use of an incredible structure. This perfect arrangement of petals is why the small, compact rose bud grows into such a beautiful flower.

Did you know? boxes contain interesting facts showing how Mathematics relates to the wider world.

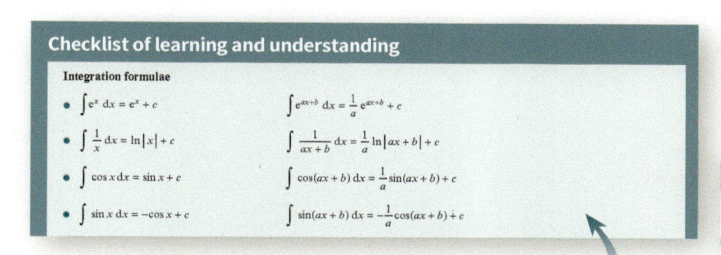

> **Checklist of learning and understanding**
> **Integration formulae**
> - $\int e^x \, dx = e^x + c$
> - $\int e^{ax+b} \, dx = \frac{1}{a} e^{ax+b} + c$
> - $\int \frac{1}{x} \, dx = \ln|x| + c$
> - $\int \frac{1}{ax+b} \, dx = \frac{1}{a} \ln|ax+b| + c$
> - $\int \cos x \, dx = \sin x + c$
> - $\int \cos(ax+b) \, dx = \frac{1}{a} \sin(ax+b) + c$
> - $\int \sin x \, dx = -\cos x + c$
> - $\int \sin(ax+b) \, dx = -\frac{1}{a} \cos(ax+b) + c$

At the end of each chapter there is a **Checklist of learning and understanding**. The checklist contains a summary of the concepts that were covered in the chapter. You can use this to quickly check that you have covered the main topics.

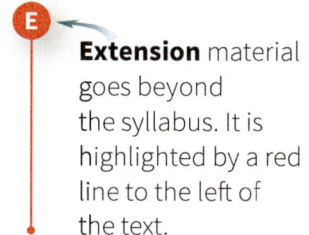

Extension material goes beyond the syllabus. It is highlighted by a red line to the left of the text.

Web link boxes contain links to useful resources on the internet.

> **WEB LINK**
> Try the *Equation or identity? (II)* resource on the Underground Mathematics website

Throughout each chapter there are multiple exercises containing practice questions. The questions are coded:

- **PS** These questions focus on problem-solving.
- **P** These questions focus on proofs.
- **M** These questions focus on modelling.
- 🚫 You should not use a calculator for these questions.
- 📘 These questions are taken from past examination papers.

The **End-of-chapter review** contains exam-style questions covering all topics in the chapter. You can use this to check your understanding of the topics you have covered. The number of marks gives an indication of how long you should be spending on the question. You should spend more time on questions with higher mark allocations; questions with only one or two marks should not need you to spend time doing complicated calculations or writing long explanations.

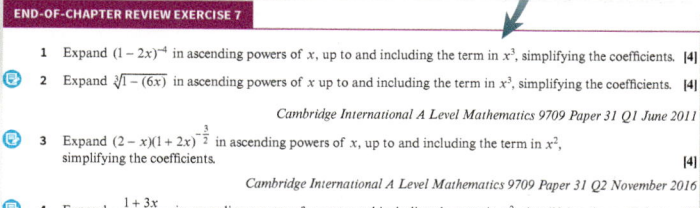

> **END-OF-CHAPTER REVIEW EXERCISE 7**
> 1. Expand $(1 - 2x)^{-4}$ in ascending powers of x, up to and including the term in x^3, simplifying the coefficients. [4]
> 2. Expand $\sqrt[3]{1 - (6x)}$ in ascending powers of x up to and including the term in x^3, simplifying the coefficients. [4]
> *Cambridge International A Level Mathematics 9709 Paper 31 Q1 June 2011*
> 3. Expand $(2 - x)(1 + 2x)^{-\frac{3}{2}}$ in ascending powers of x, up to and including the term in x^2, simplifying the coefficients. [4]
> *Cambridge International A Level Mathematics 9709 Paper 31 Q2 November 2016*
> 4. Expand $\frac{1 + 3x}{\sqrt{(1 + 2x)}}$ in ascending powers of x up to and including the term in x^2, simplifying the coefficients. [4]
> *Cambridge International A Level Mathematics 9709 Paper 31 Q2 June 2013*
> 5. Expand $\frac{1}{\sqrt{1 + 3x}}$ in ascending powers of x, up to and including the term in x^3, simplifying the coefficients. [4]

> **CROSS-TOPIC REVIEW EXERCISE 4**
> **P3** This exercise is for Pure Mathematics 3 students only.
> 1. Relative to the origin O, the position vectors of the points A and B are given by
> $\overrightarrow{OA} = \begin{pmatrix} -5 \\ 0 \\ 3 \end{pmatrix}$ and $\overrightarrow{OB} = \begin{pmatrix} 1 \\ 7 \\ 2 \end{pmatrix}$
> a. Find a vector equation of the line AB. [3]
> b. The line AB is perpendicular to the line L with vector equation:
> $\mathbf{r} = \begin{pmatrix} 4 \\ 2 \\ -3 \end{pmatrix} + \mu \begin{pmatrix} m \\ 3 \\ 9 \end{pmatrix}$

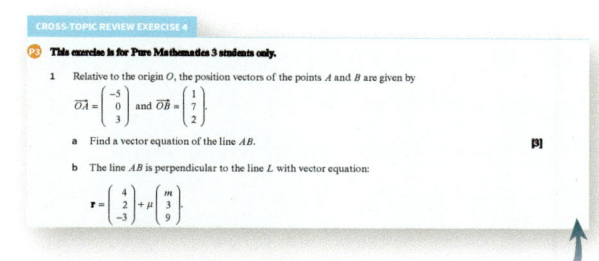

Cross-topic review exercises appear after several chapters, and cover topics from across the preceding chapters.

Acknowledgements

The authors and publishers acknowledge the following sources of copyright material and are grateful for the permissions granted. While every effort has been made, it has not always been possible to identify the sources of all the material used, or to trace all copyright holders. If any omissions are brought to our notice, we will be happy to include the appropriate acknowledgements on reprinting.

Past examination questions throughout are reproduced by permission of Cambridge Assessment International Education.

Thanks to the following for permission to reproduce images:

Cover image Mint Images – Frans Lanting/Getty Images

Inside (*in order of appearance*) sakkmesterke/Getty Images, Don Johnston/Getty Images, Photo by marianna armata/Getty Images, DEA PICTURE LIBRARY/Getty Images, malcolm park/Getty Images, agsandrew/Shutterstock, Stephan Snyder/Shutterstock, alengo/Getty Images, Moritz Haisch/EyeEm/Getty Images, Rosemary Calvert/Getty Images, bluecrayola/Shutterstock, PinkCat/Shutterstock, Maurice Rivera/EyeEm/Getty Images, PASIEKA/Getty Images, Alfred Pasieka/Science Photo Library/Getty Images

Chapter 1
Algebra

In this chapter you will learn how to:

- understand the meaning of $|x|$, sketch the graph of $y = |ax + b|$ and use relations such as $|a| = |b| \Leftrightarrow a^2 = b^2$ and $|x - a| < b \Leftrightarrow a - b < x < a + b$ in the course of solving equations and inequalities
- divide a polynomial, of degree not exceeding 4, by a linear or quadratic polynomial, and identify the quotient and remainder (which may be zero)
- use the factor theorem and the remainder theorem.

Cambridge International AS & A Level Mathematics: Pure Mathematics 2 & 3

PREREQUISITE KNOWLEDGE

Where it comes from	What you should be able to do	Check your skills

IGCSE® / O Level Mathematics	Perform long division on numbers and find the remainder where necessary.	1 Using long division, calculate:
		  a $4998 \div 14$
		  b $10287 \div 27$
		  c $4283 \div 32$

IGCSE / O Level Mathematics	Sketch straight-line graphs.	2 Sketch the graph of $y = 2x - 5$.

Why do we need to study algebra?

You have previously learnt the general formula, $x = \dfrac{-b \pm \sqrt{b^2 - 4ac}}{2a}$, for solving quadratic equations. Versions of this formula were known and used by the Babylonians nearly 4000 years ago.

It took until the 16th century for mathematicians to discover formulae for solving cubic equations, $ax^3 + bx^2 + cx + d = 0$, and quartic equations, $ax^4 + bx^3 + cx^2 + dx + e = 0$. (These formulae are very complicated and there is insufficient space to include them here but you might wish to research them on the internet.)

Mathematicians then spent many years trying to discover a general formula for solving quintic equations, $ax^5 + bx^4 + cx^3 + dx^2 + ex + f = 0$. Eventually, in 1824, a mathematician managed to prove that no general formula for this exists.

In this chapter you will develop skills for factorising and solving cubic and quartic equations. These types of equations have various applications in the real world. For example, cubic equations are used in thermodynamics and fluid mechanics to model the pressure/volume/temperature behaviour of gases and fluids.

You will also learn about a new type of function, called the modulus function.

> **WEB LINK**
>
> Explore the *Polynomials and rational functions* station on the Underground Mathematics website.

1.1 The modulus function

The **modulus** of a number is the magnitude of the number without a sign attached.

The modulus of 3 is written $|3|$.

$|3| = 3$ and $|-3| = 3$.

It is important to note that the modulus of any number (positive or negative) is always a positive number.

The modulus of a number is also called the **absolute value**.

The modulus of x, written as $|x|$, is defined as:

$$|x| = \begin{cases} x & \text{if } x \geq 0 \\ -x & \text{if } x < 0 \end{cases}$$

Chapter 1: Algebra

> **EXPLORE 1.1**
>
> You are given these eight statements:
>
> 1. $|a+b| = |a| + |b|$
> 2. $|a-b| = |a| - |b|$
> 3. $|ab| = |a| \times |b|$
> 4. $\left|\dfrac{a}{b}\right| = |a| \div |b|$, if $b \neq 0$
> 5. $|a|^2 = a^2$
> 6. $|a|^n = a^n$, where n is a positive integer
> 7. $|a+b| \leqslant |a| + |b|$
> 8. $|a-b| \leqslant |a-c| + |c-b|$
>
> You must decide whether a statement is
>
> - Always true
> - Sometimes true
> - Never true
>
> - If you think that a statement is either **always true** or **never true**, you must give a clear explanation to justify your answer.
>
> - If you think that a statement is **sometimes true** you must give an example of when it is true and an example of when it is not true.

The statement $|x| = k$, where $k \geqslant 0$, means that $x = k$ or $x = -k$.

This property is used to solve equations that involve modulus functions.

If you are solving equations of the form $|ax+b| = k$, you can solve the equations using

$ax + b = k$ and $ax + b = -k$

If you are solving harder equations of the form $|ax+b| = cx+d$, you can solve the equations using

$ax + b = cx + d$ and $ax + b = -(cx+d)$

When solving these more complicated equations you must always check that your answers satisfy the original equation.

▶▶ FAST FORWARD

You will see why it is necessary to check your answers when you learn how to sketch graphs of modulus functions in Section 1.2.

WORKED EXAMPLE 1.1

Solve:

a $|2x - 1| = 3$

b $|x - 4| = 2x + 1$

Answer

a $|2x - 1| = 3$

$2x - 1 = 3$ or $2x - 1 = -3$

$2x = 4 \qquad\quad 2x = -2$

$x = 2 \qquad\quad\ x = -1$

CHECK: $|2 \times 2 - 1| = 3$ ✓ and $|2 \times (-1) - 1| = 3$ ✓

The solution is: $x = -1$ or $x = 2$

b $|x - 4| = 2x + 1$

$x - 4 = 2x + 1$ or $x - 4 = -(2x + 1)$

$x = -5 \qquad\qquad\quad 3x = 3$

$\qquad\qquad\qquad\qquad\quad x = 1$

CHECK: $|-5 - 4| = 2 \times (-5) + 1$ ✗ and $|1 - 4| = 2 \times 1 + 1$ ✓

The solution is: $x = 1$

EXPLORE 1.2

Using $|a|^2 = a^2$ you can write $a^2 - b^2$ as $|a|^2 - |b|^2$.

Using the difference of two squares you can now write:

$$a^2 - b^2 = (|a| - |b|)(|a| + |b|)$$

Using the previous statement, explain how these three important results can be obtained:

(The symbol ⇔ means 'is equivalent to'.)

- $|a| = |b| \Leftrightarrow a^2 = b^2$
- $|a| > |b| \Leftrightarrow a^2 > b^2$
- $|a| < |b| \Leftrightarrow a^2 < b^2$, if $b \neq 0$.

Worked example 1.2 shows you how to solve equations of the form $|cx + d| = |ex + f|$.

KEY POINT 1.1

To solve equations of the form $|cx + d| = |ex + f|$ we can use the rule:

$|a| = |b| \Leftrightarrow a^2 = b^2$

Chapter 1: Algebra

WORKED EXAMPLE 1.2

Solve the equation $|3x+4| = |x+5|$.

Answer

Method 1

$|3x+4| = |x+5|$

$3x + 4 = x + 5$ or $3x + 4 = -(x+5)$

$\quad 2x = 1 \quad$ or $\quad 4x = -9$

$\quad x = \dfrac{1}{2} \qquad\qquad x = -\dfrac{9}{4}$

CHECK: $\left|3 \times \dfrac{1}{2} + 4\right| = \left|\dfrac{1}{2} + 5\right|$ ✓ , $\left|3 \times \left(-\dfrac{9}{4}\right) + 4\right| = \left|-\dfrac{9}{4} + 5\right|$ ✓

Solution is: $x = \dfrac{1}{2}$ or $x = -\dfrac{9}{4}$

Method 2

$|3x+4| = |x+5|$ .... Use $|a| = |b| \Leftrightarrow a^2 = b^2$.

$(3x+4)^2 = (x+5)^2$

$9x^2 + 24x + 16 = x^2 + 10x + 25$

$8x^2 + 14x - 9 = 0$ .... Factorise.

$(2x - 1)(4x + 9) = 0$

$x = \dfrac{1}{2}$ or $x = -\dfrac{9}{4}$

WORKED EXAMPLE 1.3

Solve $|x+3| + |x+5| = 10$.

Answer

$|x+3| + |x+5| = 10$ .... Subtract $|x+5|$ from both sides.

$|x+3| = 10 - |x+5|$ .... Split the equation into two parts.

$\quad x + 3 = 10 - |x+5|$ ---------- (1)

$\quad x + 3 = |x+5| - 10$ ---------- (2)

Using equation (1):

$-7 + x + |x+5| = 0$

$\quad |x+5| = 7 - x$ .... Split this equation into two parts.

$x + 5 = 7 - x$ or $x + 5 = -(7 - x)$

$\quad 2x = 2 \quad$ or $\quad 0 = -12$ .... $0 = -12$ is false.

$\quad x = 1$

Using equation (2):

$|x + 5| = x + 13$ .... Split this equation into two parts.

$x + 5 = x + 13$ or $x + 5 = -(x + 13)$

$0 = 8$ or $2x = -18$ .... $0 = 8$ is false.

$x = -9$

The solution is $x = 1$ or $x = -9$.

EXERCISE 1A

1 Solve.
   a $|4x - 3| = 7$
   b $|1 - 2x| = 5$
   c $\left|\dfrac{3x - 2}{5}\right| = 4$
   d $\left|\dfrac{x}{3} + 2\right| = 3$
   e $\left|\dfrac{x + 2}{3} - \dfrac{2x}{5}\right| = 2$
   f $|2x + 7| = 3x$

2 Solve.
   a $\left|\dfrac{2x + 1}{x - 2}\right| = 5$
   b $\left|\dfrac{3x - 1}{x + 5}\right| = 1$
   c $\left|2 - \dfrac{x + 2}{x - 3}\right| = 5$
   d $|3x - 5| = x + 1$
   e $x + |x + 4| = 8$
   f $8 - |1 - 2x| = x$

3 Solve.
   a $|2x + 1| = |x|$
   b $|3 - 2x| = |3x|$
   c $|2x - 5| = |1 - x|$
   d $|3x + 5| = |1 + 2x|$
   e $|x - 5| = 2|x + 1|$
   f $3|2x - 1| = \left|\dfrac{1}{2}x - 3\right|$

4 Solve.
   a $|x^2 - 2| = 7$
   b $|5 - x^2| = 3 - x$
   c $|x^2 + 2x| = x + 2$
   d $|x^2 - 3| = 2x + 1$
   e $|2x^2 - 5x| = 4 - x$
   f $|x^2 - 7x + 6| = 6 - x$

5 Solve the simultaneous equations.
   a $x + 2y = 8$
      $|x + 2| + y = 6$
   b $3x + y = 0$
      $y = |2x^2 - 5|$

6 Solve the equation $5|x - 1|^2 + 9|x - 1| - 2 = 0$.

7 a Solve the equation $x^2 - 5|x| + 6 = 0$.
   b Use graphing software to draw the graph of $y = x^2 - 5|x| + 6$.
   c Name the equation of the line of symmetry of the curve.

8 Solve the equation $|2x + 1| + |2x - 1| = 3$.

PS 9 Solve the equation $|3x - 2y - 11| + 2\sqrt{31 - 8x + 5y} = 0$.

1.2 Graphs of $y=|f(x)|$ where $f(x)$ is linear

Consider drawing the graph of $y = |x|$.

First draw the line $y = x$.

Then reflect, in the x-axis, the part of the line that is below the x-axis.

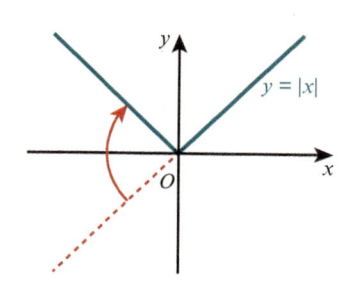

WORKED EXAMPLE 1.4

Sketch the graph of $y = \left|\dfrac{1}{2}x - 1\right|$, showing the points where the graph meets the axes.

Use your graph to express $\left|\dfrac{1}{2}x - 1\right|$ in an alternative form.

Answer

First sketch the graph of $y = \dfrac{1}{2}x - 1$.

The line has gradient $\dfrac{1}{2}$ and a y-intercept of -1.

You then reflect in the x-axis the part of the line that is below the x-axis.

The graph shows that $\left|\dfrac{1}{2}x - 1\right|$ can be written as:

$$\left|\dfrac{1}{2}x - 1\right| = \begin{cases} \dfrac{1}{2}x - 1 & \text{if } x \geq 2 \\ -\left(\dfrac{1}{2}x - 1\right) & \text{if } x < 2 \end{cases}$$

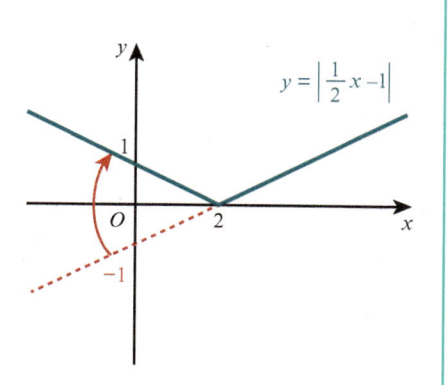

In Worked example 1.1, we found that there were two roots, $x = -1$ and $x = 2$, to the equation $|2x - 1| = 3$.

These can also be found graphically by finding the x-coordinates of the points of intersection of the graphs of $y = |2x - 1|$ and $y = 3$ as shown.

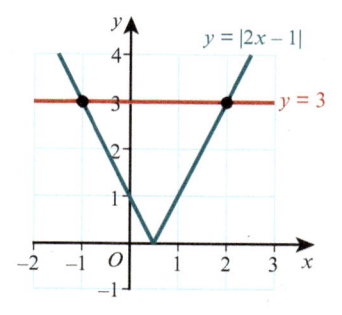

Also in Worked example 1.1, we found that there was only one root, $x = 1$, to the equation $|x - 4| = 2x + 1$.

This root can be found graphically by finding the x-coordinates of the points of intersection of the graphs of $y = |x - 4|$ and $y = 2x + 1$ as shown.

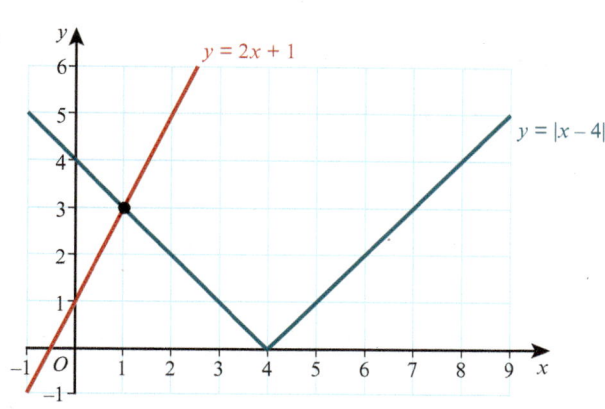

EXERCISE 1B

1. Sketch the graphs of each of the following functions showing the coordinates of the points where the graph meets the axes. Express each function in an alternative form.

   a $y = |x + 2|$
   b $y = |3 - x|$
   c $y = \left|5 - \frac{1}{2}x\right|$

2. a Complete the table of values for $y = |x - 3| + 2$.

x	0	1	2	3	4	5	6
y	5		3				

   b Draw the graph of $y = |x - 3| + 2$ for $0 \leqslant x \leqslant 6$.

   c Describe the transformation that maps the graph of $y = |x|$ onto the graph of $y = |x - 3| + 2$.

3. Describe fully the transformation (or combination of transformations) that maps the graph of $y = |x|$ onto each of these functions.

   a $y = |x + 1| + 2$
   b $y = |x - 5| - 2$
   c $y = 2 - |x|$
   d $y = |2x| - 3$
   e $y = 1 - |x + 2|$
   f $y = 5 - 2|x|$

4. Sketch the graphs of each of the functions in question 3. For each graph, state the coordinates of the vertex.

5. $f(x) = |5 - 2x| + 3$ for $2 \leqslant x \leqslant 8$

   Find the range of function f.

6. a Sketch the graph of $y = 2|x - 2| + 1$ for $-2 < x < 6$, showing the coordinates of the vertex and the y-intercept.

   b On the same diagram, sketch the graph of $y = x + 2$.

   c Use your graph to solve the equation $2|x - 2| + 1 = x + 2$.

7. a Sketch the graph of $y = |x - 2|$ for $-3 < x < 6$, showing the coordinates of the vertex and the y-intercept.

   b On the same diagram, sketch the graph of $y = |1 - 2x|$.

   c Use your graph to solve the equation $|x - 2| = |1 - 2x|$.

8. a Sketch the graph of $y = |x + 1| + |x - 1|$.

   b Use your graph to solve the equation $|x + 1| + |x - 1| = 4$.

1.3 Solving modulus inequalities

Two useful properties that can be used when solving modulus inequalities are:

$|a| \leqslant b \Leftrightarrow -b \leqslant a \leqslant b$ and $|a| \geqslant b \Leftrightarrow a \leqslant -b$ or $a \geqslant b$

The following examples illustrate the different methods that can be used when solving modulus inequalities.

WORKED EXAMPLE 1.5

Solve $|2x - 5| < 3$.

Answer

Method 1: Using algebra

$|2x - 5| < 3$
$-3 < 2x - 5 < 3$
$2 < 2x < 8$
$1 < x < 4$

Method 2: Using a graph

The graphs of $y = |2x - 5|$ and $y = 3$ intersect at the points A and B.

$|2x - 5| = \begin{cases} 2x - 5 & \text{if } x \geq 2\frac{1}{2} \\ -(2x - 5) & \text{if } x < 2\frac{1}{2} \end{cases}$

At A, the line $y = -(2x - 5)$ intersects the line $y = 3$.
$-(2x - 5) = 3$
$-2x + 5 = 3$
$2x = 2$
$x = 1$

At B, the line $y = 2x - 5$ intersects the line $y = 3$.
$2x - 5 = 3$
$2x = 8$
$x = 4$

To solve the inequality $|2x - 5| < 3$ you must find where the graph of the function $y = |2x - 5|$ is below the graph of $y = 3$.

Hence, $1 < x < 4$.

WORKED EXAMPLE 1.6

Solve the inequality $|2x - 1| \geq |3 - x|$.

Answer

Method 1: Using algebra

$|2x - 1| \geq |3 - x|$ .... Use $|a| \geq |b| \Leftrightarrow a^2 \geq b^2$.
$(2x - 1)^2 \geq (3 - x)^2$
$4x^2 - 4x + 1 \geq 9 - 6x + x^2$
$3x^2 + 2x - 8 \geq 0$ .... Factorise.
$(3x - 4)(x + 2) \geq 0$

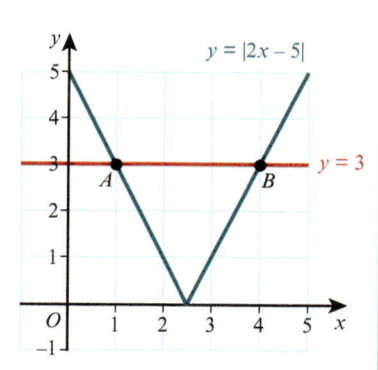

Critical values are -2 and $\dfrac{4}{3}$.

Hence, $x \leqslant -2$ or $x \geqslant \dfrac{4}{3}$.

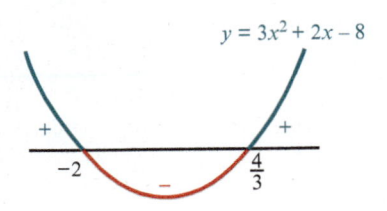

Method 2: Using a graph

The graphs of $y = |2x - 1|$ and $y = |3 - x|$ intersect at the points A and B.

$$|2x - 1| = \begin{cases} 2x - 1 & \text{if } x \geqslant \dfrac{1}{2} \\ -(2x - 1) & \text{if } x < \dfrac{1}{2} \end{cases}$$

$$|3 - x| = |x - 3| = \begin{cases} x - 3 & \text{if } x \geqslant 3 \\ -(x - 3) & \text{if } x < 3 \end{cases}$$

At A, the line $y = -(2x - 1)$ intersects the line $y = -(x - 3)$.

$2x - 1 = x - 3$

$\quad x = -2$

At B, the line $y = 2x - 1$ intersects the line $y = -(x - 3)$.

$2x - 1 = -(x - 3)$

$\quad 3x = 4$

$\quad x = \dfrac{4}{3}$

To solve the inequality $|2x - 1| \geqslant |3 - x|$ you must find where the graph of the function $y = |2x - 1|$ is above the graph of $y = |3 - x|$.

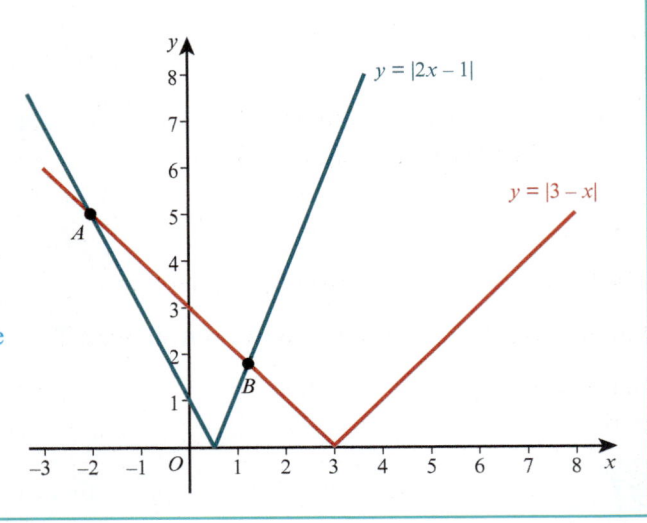

Hence, $x \leqslant -2$ or $x \geqslant \dfrac{4}{3}$.

EXERCISE 1C

1

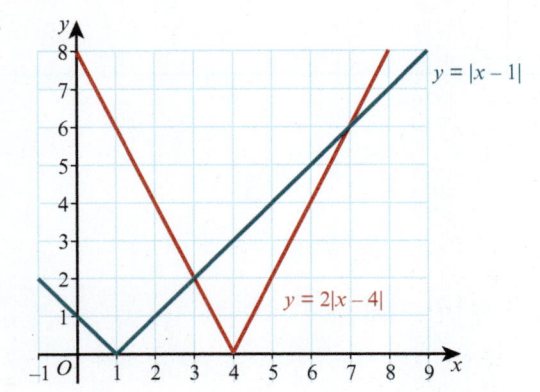

The graphs of $y = |x - 1|$ and $y = 2|x - 4|$ are shown on the grid.

Write down the solution to the inequality $|x - 1| > 2|x - 4|$.

2 a On the same axes, sketch the graphs of $y = |2x - 1|$ and $y = 4 - |x - 1|$.

   b Solve the inequality $|2x - 1| > 4 - |x - 1|$.

3 Solve. (You may use either an algebraic method or a graphical method.)

   a $|2x+1| \leqslant 3$
   b $|2-x| < 4$
   c $|3x-2| \geqslant 7$

4 Solve. (You may use either an algebraic method or a graphical method.)

   a $|2x-5| \leqslant x-2$
   b $|3+x| > 4-2x$
   c $|x-7| - 2x \leqslant 4$

5 Solve. (You may use either an algebraic method or a graphical method.)

   a $|x+4| < |2x|$
   b $|2x-5| > |3-x|$
   c $|3x-2| \leqslant |1-3x|$
   d $|x-1| > 2|x-4|$
   e $3|3-x| \geqslant |2x-1|$
   f $|3x-5| < 2|2-x|$

6 Solve $|2x+1| + |2x-1| > 5$.

1.4 Division of polynomials

A **polynomial** is an expression of the form

$$a_n x^n + a_{n-1} x^{n-1} + a_{n-2} x^{n-2} + \ldots + a_2 x^2 + a_1 x^1 + a_0$$

where:

- x is a variable
- n is a non-negative integer
- the coefficients a_n, a_{n-1}, a_{n-2}, ..., a_2, a_1, a_0 are constants
- a_n is called the leading coefficient and $a_n \neq 0$
- a_0 is called the constant term.

The highest power of x in the polynomial is called the **degree** of the polynomial.

For example, $5x^3 - 6x^2 + 2x - 6$ is a polynomial of degree 3.

In this section you will learn how to use long division to divide a polynomial by another polynomial. Alternative methods are possible, this is shown in Worked example 1.10.

> **DID YOU KNOW?**
>
> Polynomial functions can be used to predict the mass of an animal from its length.

WORKED EXAMPLE 1.7

Divide $x^3 + 2x^2 - 11x + 6$ by $x - 2$.

Answer

Step 1:

$$\begin{array}{r} x^2 \\ x-2 \overline{\smash{)}x^3 + 2x^2 - 11x + 6} \\ \underline{x^3 - 2x^2} \\ 4x^2 - 11x \end{array}$$

Divide the first term of the polynomial by x: $x^3 \div x = x^2$.

Multiply $(x-2)$ by x^2: $x^2(x-2) = x^3 - 2x^2$.

Subtract: $(x^3 + 2x^2) - (x^3 - 2x^2) = 4x^2$.

Bring down the $-11x$ from the next column.

Step 2: Repeat the process

$$\begin{array}{r} x^2 + 4x \\ x-2\overline{\smash{\big)}x^3 + 2x^2 - 11x + 6} \\ \underline{x^3 - 2x^2} \\ 4x^2 - 11x \\ \underline{4x^2 - 8x} \\ -3x + 6 \end{array}$$

Divide $4x^2$ by x: $4x^2 \div x = 4x$.

Multiply $(x-2)$ by $4x$: $4x(x-2) = 4x^2 - 8x$.

Subtract: $(4x^2 - 11x) - (4x^2 - 8x) = -3x$.

Bring down the 6 from the next column.

Step 3: Repeat the process

$$\begin{array}{r} x^2 + 4x - 3 \\ x-2\overline{\smash{\big)}x^3 + 2x^2 - 11x + 6} \\ \underline{x^3 - 2x^2} \\ 4x^2 - 11x \\ \underline{4x^2 - 8x} \\ -3x + 6 \\ \underline{-3x + 6} \\ 0 \end{array}$$

Divide $-3x$ by x: $-3x \div x = -3$.

Multiply $(x-2)$ by -3: $-3(x-2) = -3x + 6$.

Subtract: $(-3x + 6) - (-3x + 6) = 0$.

Hence, $(x^3 + 2x^2 - 11x + 6) \div (x - 2) = x^2 + 4x - 3$.

Each part in the calculation in Worked example 1.7 has a special name:

$$(x^3 + 2x^2 - 11x + 6) \div (x - 2) = (x^2 + 4x - 3)$$

      dividend   **divisor**   **quotient**

There was no **remainder** in this calculation so we say that $x - 2$ is a **factor** of $x^3 + 2x^2 - 11x + 6$.

WORKED EXAMPLE 1.8

Find the remainder when $2x^3 - x + 52$ is divided by $x + 3$.

Answer

There is no x^2 term in $2x^3 - x + 52$ so we write it as $2x^3 + 0x^2 - x + 52$.

Step 1:

$$\begin{array}{r} 2x^2 \\ x+3\overline{\smash{\big)}2x^3 + 0x^2 - x + 52} \\ \underline{2x^3 + 6x^2} \\ -6x^2 - x \end{array}$$

Divide the first term of the polynomial by x: $2x^3 \div x = 2x^2$.

Multiply $(x + 3)$ by $2x^2$: $2x^2(x + 3) = 2x^3 + 6x^2$.

Subtract: $(2x^3 + 0x^2) - (2x^3 + 6x^2) = -6x^2$.

Bring down the $-x$ from the next column.

Step 2: Repeat the process

$$\begin{array}{r} 2x^2 - 6x \\ x+3 \overline{\smash{)}2x^3 + 0x^2 - x + 52} \\ \underline{2x^3 + 6x^2 } \\ -6x^2 - x \\ \underline{-6x^2 - 18x } \\ 17x + 52 \end{array}$$

Divide $-6x^2$ by x: $-6x^2 \div x = -6x$.

Multiply $(x+3)$ by $-6x$: $-6x(x+3) = -6x^2 - 18x$.

Subtract: $(-6x^2 - x) - (-6x^2 - 18x) = 17x$.

Bring down the 52 from the next column.

Step 3: Repeat the process

$$\begin{array}{r} 2x^2 - 6x + 17 \\ x+3 \overline{\smash{)}2x^3 + 0x^2 - x + 52} \\ \underline{2x^3 + 6x^2 } \\ -6x^2 - x \\ \underline{-6x^2 - 18x } \\ 17x + 52 \\ \underline{17x + 51} \\ 1 \end{array}$$

Divide $17x$ by x: $17x \div x = 17$.

Multiply $(x+3)$ by 17: $17(x+3) = 17x + 51$.

Subtract: $(17x + 52) - (17x + 51) = 1$.

The remainder is 1.

The calculation in Worked example 1.8 can be written as:

$(2x^3 - x + 52) = (x+3) \times (2x^2 - 6x + 17) + 1$

↑ dividend ↑ divisor ↑ quotient ↑ remainder

KEY POINT 1.2

The **division algorithm for polynomials** is:

dividend = divisor × quotient + remainder

EXERCISE 1D

1. Simplify each of the following.
   a $(x^3 + 5x^2 + 5x - 2) \div (x + 2)$
   b $(x^3 - 8x^2 + 22x - 21) \div (x - 3)$
   c $(3x^3 - 7x^2 + 6x - 2) \div (x - 1)$
   d $(2x^3 - 3x^2 + 8x + 5) \div (2x + 1)$
   e $(5x^3 - 13x^2 + 10x - 8) \div (2 - x)$
   f $(6x^4 + 13x - 19) \div (1 - x)$

2. Find the quotient and remainder for each of the following.
   a $(x^3 + 2x^2 + 5x - 4) \div (x + 1)$
   b $(6x^3 + 7x^2 - 6) \div (x - 2)$
   c $(8x^3 + x^2 - 2x + 1) \div (2x - 1)$
   d $(2x^3 - 9x^2 - 9) \div (3 - x)$
   e $(x^3 + 5x^2 - 3x - 1) \div (x^2 + 2x + 1)$
   f $(5x^4 - 2x^2 - 13x + 8) \div (x^2 + 1)$

3. Use algebraic division to show these two important factorisations.
   a $x^3 - y^3 = (x - y)(x^2 + xy + y^2)$
   b $x^3 + y^3 = (x + y)(x^2 - xy + y^2)$

DID YOU KNOW?

The number 1729 is the smallest number that can be expressed as the sum of two cubes in two different ways.

4   a   Use algebraic division to show that $x - 2$ is a factor of $2x^3 + 7x^2 - 43x + 42$.

    b   Hence, factorise $2x^3 + 7x^2 - 43x + 42$ completely.

5   a   Use algebraic division to show that $2x + 1$ is a factor of $12x^3 + 16x^2 - 3x - 4$.

    b   Hence, solve the equation $12x^3 + 16x^2 - 3x - 4 = 0$.

6   a   Use algebraic division to show that $x + 1$ is a factor of $x^3 - x^2 + 2x + 4$.

    b   Hence, show that there is only one real root for the equation $x^3 - x^2 + 2x + 4 = 0$.

7   The general formula for solving the quadratic equation $ax^2 + bx + c = 0$ is $x = \dfrac{-b \pm \sqrt{b^2 - 4ac}}{2a}$.

There is a general formula for solving a cubic equation. Find out more about this formula and use it to solve $2x^3 - 5x^2 - 28x + 15 = 0$.

1.5 The factor theorem

In Worked example 1.7 you found that $x^3 + 2x^2 - 11x + 6$ is exactly divisible by $x - 2$.

$$(x^3 + 2x^2 - 11x + 6) \div (x - 2) = x^2 + 4x - 3$$

This can also be written as:

$$x^3 + 2x^2 - 11x + 6 = (x - 2)(x^2 + 4x - 3)$$

If a polynomial P(x) divides exactly by a linear factor $x - c$ to give the polynomial Q(x), then

$$P(x) = (x - c)\,Q(x)$$

Substituting $x = c$ into this formula gives $P(c) = 0$. This is known as the **factor theorem**.

KEY POINT 1.3

If for a polynomial P(x), $P(c) = 0$ then $x - c$ is a factor of P(x).

For example, when $x = 2$,

$$x^3 + 2x^2 - 11x + 6 = (2)^3 + 2(2)^2 - 11(2) + 6 = 8 + 8 - 22 + 6 = 0$$

Therefore $(x - 2)$ is a factor of $x^3 + 2x^2 - 11x + 6$.

The factor theorem can be extended to:

KEY POINT 1.4

If for a polynomial P(x), $P\left(\dfrac{b}{a}\right) = 0$ then $ax - b$ is a factor of P(x).

For example, when $x = \dfrac{2}{3}$,

$$3x^3 - 2x^2 - 3x + 2 = 3\left(\dfrac{2}{3}\right)^3 - 2\left(\dfrac{2}{3}\right)^2 - 3\left(\dfrac{2}{3}\right) + 2 = \dfrac{8}{9} - \dfrac{8}{9} - 2 + 2 = 0$$

Hence $3x - 2$ is a factor of $3x^3 - 2x^2 - 3x + 2$.

Chapter 1: Algebra

WORKED EXAMPLE 1.9

Use the factor theorem to show that $x - 3$ is a factor of $x^3 + 2x^2 - 16x + 3$.

Answer

Let $f(x) = x^3 + 2x^2 - 16x + 3$ .... If $f(3) = 0$, then $(x - 3)$ is a factor.

$f(3) = (3)^3 + 2(3)^2 - 16(3) + 3$

$= 27 + 18 - 48 + 3$

$= 0$

Hence $x - 3$ is a factor of $x^3 + 2x^2 - 16x + 3$.

EXPLORE 1.3

A $x^3 + x^2 - 3x + 1$

B $2x^3 + 7x^2 + x - 10$

C $3x^3 + 7x^2 - 4$

D $6x^3 + x^2 - 17x + 10$

E $6x^4 + x^3 - 8x^2 - x + 2$

F $3x^3 + x^2 - 10x - 8$

Discuss with your classmates which of the expressions have:

1 $x - 1$ as a factor

2 $x + 2$ as a factor

3 $2x - 1$ as a factor.

WORKED EXAMPLE 1.10

a Given that $3x^2 + 5x - 2$ is a factor of $3x^3 + 17x^2 + ax + b$, find the value of a and the value of b.

b Hence, factorise the polynomial completely.

Answer

a Let $f(x) = 3x^3 + 17x^2 + ax + b$.

Factorising $3x^2 + 5x - 2$ gives $(3x - 1)(x + 2)$.

Hence, $3x - 1$ and $x + 2$ are also factors of $f(x)$.

Using the factor theorem, $f\left(\dfrac{1}{3}\right) = 0$ and $f(-2) = 0$.

$f\left(\dfrac{1}{3}\right) = 0$ gives $3\left(\dfrac{1}{3}\right)^3 + 17\left(\dfrac{1}{3}\right)^2 + a\left(\dfrac{1}{3}\right) + b = 0$

$\dfrac{1}{9} + \dfrac{17}{9} + \dfrac{a}{3} + b = 0$

$a + 3b = -6$ -------------- (1)

$f(-2) = 0$ gives $3(-2)^3 + 17(-2)^2 + a(-2) + b = 0$
$$-24 + 68 - 2a + b = 0$$
$$2a - b = 44 \text{ ------------- (2)}$$

$(1) + 3 \times (2)$ gives $7a = 126$
$$a = 18$$

Substituting in (2) gives $b = -8$.

Hence, $a = 18$ and $b = -8$.

b Method 1: Using long division

$$\begin{array}{r} x + 4 \\ 3x^2 + 5x - 2 \overline{\smash{\big)}\, 3x^3 + 17x^2 + 18x - 8} \\ \underline{3x^3 + 5x^2 - 2x } \\ 12x^2 + 20x - 8 \\ \underline{12x^2 + 20x - 8} \\ 0 \end{array}$$

Hence, $3x^3 + 17x^2 + 18x - 8 = (3x^2 + 5x - 2)(x + 4)$
$$= (3x - 1)(x + 2)(x + 4)$$

Method 2: Equating coefficients

If $3x^2 + 5x - 2$ is a factor, $3x^3 + 17x^2 + 18x - 8$ can be written as

$$3x^3 + 17x^2 + 18x - 8 = (3x^2 + 5x - 2)(cx + d)$$

| coefficient of x^3 is 3, so $c = 1$ since $3 \times 1 = 3$ | constant term is -8, so $d = 4$ since $-2 \times 4 = -8$ |

Hence, $3x^3 + 17x^2 + 18x - 8 = (3x - 1)(x + 2)(x + 4)$

WORKED EXAMPLE 1.11

Solve $x^3 + 4x^2 - 19x + 14 = 0$.

Answer

Let $f(x) = x^3 + 4x^2 - 19x + 14$.

The positive and negative factors of 14 are ± 1, ± 2, ± 7 and ± 14.

$f(1) = (1)^3 + 4(1)^2 - 19(1) + 14 = 0$, so $x - 1$ is a factor of $f(x)$.

The other factors of $f(x)$ can be found by any of the following methods.

Method 1: Trial and error

$f(2) = (2)^3 + 4(2)^2 - 19(2) + 14 = 0$, so $x - 2$ is a factor of $f(x)$

$f(-7) = (-7)^3 + 4(-7)^2 - 19(-7) + 14 = 0$, so $x + 7$ is a factor of $f(x)$.

Hence, $f(x) = (x - 1)(x - 2)(x + 7)$.

Method 2: Long division

$$\begin{array}{r}
x^2 + 5x - 14 \\
x - 1 \overline{\smash{)} x^3 + 4x^2 - 19x + 14} \\
\underline{x^3 - x^2 } \\
5x^2 - 19x \\
\underline{5x^2 - 5x } \\
-14x + 14 \\
\underline{-14x + 14} \\
0
\end{array}$$

Hence, $f(x) = (x - 1)(x^2 + 5x - 14)$
$ = (x - 1)(x - 2)(x + 7)$

Method 3: Equating coefficients

Since $x - 1$ is a factor, $x^3 + 4x^2 - 19x + 14$ can be written as:

$$x^3 + 4x^2 - 19x + 14 = (x - 1)(ax^2 + bx + c)$$

| coefficient of x^3 is 1, so $a = 1$ since $1 \times 1 = 1$ | constant term is 14, so $c = -14$ since $-1 \times -14 = 14$ |

$x^3 + 4x^2 - 19x + 14 = (x - 1)(x^2 + bx - 14)$

Equating the coefficients of x^2 gives $4 = -1 + b$, so $b = 5$.

Hence, $f(x) = (x - 1)(x^2 + 5x - 14)$
$ = (x - 1)(x - 2)(x + 7)$

The solutions of the equation $x^3 + 4x^2 - 19x + 14 = 0$ are: $x = 1$, $x = 2$ and $x = -7$.

EXERCISE 1E

1. Use the factor theorem to show that $x + 1$ is a factor of $x^4 - 3x^3 - 4x^2 + 5x + 5$.

2. Use the factor theorem to show that $2x - 5$ is a factor of $2x^3 - 7x^2 + 9x - 10$.

3. Given that $x + 4$ is a factor of $x^3 + ax^2 - 29x + 12$, find the value of a.

4. Given that $x - 3$ is a factor of $x^3 + ax^2 + bx - 30$, express a in terms of b.

5. Given that $2x^2 + x - 1$ is a factor of $2x^3 - x^2 + ax + b$, find the value of a and the value of b.

(PS) 6 Given that $x - 3$ and $2x + 1$ are factors of $2x^3 + px^2 + (2q - 1)x + q$:

    a Find the value of p and the value of q.

    b Explain why $x + 2$ is also a factor of the expression.

(PS) 7 It is given that $x + a$ is a factor of $x^3 + 4x^2 + 7ax + 4a$. Find the possible values of a.

(PS) 8 **a** It is given that $x + 1$ is a common factor of $x^3 + px + q$ and $x^3 + (1 - p)x^2 + 19x - 2q$.

    Find the value of p and the value of q.

    b When p and q have these values, factorise $x^3 + px + q$ and $x^3 + (1 - p)x^2 + 19x - 2q$ completely.

(PS) 9 Given that $x - 1$ and $x + 2$ are factors of $x^4 - x^3 + px^2 - 11x + q$:

    a Find the value of p and the value of q.

    b When p and q have these values, factorise $x^4 - x^3 + px^2 - 11x + q$ completely.

10 Solve.

    a $x^3 - 5x^2 - 4x + 20 = 0$ **b** $x^3 + 5x^2 - 17x - 21 = 0$

    c $2x^3 - 5x^2 - 13x + 30 = 0$ **d** $3x^3 + 17x^2 + 18x - 8 = 0$

    e $x^4 + 2x^3 - 7x^2 - 8x + 12 = 0$ **f** $2x^4 - 11x^3 + 12x^2 + x - 4 = 0$

(P) 11 You are given that the equation $x^3 + ax^2 + bx + c = 0$ has three real roots and that these roots are consecutive terms in an arithmetic progression. Show that $2a^3 + 27c = 9ab$.

(PS) 12 Find the set of values for k for which the equation $3x^4 + 4x^3 - 12x^2 + k = 0$ has four real roots.

1.6 The remainder theorem

Consider $f(x) = 2x^3 - 5x^2 - 2x + 12$.

Substituting $x = 2$ in the polynomial gives $f(2) = 2(2)^3 - 5(2)^2 - 2(2) + 12 = 4$.

When $2x^3 - 5x^2 - 2x + 12$ is divided by $x - 2$, there is a remainder:

$$
\begin{array}{r}
2x^2 - x - 4 \\
x - 2 \overline{\smash{)}2x^3 - 5x^2 - 2x + 12} \\
\underline{2x^3 - 4x^2 } \\
- x^2 - 2x \\
\underline{- x^2 + 2x } \\
- 4x + 12 \\
\underline{- 4x + 8} \\
4
\end{array}
$$

The remainder is 4. This is the same value as f(2).

$f(x) = 2x^3 - 5x^2 - 2x + 12$ can be written as:

$f(x) = (x - 2)(2x^2 - x - 4) + 4$

In general:

If a polynomial P(x) is divided by $x - c$ to give the polynomial Q(x) and a remainder R, then:

$P(x) = (x - c)Q(x) + R$

Substituting $x = c$ into this formula gives $P(c) = R$.

This leads to the **remainder theorem**.

> **KEY POINT 1.5**
>
> The remainder theorem:
>
> If a polynomial $P(x)$ is divided by $x - c$, the remainder is $P(c)$.

The remainder theorem can be extended to:

> **KEY POINT 1.6**
>
> If a polynomial $P(x)$ is divided by $ax - b$, the remainder is $P\left(\dfrac{b}{a}\right)$.

WORKED EXAMPLE 1.12

Use the factor theorem to find the remainder when $4x^3 - 2x^2 + 5x - 8$ is divided by $x + 1$.

Answer

Let $f(x) = 4x^3 - 2x^2 + 5x - 8$

$\begin{aligned}\text{Remainder} &= f(-1) \\ &= 4(-1)^3 - 2(-1)^2 + 5(-1) - 8 \\ &= -4 - 2 - 5 - 8 \\ &= -19\end{aligned}$

WORKED EXAMPLE 1.13

$f(x) = 6x^3 + ax^2 + bx - 4$

When $f(x)$ is divided by $x - 1$, the remainder is 3.

When $f(x)$ is divided by $3x + 2$, the remainder is -2.

Find the value of a and the value of b.

Answer

$f(x) = 6x^3 + ax^2 + bx - 4$

When $f(x)$ is divided by $x - 1$ the remainder is 3, hence $f(1) = 3$.

$6(1)^3 + a(1)^2 + b(1) - 4 = 3$

$a + b = 1$ ------------ (1)

When $f(x)$ is divided by $3x + 2$ the remainder is -2, hence $f\left(-\dfrac{2}{3}\right) = -2$.

$$6\left(-\frac{2}{3}\right)^3 + a\left(-\frac{2}{3}\right)^2 + b\left(-\frac{2}{3}\right) - 4 = -2$$

$$2a - 3b = 17 \text{ ----------- (2)}$$

$2 \times (1) - (2)$ gives $5b = -15$

$$b = -3$$

Substituting $b = -3$ in equation (1) gives $a = 4$.

$a = 4$ and $b = -3$.

EXERCISE 1F

1 Find the remainder when:

  a $6x^3 + 3x^2 - 5x + 2$ is divided by $x - 1$

  b $x^3 + x^2 - 11x + 12$ is divided by $x + 4$

  c $x^4 + 2x^3 - 5x^2 - 2x + 8$ is divided by $x + 1$

  d $4x^3 - x^2 - 18x + 1$ is divided by $2x - 1$

2 a When $x^3 - 3x^2 + ax - 7$ is divided by $x + 2$, the remainder is -37. Find the value of a.

  b When $9x^3 + bx - 5$ is divided by $3x + 2$, the remainder is -13. Find the value of b.

3 $f(x) = x^3 + ax^2 + bx - 5$

  $f(x)$ has a factor of $x - 1$ and leaves a remainder of -6 when divided by $x + 1$.

  Find the value of a and the value of b.

4 The polynomial $3x^3 + ax^2 + bx + 8$, where a and b are constants, is denoted by $f(x)$. It is given that $x + 2$ is a factor of $f(x)$, and that when $f(x)$ is divided by $x - 1$ the remainder is 15. Find the value of a and the value of b.

5 The function $f(x) = ax^3 + 7x^2 + bx - 8$, where a and b are constants, is such that $2x + 1$ is a factor. The remainder when $f(x)$ is divided by $x + 1$ is 7.

  a Find the value of a and the value of b.

  b Factorise $f(x)$ completely.

6 The polynomial $x^3 + 6x^2 + px - 3$ leaves a remainder of R when divided by $x + 1$ and a remainder of $-10R$ when divided by $x - 3$.

  a Find the value of p.

  b Hence find the remainder when the expression is divided by $x - 2$.

7 The polynomial $x^3 + ax^2 + bx + 2$, where a and b are constants, is denoted by $f(x)$. It is given that $x - 2$ is a factor of $f(x)$, and that when $f(x)$ is divided by $x + 1$ the remainder is 21.

  a Find the value of a and the value of b.

  b Solve the equation $f(x) = 0$, giving the roots in exact form.

8 The polynomial $2x^3 + ax^2 + bx + c$ is denoted by f(x). The roots of f(x) = 0 are $-1, 2$ and k. When f(x) is divided by $x - 1$, the remainder is 6.

   a Find the value of k.

   b Find the remainder when f(x) is divided by $x + 2$.

PS 9   P(x) = $2x + 4x^2 + 6x^3 + \ldots + 100x^{50}$

    Find the remainder when P(x) is divided by $(x - 1)$.

PS 10   P(x) = $3(x+1)(x+2)(x+3) + a(x+1)(x+2) + b(x+1) + c$

    It is given that when P(x) is divided by each of $x + 1, x + 2$ and $x + 3$ the remainders are $-2, 2$ and 10, respectively. Find the value of a, the value of b and the value of c.

Checklist of learning and understanding

The modulus function

- The modulus function $|x|$ is defined by:

$$|x| = \begin{cases} x & \text{if } x \geq 0 \\ -x & \text{if } x < 0 \end{cases}$$

The division algorithm for polynomials

-    P(x)   =   D(x)   ×   Q(x)   +   R(x)
     ↑   ↑   ↑   ↑
  dividend = divisor × quotient + remainder

The factor theorem

- If for a polynomial P(x), P(c) = 0 then $x - c$ is a factor of P(x).
- If for a polynomial P(x), P$\left(\dfrac{b}{a}\right)$ = 0 then $ax - b$ is a factor of P(x).

The remainder theorem

- If a polynomial P(x) is divided by $x - c$, the remainder is P(c).
- If a polynomial P(x) is divided by $ax - b$, the remainder is P$\left(\dfrac{b}{a}\right)$.

END-OF-CHAPTER REVIEW EXERCISE 1

1. Solve the equation $|2x - 3| = |5x + 1|$. [3]
2. Solve the inequality $|5x - 3| \geq 7$. [3]
3. Solve the inequality $|2x - 3| < |2 - x|$. [3]
4. Solve the equation $|x^2 - 14| = 11$. [4]
5. The polynomial $ax^3 - 13x^2 - 41x - 2a$, where a is a constant, is denoted by p(x).
   a. Given that $(x - 4)$ is a factor of p(x), find the value of a. [2]
   b. When a has this value, factorise p(x) completely. [3]
6. The polynomial $6x^3 - 23x^2 - 38x + 15$ is denoted by f(x).
   a. Show that $(x - 5)$ is a factor of f(x) and hence factorise f(x) completely. [4]
   b. Write down the roots of $f(|x|) = 0$. [1]
7. The polynomial $x^3 - 5x^2 + ax + b$ is denoted by f(x). It is given that $(x + 2)$ is a factor of f(x) and that when f(x) is divided by $(x - 1)$ the remainder is -6. Find the value of a and the value of b. [5]
8. The polynomial $x^3 - 5x^2 + 7x - 3$ is denoted by p(x).
   a. Find the quotient and remainder when p(x) is divided by $(x^2 - 2x - 1)$. [4]
   b. Use the factor theorem to show that $(x - 3)$ is a factor of p(x). [2]
9. The polynomial $4x^4 + 4x^3 - 7x^2 - 4x + 8$ is denoted by p(x).
   a. Find the quotient and remainder when p(x) is divided by $(x^2 - 1)$. [3]
   b. Hence solve the equation $4x^4 + 4x^3 - 7x^2 - 4x + 3 = 0$. [3]
10. The polynomial $x^4 - 48x^2 - 21x - 2$ is denoted by f(x).
    a. Find the value of the constant k for which $f(x) = (x^2 + kx + 2)(x^2 - kx - 1)$. [3]
    b. Hence solve the equation $f(x) = 0$. Give your answers in exact form. [3]
11. The polynomial $2x^4 + 3x^3 - 12x^2 - 7x + a$ is denoted by p(x).
    a. Given that $(2x - 1)$ is a factor of p(x), find the value of a. [2]
    b. When a has this value, verify that $(x + 3)$ is also a factor of p(x) and hence factorise p(x) completely. [4]
12. The polynomial $3x^3 + ax^2 - 36x + 20$ is denoted by p(x).
    a. Given that $(x - 2)$ is a factor of p(x), find the value of a. [2]
    b. When a has this value, solve the equation $p(x) = 0$. [4]
13. The polynomial $2x^3 + 5x^2 - 7x + 11$ is denoted by f(x).
    a. Find the remainder when f(x) is divided by $(x - 2)$. [2]
    b. Find the quotient and remainder when f(x) is divided by $(x^2 - 4x + 2)$. [4]
14. The polynomial $ax^3 + bx^2 - x + 12$ is denoted by p(x).
    a. Given that $(x - 3)$ and $(x + 1)$ are factors of p(x), find the value of a and the value of b. [4]
    b. When a and b take these values, find the other linear factor of p(x). [2]

Chapter 1: Algebra

15 The polynomial $6x^3 + x^2 + ax - 10$, where a is a constant, is denoted by P(x). It is given that when P(x) is divided by $(x + 2)$ the remainder is -12.

    a Find the value of a and hence verify that $(2x + 1)$ is a factor of P(x). [3]

    b When a has this value, solve the equation P(x) = 0. [4]

16 The polynomial $2x^3 + ax^2 + bx + 6$ is denoted by p(x).

    a Given that $(x + 2)$ and $(x - 3)$ are factors of p(x), find the value of a and the value of b. [4]

    b When a and b take these values, factorise p(x) completely. [3]

17 The polynomials P(x) and Q(x) are defined as:

$$P(x) = x^3 + ax^2 + b \quad \text{and} \quad Q(x) = x^3 + bx^2 + a.$$

    It is given that $(x - 2)$ is a factor of P(x) and that when Q(x) is divided by $(x + 1)$ the remainder is -15.

    a Find the value of a and the value of b. [5]

    b When a and b take these values, find the least possible value of P(x) – Q(x) as x varies. [2]

18 The polynomial $5x^3 - 13x^2 + 17x - 7$ is denoted by p(x).

    a Find the quotient when p(x) is divided by $(x - 1)$, and show that the remainder is 2. [4]

    b Hence show that the polynomial $5x^3 - 13x^2 + 17x - 9$ has exactly one real root. [3]

19 The polynomial $4x^3 + kx^2 - 65x + 18$ is denoted by f(x).

    a Given that $(x + 2)$ is a factor of f(x), find the value of k. [2]

    b When k has this value, solve the equation f(x) = 0. [4]

    c Write down the roots of f(x^2) = 0. [1]

20 The polynomial $2x^3 - 5x^2 + ax + b$, where a and b are constants, is denoted by f(x). It is given that when f(x) is divided by $(x + 2)$ the remainder is 8 and that when f(x) is divided by $(x - 1)$ the remainder is 50.

    a Find the value of a and the value of b. [5]

    b When a and b have these values, find the quotient and remainder when f(x) is divided by $x^2 - x + 2$. [3]

21 The polynomial $2x^3 - 9x^2 + ax + b$, where a and b are constants, is denoted by f(x). It is given that $(x + 2)$ is a factor of f(x), and that when f(x) is divided by $(x + 1)$ the remainder is 30.

    a Find the value of a and the value of b. [5]

    b When a and b have these values, solve the equation f(x) = 0. [4]

22 The polynomial $x^3 + 3x^2 + 4x + 2$ is denoted by f(x).

    i Find the quotient and remainder when f(x) is divided by $x^2 + x - 1$. [4]

    ii Use the factor theorem to show that $(x + 1)$ is a factor of f(x). [2]

Cambridge International AS & A Level Mathematics 9709 Paper 21 Q4 June 2010

23 The polynomial $4x^3 + ax^2 + 9x + 9$, where a is a constant, is denoted by $p(x)$. It is given that when $p(x)$ is divided by $(2x - 1)$ the remainder is 10.

   i Find the value of a and hence verify that $(x - 3)$ is a factor of $p(x)$. [3]

   ii When a has this value, solve the equation $p(x) = 0$. [4]

Cambridge International AS & A Level Mathematics 9709 Paper 21 Q5 November 2011

24 The polynomial $ax^3 - 5x^2 + bx + 9$, where a and b are constants, is denoted by $p(x)$. It is given that $(2x + 3)$ is a factor of $p(x)$, and that when $p(x)$ is divided by $(x + 1)$ the remainder is 8.

   i Find the values of a and b. [5]

   ii When a and b have these values, factorise $p(x)$ completely. [3]

Cambridge International AS & A Level Mathematics 9709 Paper 21 Q4 June 2013

Chapter 2
Logarithmic and exponential functions

In this chapter you will learn how to:

- understand the relationship between logarithms and indices, and use the laws of logarithms (excluding change of base)
- understand the definition and properties of e^x and $\ln x$, including their relationship as inverse functions and their graphs
- use logarithms to solve equations and inequalities in which the unknown appears in indices
- use logarithms to transform a given relationship to linear form, and hence determine unknown constants by considering the gradient and/or intercept.

Cambridge International AS & A Level Mathematics: Pure Mathematics 2 & 3

PREREQUISITE KNOWLEDGE

Where it comes from	What you should be able to do	Check your skills
IGCSE / O Level Mathematics	Use and interpret positive, negative, fractional and zero indices.	1 Evaluate. a 5^{-2} b $8^{\frac{2}{3}}$ c 7^0 2 Solve $32^x = 2$.
IGCSE / O Level Mathematics	Use the rules of indices.	3 Simplify. a $3x^{-4} \times \frac{2}{3} x^{\frac{1}{2}}$ b $\frac{2}{5} x^{\frac{1}{2}} \div 2x^{-2}$ c $\left(\frac{2x^5}{3}\right)^3$

What is a logarithm?

At IGCSE / O Level we learnt how to solve some simple exponential equations. For example, $32^x = 2$ can be solved by first writing 32 as a power of 2. In this chapter we will learn how to solve more difficult exponential equations such as $a^x = b$ where there is no obvious way of writing the numbers a and b in index form with the same base number. In order to be able to do this we will first learn about a new type of function called the logarithmic function. A logarithmic function is the inverse of an exponential function.

At IGCSE / O Level we learnt about exponential growth and decay and its applications to numerous real-life problems such as investments and population sizes. Logarithmic functions also have real-life applications such as the amount of energy released during an earthquake (the Richter scale). One of the most fascinating occurrences of logarithms in nature is that of the logarithmic spiral that can be observed in a nautilus shell. We also see logarithmic spiral shapes in spiral galaxies, and in plants such as sunflowers.

▶▶| **FAST FORWARD**

You will encounter logarithmic functions again when studying calculus in Chapters 4, 5 and 8.

🌐 **WEB LINK**

Explore the *Exponentials and logarithms* station on the Underground Mathematics website.

2.1 Logarithms to base 10

Consider the exponential function $f(x) = 10^x$.

To solve the equation $10^x = 6$, we could make an estimate by noting that $10^0 = 1$ and $10^1 = 10$, hence $0 < x < 1$. We could then try values of x between 0 and 1 to find a more accurate value for x.

The graph of $y = 10^x$ could also be used to give a more accurate value for x when $10^x = 6$.

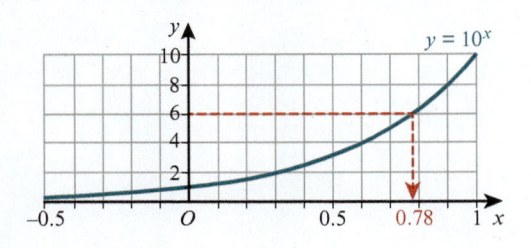

From the graph, $x \approx 0.78$.

There is, however, a function that gives the value of x directly:

If $10^x = 6$ then $x = \log_{10} 6$.

$\log_{10} 6$ is read as 'log 6 to base 10'.

log is short for **logarithm**.

On a calculator, for logs to the base 10, we use the log or lg key.

So if $10^x = 6$

then $\quad x = \log_{10} 6$

$\quad x = 0.778$ to 3 significant figures.

> **TIP**
>
> $\log_{10} 6$ can also be written as log 6 or lg 6.

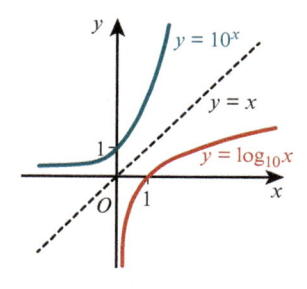

KEY POINT 2.1

The rule for base 10 is:

If $y = 10^x$ then $x = \log_{10} y$

This rule can be described in words as:

$\log_{10} y$ is the power that 10 must be raised to in order to obtain y.

For example, $\log_{10} 100 = 2$ since $100 = 10^2$.

$y = 10^x$ and $y = \log_{10} x$ are inverse functions.

We say that $y = 10^x \Leftrightarrow x = \log_{10} y$.

This is read as: $y = 10^x$ is equivalent to saying that $x = \log_{10} y$.

EXPLORE 2.1

1 Discuss with your classmates why each of these two statements is true.

$\log_{10} 10^x = x$ for $x \in \mathbb{R}$ $\qquad$ $10^{\log_{10} x} = x$ for $x > 0$

2 a Discuss with your classmates why each of these two statements is true.

$\log_{10} 10 = 1$ $\qquad$ $\log_{10} 1 = 0$

b Discuss what the missing numbers should be in this table of values, giving reasons for your answers.

$\log_{10} 0.01$	$\log_{10} 0.1$	$\log_{10} 1$	$\log_{10} 10$	$\log_{10} 100$	$\log_{10} 1000$
.....	.....	0	1	.....	.....

WORKED EXAMPLE 2.1

a Convert $10^x = 58$ to logarithmic form.

b Solve $10^x = 58$ giving your answer correct to 3 significant figures.

Answer

a Method 1

$10^x = 58$

Step 1: Identify the base and index ⟶ base = 10, index = x

Step 2: Start to write in log form ⟶ $x = \log_{10} ?$

Step 3: Complete the log form ⟶ $x = \log_{10} 58$

$10^x = 58 \Leftrightarrow x = \log_{10} 58$

Method 2

$10^x = 58$

$\log_{10} 10^x = \log_{10} 58$ .... Take logs to base 10 of both sides.

$x = \log_{10} 58$ .... Use $\log_{10} 10^x = x$.

b $10^x = 58$

$x = \log_{10} 58$

$x \approx 1.76$

> **TIP**
>
> In log form the index is always on its own and the base goes at the base of the logarithm.

WORKED EXAMPLE 2.2

a Convert $\log_{10} x = 3.5$ to exponential form.

b Solve $\log_{10} x = 3.5$ giving your answer correct to 3 significant figures.

Answer

a Method 1

$\log_{10} x = 3.5$

Step 1: Identify the base and index ⟶ base = 10, index = 3.5

Step 2: Start to write in exponential form ⟶ $10^{3.5} = ?$

> **TIP**
>
> In log form the index is always on its own.

Chapter 2: Logarithmic and exponential functions

Step 3: Complete the exponential form $\longrightarrow x = 10^{3.5}$

$\log_{10} x = 3.5 \Leftrightarrow x = 10^{3.5}$

Method 2

$\log_{10} x = 3.5$ .... Write each side as an exponent of 10.

$10^{\log_{10} x} = 10^{3.5}$

$x = 10^{3.5}$ .... Use $10^{\log_{10} x} = x$.

b $\log_{10} x = 3.5$

$x = 10^{3.5}$

$x \approx 3160$

WORKED EXAMPLE 2.3

Find the value of:

a $\log_{10} 10\,000$ **b** $\log_{10} 0.0001$ **c** $\log_{10} 1000\sqrt{10}$.

Answer

a $\log_{10} 10\,000 = \log_{10} 10^4$ .... Write 10 000 as a power of 10, $10\,000 = 10^4$.

$= 4$

b $\log_{10} 0.0001 = \log_{10} 10^{-4}$ .... Write 0.0001 as a power of 10, $0.0001 = 10^{-4}$.

$= -4$

c $\log_{10} 1000\sqrt{10} = \log_{10} 10^{3.5}$ .... Write $1000\sqrt{10}$ as a power of 10.

$= 3.5$ $1000\sqrt{10} = 10^3 \times 10^{0.5} = 10^{3.5}$.

EXERCISE 2A

1 Convert from exponential form to logarithmic form.

     a $10^2 = 100$ **b** $10^x = 200$ **c** $10^x = 0.05$

2 Solve each of these equations, giving your answers correct to 3 significant figures.

     a $10^x = 52$ **b** $10^x = 250$ **c** $10^x = 0.48$

3 Convert from logarithmic form to exponential form.

     a $\log_{10} 10\,000 = 4$ **b** $\log_{10} x = 1.2$ **c** $\log_{10} x = -0.6$

4 Solve each of these equations, giving your answers correct to 3 significant figures.

     a $\log_{10} x = 1.88$ **b** $\log_{10} x = 2.76$ **c** $\log_{10} x = -1.4$

5 Without using a calculator, find the value of:

    a $\log_{10} 100$ b $\log_{10} 0.0001$ c $\log_{10}(10\sqrt{10})$

    d $\log_{10}(\sqrt[3]{10})$ e $\log_{10}(100\sqrt[3]{10})$ f $\log_{10}\left(\dfrac{100}{\sqrt{1000}}\right)$

WEB LINK

Try the *See the power* resource on the Underground Mathematics website.

6 Given that the function f is defined as $f : x \mapsto 10^x - 3$ for $x \in \mathbb{R}$, find an expression for $f^{-1}(x)$.

7 Given that p and q are positive and that $4(\log_{10} p)^2 + 2(\log_{10} q)^2 = 9$, find the greatest possible value of p.

2.2 Logarithms to base a

In the last section you learnt about logarithms to the base of 10.

The same principles can be applied to define logarithms in other bases.

KEY POINT 2.2

If $y = a^x$ then $x = \log_a y$

$\log_a a = 1$ $\log_a 1 = 0$

$\log_a a^x = x$ $a^{\log_a x} = x$

The conditions for $\log_a x$ to be defined are:

- $a > 0$ and $a \neq 1$
- $x > 0$

We say that $y = a^x \Leftrightarrow x = \log_a y$.

WORKED EXAMPLE 2.4

Convert $5^3 = 125$ to logarithmic form.

Answer

Method 1

$5^3 = 125$

Step 1: Identify the base and index   ⟶   base = 5, index = 3

Step 2: Start to write in log form   ⟶ $3 = \log_5 ?$

Step 3: Complete the log form   ⟶ $3 = \log_5 125$

$5^3 = 125 \Leftrightarrow 3 = \log_5 125$

Method 2

$5^3 = 125$

$\log_5 5^3 = \log_5 125$ Take logs to base 5 of both sides.

$3 = \log_5 125$ Use $\log_5 5^x = x$.

TIP

In log form the index is always on its own and the base goes at the base of the logarithm.

Chapter 2: Logarithmic and exponential functions

WORKED EXAMPLE 2.5

Convert $\log_3 x = 2.5$ to exponential form.

Answer

Method 1
$\log_3 x = 2.5$
Step 1: Identify the base and index $\longrightarrow$ base = 3, index = 2.5
Step 2: Start to write in exponential form $\longrightarrow$ $3^{2.5} = ?$
Step 3: Complete the exponential form $\longrightarrow$ $3^{2.5} = x$
$\log_3 x = 2.5 \Leftrightarrow 3^{2.5} = x$

Method 2
$\log_3 x = 2.5$ .... Write each side as an exponent of 3.
$3^{\log_3 x} = 3^{2.5}$ .... Use $3^{\log_3 x} = x$
$x = 3^{2.5}$

> **TIP**
>
> In log form the index is always on its own.

WORKED EXAMPLE 2.6

Find the value of:

a $\log_2 16$

b $\log_3 \dfrac{1}{9}$

Answer

a $\log_2 16 = \log_2 2^4$ .... Write 16 as a power of 2, $16 = 2^4$.
   $= 4$

b $\log_3 \dfrac{1}{9} = \log_3 3^{-2}$ .... Write $\dfrac{1}{9}$ as a power of 3, $\dfrac{1}{9} = 3^{-2}$.
   $= -2$

WORKED EXAMPLE 2.7

Simplify $\log_x\left(\dfrac{\sqrt[3]{x}}{x}\right)$.

Answer

$\log_x\left(\dfrac{\sqrt[3]{x}}{x}\right) = \log_x\left(x^{-\frac{2}{3}}\right)$ .... Write $\dfrac{\sqrt[3]{x}}{x}$ as a power of x.

$= -\dfrac{2}{3}$

EXERCISE 2B

1. Convert from exponential form to logarithmic form.
   - a $5^2 = 25$
   - b $2^4 = 16$
   - c $3^{-5} = \dfrac{1}{243}$
   - d $2^{-10} = \dfrac{1}{1024}$
   - e $8^x = 15$
   - f $x^y = 6$
   - g $a^b = c$
   - h $x^{5y} = 7$

2. Convert from logarithmic form to exponential form.
   - a $\log_2 8 = 3$
   - b $\log_3 81 = 4$
   - c $\log_8 1 = 0$
   - d $\log_{16} 4 = \dfrac{1}{2}$
   - e $\log_8 2 = \dfrac{1}{3}$
   - f $\log_2 y = 4$
   - g $\log_a 1 = 0$
   - h $\log_x 5 = y$

3. Solve.
   - a $\log_2 x = 3$
   - b $\log_3 x = 2$
   - c $\log_4 x = 0$
   - d $\log_x 49 = 2$

4. Solve.
   - a $\log_3(x + 5) = 2$
   - b $\log_2(3x - 1) = 5$
   - c $\log_y(7 - 2x) = 0$

5. Find the value of:
   - a $\log_3 27$
   - b $\log_6 36$
   - c $\log_4 2$
   - d $\log_2 0.125$
   - e $\log_4\left(\dfrac{1}{16}\right)$
   - f $\log_5(5\sqrt{5})$
   - g $\log_3\left(\dfrac{\sqrt{3}}{3}\right)$
   - h $\log_7\left(\dfrac{\sqrt[3]{7}}{7}\right)^2$

6. Simplify.
   - a $\log_x x^3$
   - b $\log_x \sqrt[3]{x}$
   - c $\log_x\left(x^2\sqrt{x}\right)$
   - d $\log_x \dfrac{1}{x^4}$
   - e $\log_x\left(\dfrac{1}{x^3}\right)^2$
   - f $\log_x\left(\sqrt{x^5}\right)$
   - g $\log_x\left(\dfrac{x}{\sqrt[3]{x}}\right)$
   - h $\log_x\left(\dfrac{x^2}{\sqrt{x}}\right)^3$

7. Given that the function f is defined as $f: x \mapsto 1 + \log_2(x - 3)$ for $x \in \mathbb{R}, x > 3$, find an expression for $f^{-1}(x)$.

PS 8. Solve.
   - a $\log_2(\log_3 x) = -1$
   - b $\log_4 2^{3-5x} = x^2$

9 $\log_3 9$ $\log_2 8$ $\log_3 20$ $\log_3 4$ $\log_2 2$ $\log_4 3$ $\log_2 3$

Without using a calculator, arrange these logarithms in ascending order.

Add some more logarithms of your own in the correct position in your ordered list.

WEB LINK

Try the *Logarithm lattice* resource on the Underground Mathematics website.

EXPLORE 2.2

1. Prya and Hamish are asked what they think $\log_{10} 12 + \log_{10} 4$ simplifies to.

   Prya says that: $\log_{10} 12 + \log_{10} 4 = \log_{10}(12 + 4) = \log_{10} 16$

   Hamish says that: $\log_{10} 12 + \log_{10} 4 = \log_{10}(12 \times 4) = \log_{10} 48$

   Use a calculator to see who is correct and check that their rule works for other pairs of numbers.

2. Can you predict what $\log_{10} 12 - \log_{10} 4$ simplifies to?

   Use a calculator to check that you have predicted correctly and check for other pairs of numbers.

2.3 The laws of logarithms

If x and y are both positive and $a > 0$ and $a \neq 1$, then the following **laws of logarithms** can be used.

KEY POINT 2.3

Multiplication law	Division law	Power law
$\log_a(xy) = \log_a x + \log_a y$	$\log_a\left(\dfrac{x}{y}\right) = \log_a x - \log_a y$	$\log_a(x)^m = m \log_a x$

Proofs:

Multiplication law

$\log_a(xy)$
$= \log_a(a^{\log_a x} \times a^{\log_a y})$
$= \log_a(a^{\log_a x + \log_a y})$
$= \log_a x + \log_a y$

Division law

$\log_a\left(\dfrac{x}{y}\right)$
$= \log_a\left(\dfrac{a^{\log_a x}}{a^{\log_a y}}\right)$
$= \log_a(a^{\log_a x - \log_a y})$
$= \log_a x - \log_a y$

Power law

$\log_a(x)^m$
$= \log_a\left((a^{\log_a x})^m\right)$
$= \log_a(a^{m \log_a x})$
$= m \log_a x$

Using the power law, $\log_a\left(\dfrac{1}{x}\right) = \log_a x^{-1}$
$= -\log_a x$

Another useful rule to remember:

> **KEY POINT 2.4**
>
> $\log_a\left(\dfrac{1}{x}\right) = -\log_a x$

WORKED EXAMPLE 2.8

Use the laws of logarithms to simplify these expressions.

   a $\log_2 3 + \log_2 5$ **b** $\log_3 8 - \log_3 4$ **c** $2\log_5 2 + \log_5 3$

Answer

  a $\log_2 3 + \log_2 5$ **b** $\log_3 8 - \log_3 4$ **c** $2\log_5 2 + \log_5 3$

      $= \log_2(3 \times 5)$ $= \log_3\left(\dfrac{8}{4}\right)$ $= \log_5 2^2 + \log_5 3$

      $= \log_2 15$ $= \log_3 2$ $= \log_5(4 \times 3)$

                                                                                                                      $= \log_5 12$

WORKED EXAMPLE 2.9

Given that $\log_4 p = x$ and $\log_4 q = y$, express in terms of x and/or y

   a $\log_4 p^5 - \log_4 q^2$ **b** $\log_4 \sqrt{p} + 5\log_4 \sqrt[3]{q}$ **c** $\log_4\left(\dfrac{64}{p}\right)$.

Answer

  a $\log_4 p^5 - \log_4 q^2$ **b** $\log_4 \sqrt{p} + 5\log_4 \sqrt[3]{q}$ **c** $\log_4\left(\dfrac{64}{p}\right)$

      $= 5\log_4 p - 2\log_4 q$ $= \dfrac{1}{2}\log_4 p + \dfrac{5}{3}\log_4 q$ $= \log_4 4^3 - \log_4 p$

      $= 5x - 2y$ $= \dfrac{1}{2}x + \dfrac{5}{3}y$ $= 3 - x$

EXERCISE 2C

  1 Write as a single logarithm.

      a $\log_2 7 + \log_2 11$ **b** $\log_6 20 - \log_6 4$ **c** $3\log_5 2 - \log_5 4$

      d $2\log_3 8 - 5\log_3 2$ **e** $1 + 2\log_2 3$ **f** $2 - \log_4 2$

  2 Write as a single logarithm, then simplify your answer.

      a $\log_2 40 - \log_2 5$ **b** $\log_6 20 + \log_6 5$ **c** $\log_2 60 - \log_2 5$

3 Simplify.

a $3\log_5 2 + \frac{1}{2}\log_5 36 - \log_5 12$

b $\frac{1}{2}\log_3 8 + \frac{1}{2}\log_3 18 - 1$

4 Express 8 and 0.25 as powers of 2 and hence simplify $\dfrac{\log_5 8}{\log_5 0.25}$.

5 Simplify.

a $\dfrac{\log_2 27}{\log_2 3}$

b $\dfrac{\log_3 128}{\log_3 16}$

c $\dfrac{\log_3 25}{\log_3 0.04}$

d $\dfrac{\log_5 1000}{\log_5 0.01}$

6 Given that $\log_8 y = \log_8(x - 2) - 2\log_8 x$, express y in terms of x.

7 Given that $\log_3(z - 1) - \log_3 z = 1 + 3\log_3 y$, express z in terms of y.

8 Given that $y = \log_5 x$, find in terms of y:

a x

b $\log_5(25x)$

c $\log_5\left(\dfrac{x\sqrt{x}}{125}\right)$

d $\log_x 125$

9 Given that $x = \log_4 p$ and $y = \log_4 q$, express in terms of x and/or y:

a $\log_4(64p)$

b $\log_4\left(\dfrac{2}{q}\right)$

c pq

d $\log_4 p^2 - \log_4\left(4\sqrt{q}\right)$

10 Given that $\log_a x = 7$ and $\log_a y = 4$, find the value of:

a $\log_a\left(\dfrac{x}{y}\right)$

b $\log_a\left(\dfrac{\sqrt{x}}{y^2}\right)$

c $\log_a\left(x^2\sqrt{y}\right)$

d $\log_a\left(\dfrac{y}{\sqrt[3]{x}}\right)$.

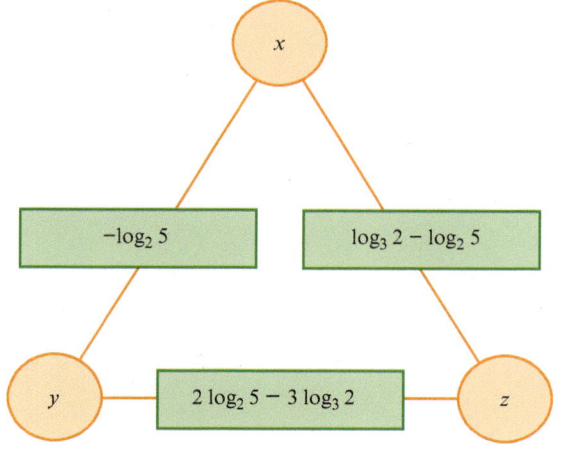

PS 11

Sides of triangle: $-\log_2 5$, $\log_3 2 - \log_2 5$, $2\log_2 5 - 3\log_3 2$

Vertices: x, y, z

The number in the rectangle on the side of the triangle is the **sum** of the numbers at the adjacent vertices.

For example, $x + y = -\log_2 5$.

Find the value of x, the value of y and the value of z.

WEB LINK

Try the *Summing to one* and the *Factorial fragments* resources on the Underground Mathematics website.

2.4 Solving logarithmic equations

We have already learnt how to solve simple logarithmic equations. In this section we will learn how to solve more complicated equations.

Since $\log_a x$ only exists for $x > 0$ and for $a > 0$ and $a \ne 1$, it is essential when solving equations involving logs that all roots are checked in the original equation.

WORKED EXAMPLE 2.10

Solve:

a $2\log_8(x+2) = \log_8(2x+19)$

b $4\log_x 2 - \log_x 4 = 2$

Answer

a $2\log_8(x+2) = \log_8(2x+19)$ — Use the power law.

$\log_8(x+2)^2 = \log_8(2x+19)$ — Use equality of logarithms.

$(x+2)^2 = 2x+19$ — Expand brackets.

$x^2 + 4x + 4 = 2x + 19$

$x^2 + 2x - 15 = 0$

$(x-3)(x+5) = 0$

$x = 3$ or $x = -5$

Check when $x = 3$: $2\log_8(x+2) = 2\log_8 5 = \log_8 25$ is defined

$\log_8(2x+19) = \log_8 25$ is defined.

So $x = 3$ is a solution, since both sides of the equation are defined and equivalent in value.

Check when $x = -5$: $2\log_8(x+2) = 2\log_8(-3)$ is not defined.

So $x = -5$ is **not** a solution of the original equation.

Hence, the solution is $x = 3$

b $4\log_x 2 - \log_x 4 = 2$ — Use the power law.

$\log_x 2^4 - \log_x 2^2 = 2$ — Use the division law.

$\log_x 2^{4-2} = 2$

$\log_x 2^2 = 2$

$\log_x 4 = 2$ — Convert to exponential form.

$x^2 = 4$

$x = \pm 2$

Since logarithms only exist for positive bases, $x = -2$ is not a solution.

Check when $x = 2$: $4\log_2 2 - \log_2 4 = 4 - 2\log_2 2$

$= 4 - 2$

$= 2$

So $x = 2$ satisfies the original equation.

Hence, the solution is $x = 2$

EXERCISE 2D

1 Solve.

    a $\log_5 x + \log_5 3 = \log_5 30$

    b $\log_3 4x - \log_3 2 = \log_3 7$

    c $\log_2(2 - x) + \log_2 9 = 2\log_2 8$

    d $\log_{10}(x - 4) = 2\log_{10} 5 + \log_{10} 2$

2 Solve.

    a $\log_{10}(2x + 9) - \log_{10} 5 = 1$

    b $\log_3 2x - \log_3(x - 1) = 2$

    c $\log_2(5 - 2x) = 3 + \log_2(x + 1)$

    d $\log_5(2x - 3) + 2\log_5 2 = 1 + \log_5(6 - 2x)$

3 Solve.

    a $\log_2 x + \log_2(x - 1) = \log_2 20$

    b $\log_5(x - 2) + \log_5(x - 5) = \log_5 4x$

    c $2\log_3 x - \log_3(x - 2) = 2$

    d $3 + 2\log_2 x = \log_2(3 - 10x)$

4 Solve.

    a $\log_x 40 - \log_x 5 = 1$

    b $\log_x 36 + \log_x 4 = 2$

    c $\log_x 25 - 2\log_x 3 = 2$

    d $2\log_x 32 = 3 + 2\log_x 4$

5 Solve.

    a $(\log_2 x)^2 - 8\log_2 x + 15 = 0$

    b $(\log_{10} x)^2 - \log_{10}(x^2) = 3$

    c $(\log_5 x)^2 + \log_5(x^3) = 10$

    d $4(\log_2 x)^2 - 2\log_2(x^2) = 3$

PS 6 Solve the simultaneous equations.

    a $xy = 81$
       $\log_x y = 3$

    b $4^x = 2^y$
       $3\log_{10} y = \log_{10} x + \log_{10} 2$

    c $\log_4(x - y) = 2\log_4 x$
       $\log_4(x + y + 9) = 0$

    d $\log_{10} x = 2\log_{10} y$
       $\log_{10}(2x - 17y) = 2$

PS 7 Given that $\log_{10}(x^2 y) = 4$ and $\log_{10}\left(\dfrac{x^4}{y^3}\right) = 18$, find the value of $\log_{10} x$ and of $\log_{10} y$.

2.5 Solving exponential equations

At IGCSE / O Level you learnt how to solve exponential equations whose terms could be converted to the same base. In this section you will learn how to solve exponential equations whose terms cannot be converted to the same base.

WORKED EXAMPLE 2.11

Solve, giving your answers correct to 3 significant figures.

a $5^{2x+1} = 7$

b $3^{2x} = 4^{x+5}$

Answer

a $\quad 5^{2x+1} = 7$ — Take logs to base 10 of both sides.

$\quad \log 5^{2x+1} = \log 7$ — Use the power rule.

$\quad (2x+1) \log 5 = \log 7$ — Expand the brackets.

$\quad 2x \log 5 + \log 5 = \log 7$ — Rearrange to find x.

$$x = \frac{\log 7 - \log 5}{2 \log 5}$$

$\quad x = 0.105$ to 3 significant figures.

b $\quad 3^{2x} = 4^{x+5}$ — Take logs to base 10 of both sides.

$\quad \log 3^{2x} = \log 4^{x+5}$ — Use the power rule.

$\quad 2x \log 3 = (x+5) \log 4$ — Expand the brackets.

$\quad 2x \log 3 = x \log 4 + 5 \log 4$ — Collect x terms on one side.

$\quad x(2 \log 3 - \log 4) = 5 \log 4$ — Divide both sides by $(2 \log 3 - \log 4)$.

$$x = \frac{5 \log 4}{2 \log 3 - \log 4}$$

$\quad x = 8.55$ to 3 significant figures.

WORKED EXAMPLE 2.12

Solve the equation $2(3^{2x}) + 7(3^x) = 15$, giving your answers correct to 3 significant figures.

Answer

$2(3^{2x}) + 7(3^x) - 15 = 0$ — Use the substitution $y = 3^x$.

$\quad 2y^2 + 7y - 15 = 0$ — Factorise.

$\quad (2y - 3)(y + 5) = 0$

$\quad y = 1.5$ or $y = -5$

When $y = 1.5$

$\quad 3^x = 1.5$ — Take logs of both sides.

$\quad x \log 3 = \log 1.5$

$$x = \frac{\log 1.5}{\log 3} = 0.369 \ldots$$

When $y = -5$

$3^x = -5$ .... There are no solutions to this equation since 3^x is always positive.

Hence, the solution is $x = 0.369$ to 3 significant figures.

EXERCISE 2E

1. Solve, giving your answers correct to 3 significant figures.
   - **a** $5^x = 18$
   - **b** $2^x = 35$
   - **c** $3^{2x} = 8$
   - **d** $2^{x+1} = 25$
   - **e** $3^{2x-5} = 20$
   - **f** $3^x = 2^{x+1}$
   - **g** $5^{x+3} = 7^{4-3x}$
   - **h** $4^{1-3x} = 3^{x-2}$
   - **i** $2^x = 2(5^x)$
   - **j** $4^x = 7(3^x)$
   - **k** $2^{x+1} = 3(5^x)$
   - **l** $5(2^{x-3}) = 4(3^{x+4})$

2.
   - **a** Show that $2^{x+1} + 6(2^{x-1}) = 12$ can be written as $2(2^x) + 3(2^x) = 12$.
   - **b** Hence solve the equation $2^{x+1} + 6(2^{x-1}) = 12$, giving your answer correct to 3 significant figures.

3. Solve, giving your answers correct to 3 significant figures.
   - **a** $2^{x+2} + 2^x = 2^2$
   - **b** $3^{x+1} = 3^{x-1} + 3^2$
   - **c** $2^{x+1} + 5(2^{x-1}) = 2^4$
   - **d** $5^x + 5^3 = 5^{x+2}$
   - **e** $4^{x-1} = 4^x - 4^3$
   - **f** $3^{x+1} - 2(3^{x-2}) = 5$

4. Use the substitution $y = 2^x$ to solve the equation $2^{2x} + 32 = 12(2^x)$.

5. Solve, giving your answers correct to 3 significant figures.
   - **a** $5^{2x} - 5(5^x) + 6 = 0$
   - **b** $2^{2x} + 5 = 6 \times 2^x$
   - **c** $3^{2x} = 6 \times 3^x + 7$
   - **d** $4^{2x} + 27 = 12(4^x)$

6. Use the substitution $u = 2^x$ to solve the equation $2^{2x} - 5(2^{x+1}) + 24 = 0$.

7. Solve, giving your answers correct to 3 significant figures.
   - **a** $2^{2x} - 2^{x+1} - 35 = 0$
   - **b** $3^{2x} - 3^{x+1} = 10$
   - **c** $2(5^{x+1}) - 5^{2x} = 16$
   - **d** $4^{2x+1} = 17(4^x) - 15$

8. Solve, giving your answers correct to 3 significant figures.
   - **a** $4^x + 2^x - 12 = 0$
   - **b** $3(16^x) - 10(4^x) + 3 = 0$
   - **c** $4^x + 15 = 4(2^{x+1})$
   - **d** $2(9^x) = 3^{x+1} + 27$

9. For each of the following equations find the value of $\dfrac{x}{y}$ correct to 3 significant figures.
   - **a** $3^x = 7^y$
   - **b** $7^x = (2.7)^y$
   - **c** $4^{2x} = 3^{5y}$

10. Solve the equation $x^{2.5} = 20x^{1.25}$, giving your answers in exact form.

11. Solve, giving your answers correct to 3 significant figures.
    - **a** $|4^x - 8| = 2$
    - **b** $|2(3^x) - 4| = 8$
    - **c** $3|2^x - 5| = 2^x$
    - **d** $|3^x + 4| = |3^x - 9|$
    - **e** $|5(2^x) + 3| = |5(2^x) - 10|$
    - **f** $|2^{x+2} + 1| = |2^x + 12|$
    - **g** $3^{2|x|} = 6(3^{|x|}) + 16$
    - **h** $4^{|x|} = 5(2^{|x|}) + 14$

Cambridge International AS & A Level Mathematics: Pure Mathematics 2 & 3

PS 12 Given that $2^{4x+1} \times 3^{2-x} = 8^x \times 3^{5-2x}$, find the value of:

  a 6^x
  b x

PS 13 Solve the equation $8(8^{x-1} - 1) = 7(4^x - 2^{x+1})$.

> **WEB LINK**
>
> Try the *To log or not to log?* resource on the Underground Mathematics website.

2.6 Solving exponential inequalities

To solve inequalities such as $2^x > 3$ we take the logarithm of both sides of the inequality. We can take logarithms to any base but care must be taken with the inequality symbol when doing so.

- If the base of the logarithm is greater than 1 then the inequality symbol remains the same.
- If the base of the logarithm is between 0 and 1 then the inequality symbol must be reversed.

This is because $y = \log_a x$ is an increasing function if $a > 1$ and it is a decreasing function when $0 < a < 1$.

Hence, when taking logarithms to base 10 of both sides of an inequality the inequality symbol is unchanged.

It is also important to remember that if we divide both sides of an inequality by a negative number then the inequality symbol must be reversed.

WORKED EXAMPLE 2.13

Solve the inequality $0.6^x < 0.7$, giving your answer in terms of base 10 logarithms.

Answer

$0.6^x < 0.7$ …… Take logs to base 10 of both sides, inequality symbol is unchanged.

$\log 0.6^x < \log 0.7$ …… Use the power rule for logs.

$x \log 0.6 < \log 0.7$ …… Divide both sides by $\log_{10} 0.6$.

$x > \dfrac{\log 0.7}{\log 0.6}$ …… log 0.6 is negative so the inequality symbol is reversed.

$x > 0.698$ (correct to 3 significant figures)

Chapter 2: Logarithmic and exponential functions

WORKED EXAMPLE 2.14

Solve the inequality $4 \times 3^{2x-1} > 5$, giving your answer in terms of base 10 logarithms.

Answer

$4 \times 3^{2x-1} > 5$ — Divide both sides by 4.

$3^{2x-1} > \dfrac{5}{4}$ — Take logs to base 10 of both sides and use the power rule.

$(2x-1)\log 3 > \log \dfrac{5}{4}$ — Expand brackets.

$(2\log 3)x - \log 3 > \log \dfrac{5}{4}$ — Rearrange.

$x > \dfrac{\log \dfrac{5}{4} + \log 3}{2\log 3}$ — Simplify.

$x > \dfrac{\log \dfrac{15}{4}}{\log 9}$

EXERCISE 2F

1. Solve the inequalities, giving your answer in terms of base 10 logarithms.

   a $2^x < 5$
   b $5^x \geqslant 7$
   c $\left(\dfrac{2}{3}\right)^x > 3$
   d $0.8^x < 0.3$

2. Solve the inequalities, giving your answer in terms of base 10 logarithms.

   a $8^{5-x} < 10$
   b $3^{2x+5} > 20$
   c $2 \times 5^{2x+1} \leqslant 3$
   d $7 \times \left(\dfrac{5}{6}\right)^{3-x} > 4$

3. Solve $5^{x^2} > 2^x$, giving your answer in terms of logarithms.

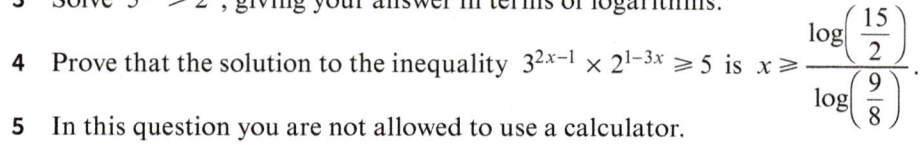

4. Prove that the solution to the inequality $3^{2x-1} \times 2^{1-3x} \geqslant 5$ is $x \geqslant \dfrac{\log\left(\dfrac{15}{2}\right)}{\log\left(\dfrac{9}{8}\right)}$.

5. In this question you are not allowed to use a calculator.

   You are given that $\log_{10} 4 = 0.60206$ correct to 5 decimal places and that $10^{0.206} < 2$.

   a Find the number of digits in the number 4^{100}.

   b Find the first digit in the number 4^{100}.

2.7 Natural logarithms

There is another type of logarithm to a special base called e.

The number e is an irrational number and $e \approx 2.718$.

The function $y = e^x$ is called the **natural exponential function**.

Logarithms to the base of e are called **natural logarithms**.

$\ln x$ is used to represent $\log_e x$.

> **FAST FORWARD**
>
> The number e is a very important number in Mathematics as it has very special properties. You will learn about these special properties in Chapters 4 and 5.

> **KEY POINT 2.5**
>
> If $y = e^x$ then $x = \ln y$

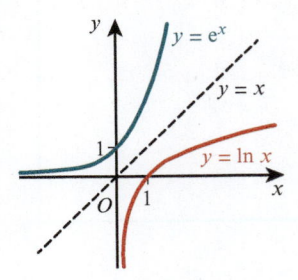

The curve $y = \ln x$ is the reflection of $y = e^x$ in the line $y = x$.

$y = \ln x$ and $y = e^x$ are inverse functions.

All the rules of logarithms that we have learnt so far also apply for natural logarithms.

We say that $y = e^x \Leftrightarrow x = \log_e y$.

> **DID YOU KNOW?**
>
> The Scottish mathematician John Napier (1550–1617) is credited with the discovery of logarithms. His original studies involved logarithms to the base of $\frac{1}{e}$.

> **DID YOU KNOW?**
>
> The number e is also known as Euler's number. It is named after Leonhard Euler (1707–1783), who devised the following formula for calculating the value of e:
>
> $$e = 1 + \frac{1}{1} + \frac{1}{1 \times 2} + \frac{1}{1 \times 2 \times 3} + \frac{1}{1 \times 2 \times 3 \times 4} + \frac{1}{1 \times 2 \times 3 \times 4 \times 5} + \ldots$$

Chapter 2: Logarithmic and exponential functions

EXERCISE 2G

1 Use a calculator to evaluate correct to 3 significant figures:

   a e^3 **b** $e^{2.7}$ **c** $e^{0.8}$ **d** e^{-2}

2 Use a calculator to evaluate correct to 3 significant figures:

   a $\ln 3$ **b** $\ln 1.4$ **c** $\ln 0.9$ **d** $\ln 0.15$

3 Without using a calculator, find the value of:

   a $e^{\ln 2}$ **b** $e^{\frac{1}{2}\ln 9}$ **c** $5e^{\ln 6}$ **d** $e^{-\ln\frac{1}{2}}$

4 Solve.

   a $e^{\ln x} = 5$ **b** $\ln e^x = 15$ **c** $e^{3\ln x} = 64$ **d** $e^{-\ln x} = 3$

5 Solve, giving your answers correct to 3 significant figures.

   a $e^x = 18$ **b** $e^{2x} = 25$ **c** $e^{x+1} = 8$ **d** $e^{2x-3} = 16$

6 Solve, giving your answers in terms of natural logarithms.

   a $e^x = 13$ **b** $e^{3x} = 7$ **c** $e^{2x-1} = 6$ **d** $e^{\frac{1}{2}x+3} = 4$

7 Solve, giving your answers in terms of natural logarithms.

   a $e^x > 10$ **b** $e^{5x-2} \leqslant 35$ **c** $5 \times e^{2x+3} < 1$

8 Solve, giving your answers correct to 3 significant figures.

   a $\ln x = 5$ **b** $\ln x = -4$ **c** $\ln(x-2) = 6$ **d** $\ln(2x+1) = -2$

9 Solve, giving your answers correct to 3 significant figures.

   a $\ln(3 - x^2) = 2\ln x$ **b** $2\ln(x+2) - \ln x = \ln(2x-1)$

   c $2\ln(x+1) = \ln(2x+3)$ **d** $\ln(2x+1) = 2\ln x + \ln 5$

   e $\ln(x+2) - \ln x = 1$ **f** $\ln(2 + x^2) = 1 + 2\ln x$

10 Express y in terms of x for each of these equations.

   a $2\ln(y+1) - \ln y = \ln(x+y)$ **b** $\ln(y+2) - \ln y = 1 + 2\ln x$

11 Solve, giving your answers in exact form.

   a $e^{2x} + 2e^x - 15 = 0$ **b** $e^{2x} - 5e^x + 6 = 0$

   c $6e^{2x} - 13e^x - 5 = 0$ **d** $e^x - 21e^{-x} = 4$

12 Given that the function f is defined as $f : x \mapsto 5e^x + 2$ for $x \in \mathbb{R}$, find an expression for $f^{-1}(x)$.

13 Solve $2\ln(3 - e^{2x}) = 1$, giving your answer correct to 3 significant figures.

14 Solve the simultaneous equations, giving your answers in exact form.

   a $2\ln x + \ln y = 1 + \ln 5$ **b** $e^{3x+4y} = 2e^{2x-y}$
   $\ln 10x - \ln y = 2 + \ln 2$ $e^{2x+y} = 8e^{x+6y}$

15 Solve $\ln(2x+1) \leqslant \ln(x+4)$.

2.8 Transforming a relationship to linear form

When we collect experimental data for two variables we often wish to find a mathematical relationship connecting the variables. When the data plots on a graph lie on a straight line the relationship can be easily found using the equation $y = mx + c$, where m is the gradient and c is the y-intercept. It is more common, however, for the data plots to lie on a curve.

Logarithms can be used to convert some curves into straight lines.

This is the case for relationships of the form $y = kx^n$ and $y = k(a^x)$ where k, a and n are constants.

Logarithms to any base can be used but it is more usual to use those that are most commonly available on calculators, which are natural logarithms or logarithms to the base 10.

WORKED EXAMPLE 2.15

Convert $y = ae^{bx}$, where a and b are constants, into the form $Y = mX + c$.

Answer

$$y = ae^{bx}$$

Taking natural logarithms of both sides gives:

$$\ln y = \ln(ae^{bx})$$
$$\ln y = \ln a + \ln e^{bx}$$
$$\ln y = \ln a + bx$$
$$\ln y = bx + \ln a$$

Now compare $\ln y = bx + \ln a$ with $Y = mX + c$:

$$\boxed{\ln y} = b \; \boxed{x} + \ln a$$
$$Y \;\; = \;\; m \;\; X \;\; + \;\; c$$

The non-linear equation $y = ae^{bx}$ becomes the linear equation:
$Y = mX + c$, where $Y = \ln y$, $X = x$, $m = b$ and $c = \ln a$.

It is important to note:

- The variables X and Y in $Y = mX + c$ must contain only the original variables x and y. (They must **not** contain the unknown constants a and b.)
- The constants m and c must contain only the original unknown constants a and b. (They must **not** contain the variables x and y.)

WORKED EXAMPLE 2.16

The variables x and y satisfy the equation $y = a\left(e^{-bx^2}\right)$, where a and b are constants. The graph of $\ln y$ against x^2 is a straight line passing through the points $(0.55, 0.84)$ and $(1.72, 0.26)$ as shown in the diagram.
Find the value of a and the value of b correct to 2 decimal places.

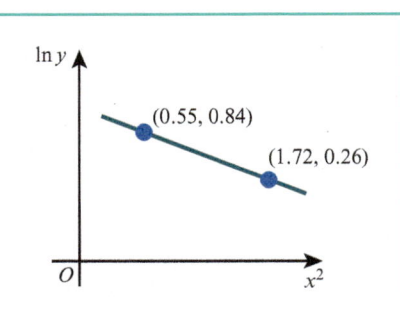

Chapter 2: Logarithmic and exponential functions

Answer

$$y = a \times e^{-bx^2}$$

Taking natural logarithms of both sides gives:

$$\ln y = \ln(a \times e^{-bx^2})$$
$$\ln y = \ln a + \ln e^{-bx^2}$$
$$\ln y = \ln a - bx^2$$
$$\ln y = -bx^2 + \ln a$$

Now compare $\ln y = -b\,x^2 + \ln a$ with $Y = mX + c$:

$$\boxed{\ln y} = -b \boxed{x^2} + \ln a$$
$$Y = m \quad X \quad + \quad c$$

Gradient $= m = \dfrac{0.26 - 0.84}{1.72 - 0.55} = -0.4957\ldots$

$\therefore b = 0.50$ to 2 decimal places

Using $Y = mX + c$, $X = 0.55$, $Y = 0.84$ and $m = -0.4957\ldots$

$$0.84 = -0.4957\ldots \times 0.55 + c$$
$$c = 1.1126\ldots$$
$$\therefore \ln a = 1.1126\ldots$$
$$a = e^{1.1126\ldots}$$

Hence $a = 3.04$ and $b = 0.50$ to 2 decimal places.

> **DID YOU KNOW?**
>
> When one variable on a graph is a log but the other variable is not a log, it is called a semi-log graph. When both variables on a graph are logs, it is called a log–log graph.

WORKED EXAMPLE 2.17

x	5	10	20	40	80
y	2593	1596	983	605	372

The table shows experimental values of the variables x and y.

a By plotting a suitable straight-line graph, show that x and y are related by the equation $y = k \times x^n$, where k and n are constants.

b Use your graph to estimate the value of k and the value of n.

Answer

a $y = k \times x^n$ .... Take natural logarithms of both sides.
$\ln y = \ln(k \times x^n)$ .... Use the multiplication law.
$\ln y = \ln k + \ln x^n$
$\ln y = n \ln x + \ln k$ .... Use the power law.

Now compare $\ln y = n \ln x + \ln k$ with $Y = mX + c$:

$$\ln y = n \ln x + \ln k$$
$$Y = m \ X + \ c$$

Hence the graph of $\ln y$ against $\ln x$ needs to be drawn where:
- gradient = n
- intercept on vertical axis = $\ln k$

Table of values is

$\ln x$	1.61	2.30	3.00	3.69	4.38
$\ln y$	7.86	7.38	6.89	6.41	5.92

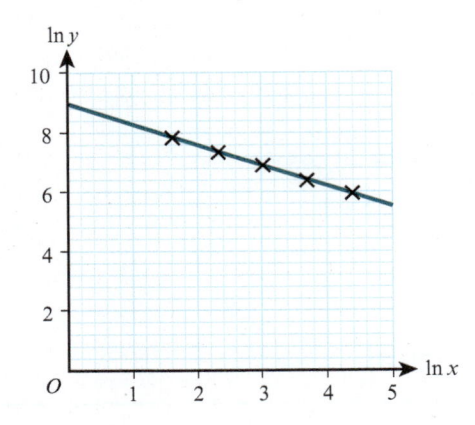

The points form an approximate straight line, so x and y are related by the equation $y = k \times x^n$.

b n = gradient $\approx \dfrac{5.92 - 7.86}{4.38 - 1.61} \approx -0.70$ to 2 significant figures.

$\ln k$ = intercept on vertical axis

$\ln k \approx 9$

$k \approx e^9$

$k \approx 8100$ to 2 significant figures.

DID YOU KNOW?

The distributions of a diverse range of natural and man-made phenomena approximately follow the power law $y = ax^b$. These include the size of craters on the moon, avalanches, species extinction, body mass, income and populations of cities.

Chapter 2: Logarithmic and exponential functions

EXERCISE 2H

1 Given that a and b are constants, use logarithms to change each of these non-linear equations into the form $Y = mX + c$. State what the variables X and Y and the constants m and c represent. (Note: there might be more than one method to do this.)

  a $y = e^{ax+b}$
  b $y = 10^{ax-b}$
  c $y = a(x^{-b})$
  d $y = a(b^x)$

  E e $a = e^{x^2+by}$
  f $x^a y^b = 8$
  g $xa^y = b$
  h $y = a(e^{-bx})$

2 The variables x and y satisfy the equation $y = a \times x^n$, where a and n are constants. The graph of $\ln y$ against $\ln x$ is a straight line passing through the points $(0.31, 4.02)$ and $(1.83, 3.22)$ as shown in the diagram. Find the value of a and the value of n correct to 2 significant figures.

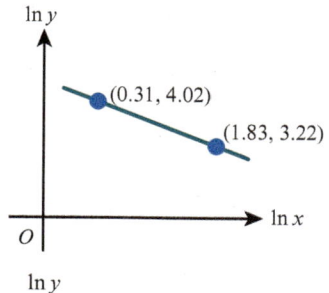

3 The variables x and y satisfy the equation $y = k \times e^{n(x-2)}$, where k and n are constants. The graph of $\ln y$ against x is a straight line passing through the points $(1, 1.84)$ and $(7, 4.33)$ as shown in the diagram. Find the value of k and the value of n correct to 2 significant figures.

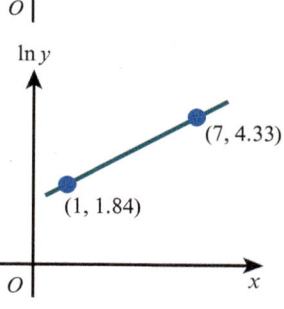

4 Variables x and y are related so that, when $\log_{10} y$ is plotted on the vertical axis and x is plotted on the horizontal axis, a straight-line graph passing through the points $(2, 5)$ and $(6, 11)$ is obtained.

  a Express $\log_{10} y$ in terms of x.
  b Express y in terms of x, giving your answer in the form $y = a \times 10^{bx}$.

5 Variables x and y are related so that, when $\ln y$ is plotted on the vertical axis and $\ln x$ is plotted on the horizontal axis, a straight-line graph passing through the points $(2, 4)$ and $(5, 13)$ is obtained.

  a Express $\ln y$ in terms of x.
  b Express y in terms of x.

6 The variables x and y satisfy the equation $5^{2y} = 3^{2x+1}$. By taking natural logarithms, show that the graph of $\ln y$ against $\ln x$ is a straight line, and find the exact value of the gradient of this line and state the coordinates of the point at which the line cuts the y-axis.

7 The mass, m grams, of a radioactive substance is given by the formula $m = m_0 e^{-kt}$, where t is the time in days after the mass was first recorded and m_0 and k are constants.

  The table below shows experimental values of t and m.

t	10	20	30	40	50
m	40.9	33.5	27.4	22.5	18.4

  a Draw the graph of $\ln m$ against t.
  b Use your graph to estimate the value of m_0 and k.
  c The half-life of a radioactive substance is the time it takes to decay to half of its original mass. Find the half-life of this radioactive substance.

8 The temperature, $T\,°C$, of a hot drink, t minutes after it is made, can be modelled by the equation $T = 25 + ke^{-nt}$, where k and n are constants. The table below shows experimental values of t and T.

t	2	4	6	8	10
T	63.3	57.7	52.8	48.7	45.2

a Convert the equation to a form suitable for drawing a straight-line graph.

b Draw the straight-line graph and use it to estimate the value of k and n.

c Estimate:

i the initial temperature of the drink

ii the time taken for the temperature to reach $28\,°C$

iii the room temperature.

Checklist of learning and understanding

The rules of logarithms

- If $y = a^x$ then $x = \log_a y$.
- $\log_a a = 1$, $\log_a 1 = 0$, $\log_a a^x = x$, $a^{\log_a x} = x$
- Product rule: $\log_a(xy) = \log_a x + \log_a y$
- Division rule: $\log_a\left(\dfrac{x}{y}\right) = \log_a x - \log_a y$
- Power rule: $\log_a(x)^m = m \log_a x$
- Special case of power rule: $\log_a\left(\dfrac{1}{x}\right) = -\log_a x$

Natural logarithms

- Logarithms to the base of e are called natural logarithms.
- $e \approx 2.718$
- $\ln x$ is used to represent $\log_e x$.
- If $y = e^x$ then $x = \ln y$.
- All the rules of logarithms apply for natural logarithms.

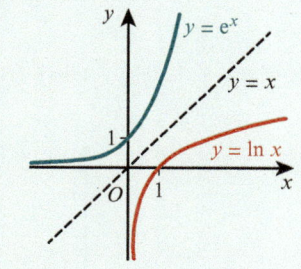

Transforming a relationship to linear form

- Logarithms can be used to convert relationships of the form $y = kx^n$ and $y = k(a^x)$, where k, a and n are constants, into straight lines of the form $Y = mX + c$, where X and Y are functions of x and of y.

END-OF-CHAPTER REVIEW EXERCISE 2

1. Solve the inequality $2^x > 7$, giving your answer in terms of logarithms. [2]

2. Given that $\ln p = 2\ln q - \ln(3+q)$ and that $q > 0$, express p in terms of q not involving logarithms. [3]

3. Solve the inequality $3 \times 2^{3x+2} < 8$, giving your answer in terms of logarithms. [4]

4. Use logarithms to solve the equation
$$5^{x+3} = 7^{x-1},$$
giving the answer correct to 3 significant figures. [4]

Cambridge International AS & A Level Mathematics 9709 Paper 21 Q1 November 2015

5. Solve the equation $6(4^x) - 11(2^x) + 4 = 0$, giving your answers for x in terms of logarithms where appropriate. [5]

6. Solve the equation $\ln(5x+4) = 2\ln x + \ln 6$. [5]

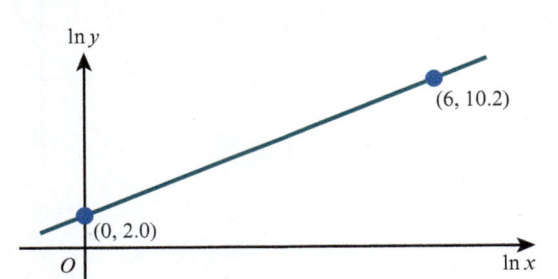

7. The variables x and y satisfy the equation $y = Kx^m$, where K and m are constants. The graph of $\ln y$ against $\ln x$ is a straight line passing through the points $(0, 2.0)$ and $(6, 10.2)$, as shown in the diagram. Find the values of K and m, correct to 2 decimal places. [5]

Cambridge International AS & A Level Mathematics 9709 Paper 21 Q3 June 2011

8. i Given that $y = 2^x$, show that the equation
$$2^x + 3(2^{-x}) = 4$$
can be written in the form
$$y^2 - 4y + 3 = 0.$$ [3]

   ii Hence solve the equation
$$2^x + 3(2^{-x}) = 4,$$
giving the values of x correct to 3 significant figures where appropriate. [3]

Cambridge International AS & A Level Mathematics 9709 Paper 21 Q5 June 2010

9. Given that $(1.2)^x = 6^y$, use logarithms to find the value of $\dfrac{x}{y}$ correct to 3 significant figures. [3]

10. The polynomial $f(x)$ is defined by
$$f(x) = 12x^3 + 25x^2 - 4x - 12.$$

    i Show that $f(-2) = 0$ and factorise $f(x)$ completely. [4]

    ii Given that $12 \times 27^y + 25 \times 9^y - 4 \times 3^y - 12 = 0$, state the value of 3^y and hence find y correct to 3 significant figures. [3]

Cambridge International A Level Mathematics 9709 Paper 31 Q4 June 2011

11 Solve the equation $|4 - 2^x| = 10$, giving your answer correct to 3 significant figures. [3]

Cambridge International A Level Mathematics 9709 Paper 31 Q1 June 2012

12 Use logarithms to solve the equation $e^x = 3^{x-2}$, giving your answer correct to 3 decimal places. [3]

Cambridge International A Level Mathematics 9709 Paper 31 Q1 November 2014

13 Using the substitution $u = 3^x$, solve the equation $3^x + 3^{2x} = 3^{3x}$ giving your answer correct to 3 significant figures. [5]

Cambridge International A Level Mathematics 9709 Paper 31 Q2 November 2015

14 The variables x and y satisfy the equation $5^y = 3^{2x-4}$.

   a By taking natural logarithms, show that the graph of y against x is a straight line and find the exact value of the gradient of this line. [3]

   b This line intersects the x-axis at P and the y-axis at Q. Find the exact coordinates of the midpoint of PQ. [3]

15

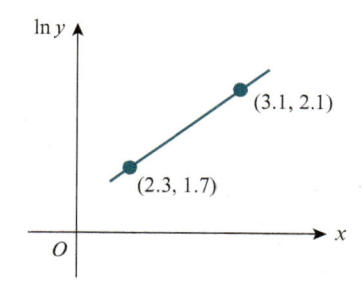

The variables x and y satisfy the equation $y = K(b^x)$, where K and b are constants. The graph of $\ln y$ against x is a straight line passing through the points $(2.3, 1.7)$ and $(3.1, 2.1)$, as shown in the diagram. Find the values of K and b, correct to 2 decimal places. [6]

Chapter 3
Trigonometry

In this chapter you will learn how to:

- understand the relationship of the secant, cosecant and cotangent functions to cosine, sine and tangent, and use properties and graphs of all six trigonometric functions for angles of any magnitude
- use trigonometrical identities for the simplification and exact evaluation of expressions and in the course of solving equations, and select an identity or identities appropriate to the context, showing familiarity in particular with the use of:
  - $\sec^2\theta = 1 + \tan^2\theta$ and $\csc^2\theta = 1 + \cot^2\theta$
  - the expansion of $\sin(A \pm B)$, $\cos(A \pm B)$ and $\tan(A \pm B)$
  - the formulae for $\sin 2A$, $\cos 2A$ and $\tan 2A$
  - the expression of $a\sin\theta + b\cos\theta$ in the forms $R\sin(\theta \pm \alpha)$ and $R\cos(\theta \pm \alpha)$.

Chapter 3: Trigonometry

PREREQUISITE KNOWLEDGE

Where it comes from	What you should be able to do	Check your skills

Pure Mathematics 1 Coursebook, Chapter 5	Sketch graphs of the sine, cosine and tangent functions.	1 Sketch the graphs of these functions for $0° \leq x \leq 360°$.
		  a $y = 2\sin x + 1$
		  b $y = \tan(x - 90°)$
		  c $y = \cos(2x + 90°)$

Pure Mathematics 1 Coursebook, Chapter 5	Find exact values of the sine, cosine and tangent of $30°$, $45°$ and $60°$ and related angles.	2 Find the exact values of:
		  a $\sin 120°$
		  b $\cos 225°$
		  c $\tan 150°$

Pure Mathematics 1 Coursebook, Chapter 5	Solve trigonometric equations using the identities $\tan\theta \equiv \dfrac{\sin\theta}{\cos\theta}$ and $\sin^2\theta + \cos^2\theta \equiv 1$.	3 Solve for $0° \leq x \leq 360°$.
		  a $5\sin x = 3\cos x$
		  b $2\cos^2 x - \sin x = 1$

Why do we study trigonometry?

In addition to the three main trigonometrical functions (sine, cosine and tangent) there are three more commonly used trigonometrical functions (cosecant, secant and cotangent) that we will learn about. These are sometimes referred to as the **reciprocal trigonometrical functions**.

We will also learn about the compound angle formulae, which form a basis for numerous important mathematical techniques. These techniques include solving equations, deriving further trigonometric identities, the addition of different sine and cosine functions and deriving rules for differentiating and integrating trigonometric functions, which we will cover later in this book.

As you know, scientists and engineers represent oscillations and waves using trigonometric functions. These functions also have further uses in navigation, engineering and physics.

> **REWIND**
>
> This chapter builds on the work we covered in the Pure Mathematics 1 Coursebook, Chapter 5.

> **WEB LINK**
>
> Explore the *Trigonometry: Compound angles* station on the Underground Mathematics website.

3.1 The cosecant, secant and cotangent ratios

There are a total of six trigonometric ratios. You have already learnt about the ratios sine, cosine and tangent. In this section you will learn about the other three ratios, which are **cosecant** (cosec), **secant** (sec) and **cotangent** (cot).

> **KEY POINT 3.1**
>
> The three reciprocal trigonometric ratios are defined as:
>
> $$\operatorname{cosec}\theta = \frac{1}{\sin\theta} \qquad \sec\theta = \frac{1}{\cos\theta} \qquad \cot\theta = \frac{1}{\tan\theta}\left(=\frac{\cos\theta}{\sin\theta}\right)$$

Consider the right-angled triangle:

$$\sin\theta = \frac{y}{r} \qquad \cos\theta = \frac{x}{r} \qquad \tan\theta = \frac{y}{x}$$

$$\operatorname{cosec}\theta = \frac{r}{y} \qquad \sec\theta = \frac{r}{x} \qquad \cot\theta = \frac{x}{y}$$

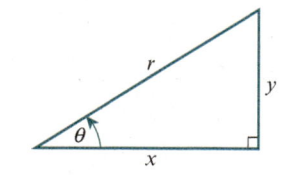

The following identities can be found from this triangle:

$x^2 + y^2 = r^2$ Divide both sides by x^2.

$1 + \left(\dfrac{y}{x}\right)^2 = \left(\dfrac{r}{x}\right)^2$ Use $\dfrac{y}{x} = \tan\theta$ and $\dfrac{r}{x} = \sec\theta$.

$1 + \tan^2\theta \equiv \sec^2\theta$

$x^2 + y^2 = r^2$ Divide both sides by y^2.

$\left(\dfrac{x}{y}\right)^2 + 1 = \left(\dfrac{r}{y}\right)^2$ Use $\dfrac{x}{y} = \cot\theta$ and $\dfrac{r}{y} = \operatorname{cosec}\theta$.

$\cot^2\theta + 1 \equiv \operatorname{cosec}^2\theta$

We will need these important identities to solve trigonometric equations later in this section.

> **KEY POINT 3.2**
>
> $1 + \tan^2\theta \equiv \sec^2\theta$
>
> $\cot^2\theta + 1 \equiv \operatorname{cosec}^2\theta$

The graphs of $y = \sin\theta$ and $y = \operatorname{cosec}\theta$

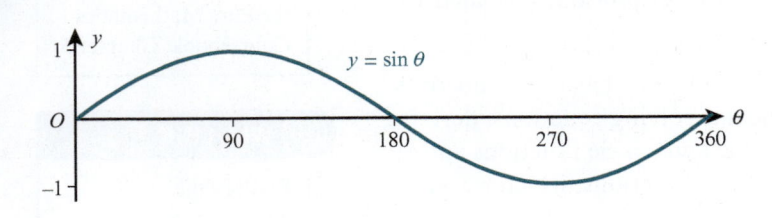

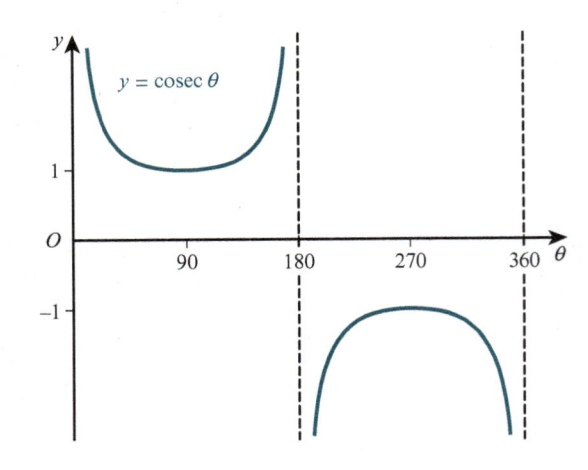

The function $y = \operatorname{cosec}\theta$ is the reciprocal of the function $y = \sin\theta$.

The function $y = \sin\theta$ is zero when $\theta = 0°, 180°, 360°, \ldots$, this means that $y = \operatorname{cosec}\theta$ is **not** defined at each of these points because $\dfrac{1}{0}$ is not defined. Hence vertical asymptotes need to be drawn at each of these points.

The graphs of $y = \cos\theta$ and $y = \sec\theta$

We can find the graph of $y = \sec\theta$ in a similar way from the graph of $y = \cos\theta$.

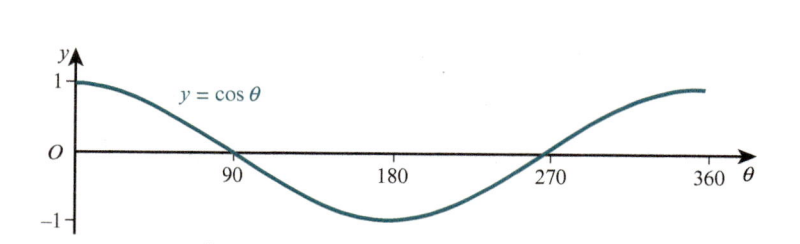

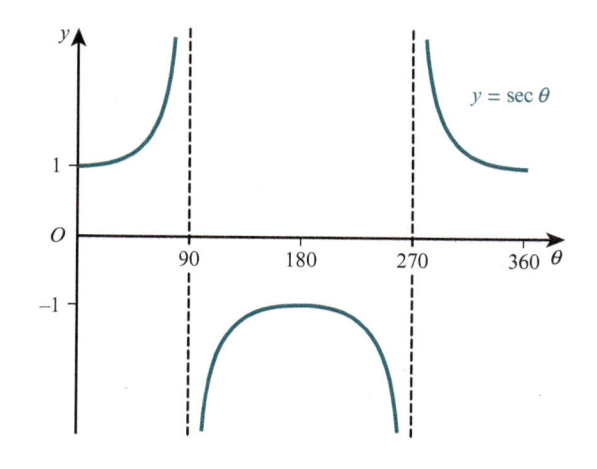

The graphs of $y = \tan\theta$ and $y = \cot\theta$

We can find the graph of $y = \cot\theta$ from the graph of $y = \tan\theta$.

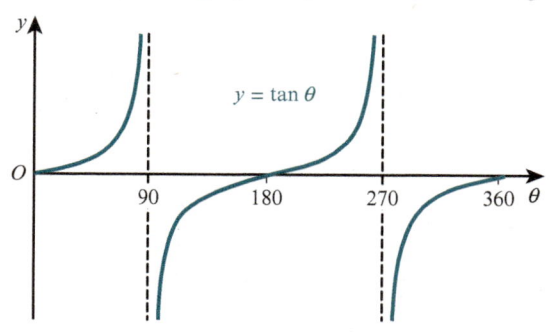

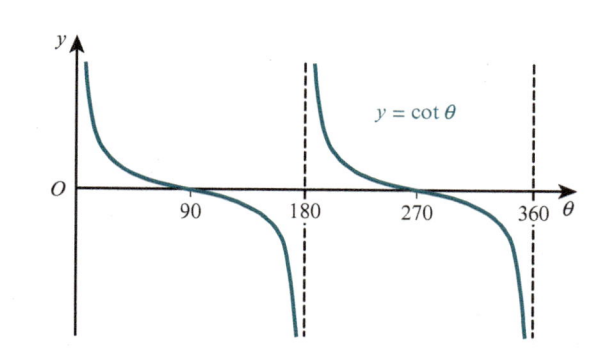

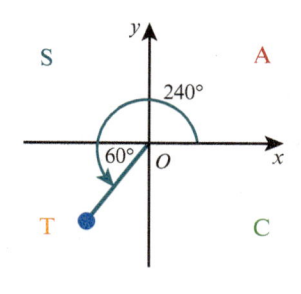

WORKED EXAMPLE 3.1

Find the exact value of $\operatorname{cosec} 240°$.

Answer

$$\operatorname{cosec} 240° = \frac{1}{\sin 240°}$$

$$= \frac{1}{-\sin 60°}$$

$$= \frac{1}{-\frac{\sqrt{3}}{2}}$$

$$= -\frac{2}{\sqrt{3}}$$

WORKED EXAMPLE 3.2

Solve $\sec^2 x - \tan x - 3 = 0$ for $0° \leqslant x \leqslant 360°$.

Answer

$\sec^2 x - \tan x - 3 = 0$ .... Use $1 + \tan^2 x \equiv \sec^2 x$.

$\tan^2 x - \tan x - 2 = 0$ .... Factorise.

$(\tan x - 2)(\tan x + 1) = 0$

$\tan x = 2$ or $\tan x = -1$

$\tan x = 2 \Rightarrow x = 63.4°$ (calculator)

or $x = 63.4° + 180° = 243.4°$

$\tan x = -1 \Rightarrow x = -45°$ (not in required range)

or $x = -45° + 180° = 135°$

or $x = 135° + 180° = 315°$

The values of x are $63.4°, 135°, 243.4°, 315°$.

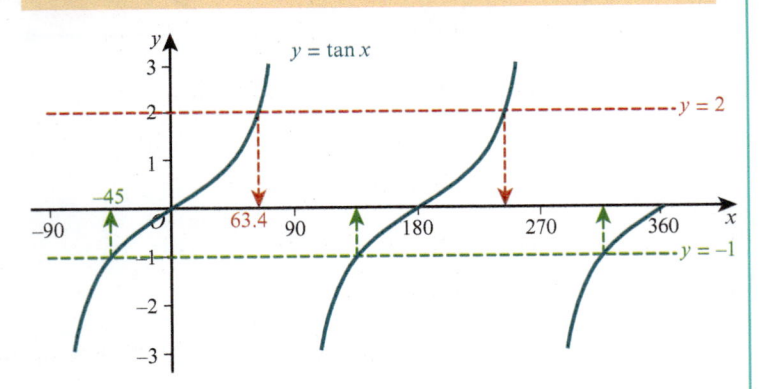

EXERCISE 3A

1 Find the exact values of:

   a $\sec 60°$ **b** $\operatorname{cosec} 45°$ **c** $\cot 120°$ **d** $\sec 300°$

   e $\operatorname{cosec} 135°$ **f** $\cot 330°$ **g** $\sec 150°$ **h** $\cot(-30°)$.

2 Find the exact values of:

   a $\operatorname{cosec} \dfrac{\pi}{6}$ **b** $\cot \dfrac{\pi}{3}$ **c** $\sec \dfrac{\pi}{4}$ **d** $\cot \dfrac{2\pi}{3}$

   e $\sec \dfrac{5\pi}{6}$ **f** $\operatorname{cosec} \dfrac{7\pi}{4}$ **g** $\cot \dfrac{4\pi}{3}$ **h** $\sec\left(-\dfrac{\pi}{6}\right)$.

3 Solve each equation for $0° \leqslant x \leqslant 360°$.

   a $\sec x = 3$ **b** $\cot x = 0.8$ **c** $\operatorname{cosec} x = -3$ **d** $3 \sec x - 4 = 0$

4 Solve each equation for $0 \leqslant x \leqslant 2\pi$.

   a $\operatorname{cosec} x = 2$ **b** $\sec x = -1$ **c** $\cot x = 2$ **d** $2 \cot x + 5 = 0$

5 Solve each equation for $0° \leqslant x \leqslant 180°$.

   a $\operatorname{cosec} 2x = 1.2$ **b** $\sec 2x = 4$ **c** $\cot 2x = 1$ **d** $2 \sec 2x = 3$

6 Solve each equation for the given domains.

   a $\operatorname{cosec}(x - 30°) = 2$ for $0° \leqslant x \leqslant 360°$ **b** $\sec(2x + 60°) = -1.5$ for $0° \leqslant x \leqslant 180°$

   c $\cot\left(x + \dfrac{\pi}{4}\right) = 2$ for $0 \leqslant x \leqslant 2\pi$ **d** $2 \operatorname{cosec}(2x - 1) = 3$ for $-\pi \leqslant x \leqslant \pi$

7 Solve each equation for $-180° \leqslant x \leqslant 180°$.

   a $\operatorname{cosec}^2 x = 4$
   b $\sec^2 x = 9$
   c $9\cot^2 \dfrac{1}{2} x = 4$
   d $\sec x = \cos x$
   e $\operatorname{cosec} x = \sec x$
   f $2\tan x = 3\operatorname{cosec} x$

8 Solve each equation for $0° \leqslant \theta \leqslant 360°$.

   a $2\tan^2 \theta - 1 = \sec \theta$
   b $3\operatorname{cosec}^2 \theta = 13 - \cot \theta$
   c $2\cot^2 \theta - \operatorname{cosec} \theta = 13$
   d $\operatorname{cosec} \theta + \cot \theta = 2\sin \theta$
   e $\tan^2 \theta + 3\sec \theta = 0$
   f $\sqrt{3} \sec^2 \theta = \operatorname{cosec} \theta$

9 Solve each equation for $0° \leqslant \theta \leqslant 180°$.

   a $\sec \theta = 3\cos \theta - \tan \theta$
   b $2\sec^2 2\theta = 3\tan 2\theta + 1$
   c $\sec^4 \theta + 2 = 6\tan^2 \theta$
   d $2\cot^2 2\theta + 7\operatorname{cosec} 2\theta = 2$

10 Solve each equation for $0 \leqslant \theta \leqslant 2\pi$.

   a $\tan^2 \theta + 3\sec \theta + 3 = 0$
   b $3\cot^2 \theta + 5\operatorname{cosec} \theta + 1 = 0$

11 a Sketch each of the following functions for the interval $0 \leqslant x \leqslant 2\pi$.

   i $y = 1 + \sec x$
   ii $y = \cot 2x$
   iii $y = 2\operatorname{cosec}\left(x - \dfrac{\pi}{2}\right)$
   iv $y = 1 - \sec x$
   v $y = 1 + \operatorname{cosec} \dfrac{1}{2} x$
   vi $y = \sec\left(2x + \dfrac{\pi}{4}\right)$

   b Write down the equation of each of the asymptotes for your graph for **part a vi**.

12 Prove each of these identities.

   a $\sin x + \cos x \cot x \equiv \operatorname{cosec} x$
   b $\operatorname{cosec} x - \sin x \equiv \cos x \cot x$
   c $\sec x \operatorname{cosec} x - \cot x \equiv \tan x$
   d $(1 + \sec x)(\operatorname{cosec} x - \cot x) \equiv \tan x$

13 Prove each of these identities.

   a $\dfrac{1}{\tan x + \cot x} \equiv \sin x \cos x$
   b $\sec^2 x + \sec x \tan x \equiv \dfrac{1}{1 - \sin x}$
   c $\dfrac{1 - \cos^2 x}{\sec^2 x - 1} \equiv 1 - \sin^2 x$
   d $\dfrac{1 + \tan^2 x}{\tan x} \equiv \sec x \operatorname{cosec} x$
   e $\dfrac{\sin x}{1 - \cos^2 x} \equiv \operatorname{cosec} x$
   f $\dfrac{1 + \sin x}{1 - \sin x} \equiv (\tan x + \sec x)^2$
   g $\dfrac{1}{1 + \cos x} + \dfrac{1}{1 - \cos x} \equiv 2\operatorname{cosec}^2 x$
   h $\dfrac{\cos x}{1 + \sin x} + \dfrac{\cos x}{1 - \sin x} \equiv 2\sec x$

14 Solve each equation for $0° \leqslant \theta \leqslant 180°$.

   a $6\sec^3 \theta - 5\sec^2 \theta - 8\sec \theta + 3 = 0$
   b $2\cot^3 \theta + 3\operatorname{cosec}^2 \theta - 8\cot \theta = 0$

3.2 Compound angle formulae

> **EXPLORE 3.1**
>
> 1. Sunita says that: $\sin(A+B) = \sin A + \sin B$.
>
>   Use a calculator with $A = 20°$ and $B = 50°$ to prove that Sunita is wrong.
>
> 2. The diagram shows $\triangle PQR$.
>
> RX is perpendicular to PQ.
>
> $\angle A$ and $\angle B$ are acute angles.
>
> $QR = p$, $PR = q$ and $RX = h$.
>
>   a. Write down an expression for h in terms of:
>
>   i. q and A ii. p and B
>
>   b. Using the formula area of triangle $= \dfrac{1}{2}ab\sin C$, we can write:
>
>   Area of $\triangle PQR = \dfrac{1}{2}pq\sin(A+B)$.
>
>   Find an expression in terms of p, q, A and B for the area of:
>
>   i. $\triangle PRX$ ii. $\triangle QRX$
>
>   c. Use your expressions for the areas of triangles PQR, PRX and QRX to prove that:
>
> $\sin(A+B) \equiv \sin A \cos B + \cos A \sin B$
>
> 3. In the identity $\sin(A+B) \equiv \sin A \cos B + \cos A \sin B$, replace B by $-B$ to show that:
>
> $\sin(A-B) \equiv \sin A \cos B - \cos A \sin B$
>
> 4. In the identity $\sin(A-B) \equiv \sin A \cos B - \cos A \sin B$, replace A by $(90° - A)$ to show that:
>
> $\cos(A+B) \equiv \cos A \cos B - \sin A \sin B$
>
> 5. In the identity $\cos(A+B) \equiv \cos A \cos B - \sin A \sin B$, replace B by $-B$ to show that:
>
> $\cos(A-B) \equiv \cos A \cos B + \sin A \sin B$
>
> 6. By writing $\tan(A+B)$ as $\dfrac{\sin(A+B)}{\cos(A+B)}$ and $\tan(A-B)$ as $\dfrac{\sin(A-B)}{\cos(A-B)}$, show that:
>
> $\tan(A+B) \equiv \dfrac{\tan A + \tan B}{1 - \tan A \tan B}$ and $\tan(A-B) \equiv \dfrac{\tan A - \tan B}{1 + \tan A \tan B}$

KEY POINT 3.3

Summarising, the six **compound angle formulae** are:

$\sin(A+B) \equiv \sin A \cos B + \cos A \sin B$

$\sin(A-B) \equiv \sin A \cos B - \cos A \sin B$

$\cos(A+B) \equiv \cos A \cos B - \sin A \sin B$

$\cos(A-B) \equiv \cos A \cos B + \sin A \sin B$

$\tan(A+B) \equiv \dfrac{\tan A + \tan B}{1 - \tan A \tan B}$

$\tan(A-B) \equiv \dfrac{\tan A - \tan B}{1 + \tan A \tan B}$

Chapter 3: Trigonometry

> **WORKED EXAMPLE 3.3**
>
> Find the exact value of:
>
>   **a** $\sin 105°$ **b** $\cos 42° \cos 12° + \sin 42° \sin 12°$
>
> **Answer**
>
>   **a** $\sin 105° = \sin(60° + 45°)$
>
> $= \sin 60° \cos 45° + \cos 60° \sin 45°$
>
> $= \left(\dfrac{\sqrt{3}}{2}\right)\left(\dfrac{1}{\sqrt{2}}\right) + \left(\dfrac{1}{2}\right)\left(\dfrac{1}{\sqrt{2}}\right)$
>
> $= \dfrac{\sqrt{3} + 1}{2\sqrt{2}}$
>
>   **b** $\cos 42° \cos 12° + \sin 42° \sin 12°$ Use $\cos(A - B) \equiv \cos A \cos B + \sin A \sin B.$
>
> $= \cos(42° - 12°)$
>
> $= \cos 30°$
>
> $= \dfrac{\sqrt{3}}{2}$

> **WORKED EXAMPLE 3.4**
>
> Given that $\sin A = \dfrac{4}{5}$ and $\cos B = \dfrac{5}{13}$, where A is obtuse and B is acute, find the value of:
>
>   **a** $\sin(A + B)$ **b** $\cos(A - B)$ **c** $\tan(A - B)$
>
> **Answer** 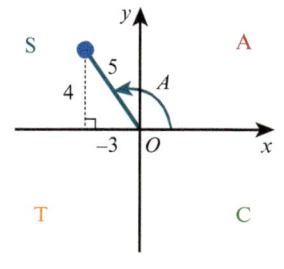 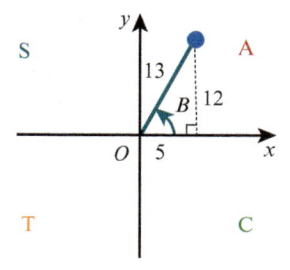
>
> $\sin A = \dfrac{4}{5}, \; \cos A = -\dfrac{3}{5}, \; \tan A = -\dfrac{4}{3}$ $\sin B = \dfrac{12}{13}, \; \cos B = \dfrac{5}{13}, \; \tan B = \dfrac{12}{5}$
>
>   **a** $\sin(A + B) = \sin A \cos B + \cos A \sin B$
>
> $= \left(\dfrac{4}{5}\right)\left(\dfrac{5}{13}\right) + \left(-\dfrac{3}{5}\right)\left(\dfrac{12}{13}\right)$
>
> $= -\dfrac{16}{65}$
>
>   **b** $\cos(A - B) = \cos A \cos B + \sin A \sin B$
>
> $= \left(-\dfrac{3}{5}\right)\left(\dfrac{5}{13}\right) + \left(\dfrac{4}{5}\right)\left(\dfrac{12}{13}\right)$
>
> $= \dfrac{33}{65}$

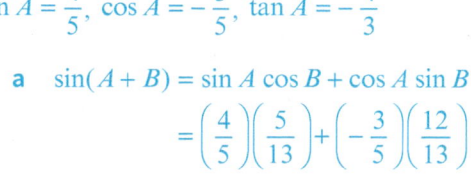

c $\tan(A-B) = \dfrac{\tan A - \tan B}{1 + \tan A \tan B}$

$= \dfrac{\left(-\dfrac{4}{3}\right) - \left(\dfrac{12}{5}\right)}{1 + \left(-\dfrac{4}{3}\right)\left(\dfrac{12}{5}\right)}$

$= \dfrac{56}{33}$

WORKED EXAMPLE 3.5

Solve the equation $\sin(60° - x) = 2\sin x$ for $0° < x < 360°$.

Answer

$\sin(60° - x) = 2\sin x$ Expand $\sin(60° - x)$.

$\sin 60° \cos x - \cos 60° \sin x = 2\sin x$ Use $\sin 60° = \dfrac{\sqrt{3}}{2}$ and $\cos 60° = \dfrac{1}{2}$.

$\dfrac{\sqrt{3}}{2}\cos x - \dfrac{1}{2}\sin x = 2\sin x$ Multiply both sides by 2.

$\sqrt{3}\cos x - \sin x = 4\sin x$ Add $\sin x$ to both sides.

$\sqrt{3}\cos x = 5\sin x$ Rearrange to find $\tan x$.

$\dfrac{\sqrt{3}}{5} = \tan x$

$x = 19.1°$ or $x = 180° + 19.1°$

The values of x are $19.1°$ and $199.1°$.

EXERCISE 3B

1. Expand and simplify $\cos(x + 30°)$.

2. Without using a calculator, find the exact value of each of the following:

   a $\sin 20° \cos 70° + \cos 20° \sin 70°$

   b $\sin 172° \cos 37° - \cos 172° \sin 37°$

   c $\cos 25° \cos 35° - \sin 25° \sin 35°$

   d $\cos 99° \cos 69° + \sin 99° \sin 69°$

   e $\dfrac{\tan 25° + \tan 20°}{1 - \tan 25° \tan 20°}$

   f $\dfrac{\tan 82° - \tan 52°}{1 + \tan 82° \tan 52°}$

3. Without using a calculator, find the exact value of:

   a $\sin 75°$ b $\tan 75°$ c $\cos 105°$ d $\tan(-15°)$

   e $\sin\dfrac{\pi}{12}$ f $\cos\dfrac{19\pi}{12}$ g $\cot\dfrac{7\pi}{12}$ h $\sin\dfrac{7\pi}{12}$

4. Given that $\cos x = \dfrac{4}{5}$ and that $0° < x < 90°$, without using a calculator find the exact value of $\cos(x - 60°)$.

5 Given that $\sin A = \dfrac{4}{5}$ and $\sin B = \dfrac{5}{13}$, where A and B are acute, show that $\sin(A+B) = \dfrac{63}{65}$.

6 Given that $\sin A = \dfrac{12}{13}$ and $\sin B = \dfrac{3}{5}$, where A is obtuse and B is acute, find the value of:

   a $\sin(A+B)$ b $\cos(A-B)$ c $\tan(A+B)$

7 Given that $\cos A = -\dfrac{4}{5}$ and $\sin B = -\dfrac{8}{17}$ and that A and B lie in the same quadrant, find the value of:

   a $\sin(A+B)$ b $\cos(A+B)$ c $\tan(A-B)$

8 Given that $\tan A = t$ and that $\tan(A-B) = 2$, find $\tan B$ in terms of t.

9 If $\cos(A-B) = 3\cos(A+B)$, find the exact value of $\tan A \tan B$.

10 a Given that $8 + \operatorname{cosec}^2 \theta = 6\cot\theta$, find the value of $\tan\theta$.

   b Hence find the exact value of $\tan(\theta + 45°)$.

11 a Given that x is acute and that $2\sec^2 x + 7\tan x = 17$, find the exact value of $\tan x$.

   b Hence find the exact value of $\tan(225° - x)$.

12 a Show that the equation $\sin(x + 30°) = 5\cos(x - 60°)$ can be written in the form $\cot x = -\sqrt{3}$.

   b Hence solve the equation $\sin(x + 30°) = 5\cos(x - 60°)$ for $-180° < x < 180°$.

13 Solve each equation for $0° \leqslant x \leqslant 360°$.

   a $\cos(x + 30°) = 2\sin x$ b $\cos(x - 60°) = 3\cos x$

   c $\sin(30° - x) = 4\sin x$ d $\cos(x + 30°) = 2\sin(x + 60°)$

14 Solve each equation for $0° \leqslant x \leqslant 180°$.

   a $2\tan(60° - x) = \tan x$ b $2\tan(45° - x) = 3\tan x$

   c $\sin(x + 60°) = 2\cos(x + 45°)$ d $\sin(x - 45°) = 2\cos(x + 60°)$

   e $\tan(x + 45°) = 6\tan x$ f $\cos(x + 225°) = 2\sin(x - 60°)$

15 Solve the equation $\tan(x - 45°) + \cot x = 2$ for $0° \leqslant x \leqslant 180°$.

P 16 Use the expansions of $\cos(5x + x)$ and $\cos(5x - x)$ to prove that $\cos 6x + \cos 4x \equiv 2\cos 5x \cos x$.

PS 17 Given that $\sin x + \sin y = p$ and $\cos x + \cos y = q$, find an expression for $\cos(x - y)$, in terms of p and q.

3.3 Double angle formulae

We can use the compound angle formulae to derive three new identities.

Consider $\sin(A + B) \equiv \sin A \cos B + \cos A \sin B$

If $B = A$, $\sin(A + A) \equiv \sin A \cos A + \cos A \sin A$

               $\therefore \sin 2A \equiv 2\sin A \cos A$

Consider $\cos(A + B) \equiv \cos A \cos B - \sin A \sin B$

If $B = A$, $\cos(A + A) \equiv \cos A \cos A - \sin A \sin A$

               $\therefore \cos 2A \equiv \cos^2 A - \sin^2 A$

Consider $\quad \tan(A+B) \equiv \dfrac{\tan A + \tan B}{1 - \tan A \tan B}$

If $B = A$, $\quad \tan(A+A) \equiv \dfrac{\tan A + \tan A}{1 - \tan A \tan A}$

$\therefore \tan 2A \equiv \dfrac{2\tan A}{1 - \tan^2 A}$

These three formulae are known as the **double angle formulae**. We can use the identity $\sin^2 A + \cos^2 A \equiv 1$ to write the identity $\cos 2A \equiv \cos^2 A - \sin^2 A$ in two alternative forms:

Using $\sin^2 A \equiv 1 - \cos^2 A$ gives: $\cos 2A \equiv 2\cos^2 A - 1$

Using $\cos^2 A \equiv 1 - \sin^2 A$ gives: $\cos 2A \equiv 1 - 2\sin^2 A$

Summarising, the double angle formulae are:

KEY POINT 3.4

$\sin 2A \equiv 2\sin A \cos A \qquad \cos 2A \equiv \cos^2 A - \sin^2 A \qquad \tan 2A \equiv \dfrac{2\tan A}{1 - \tan^2 A}$

$\equiv 2\cos^2 A - 1$

$\equiv 1 - 2\sin^2 A$

WORKED EXAMPLE 3.6

Given that $\sin x = -\dfrac{3}{5}$ and that $180° < x < 270°$, find the exact value of:

a $\sin 2x$ **b** $\cos 2x$ **c** $\tan \dfrac{x}{2}$

Answer

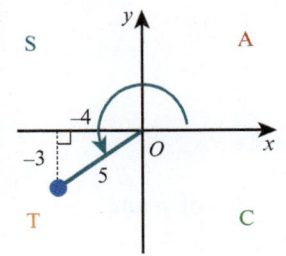

From the diagram:

$\sin x = -\dfrac{3}{5}$, $\cos x = -\dfrac{4}{5}$ and $\tan x = \dfrac{3}{4}$

a $\sin 2x = 2\sin x \cos x$

$= 2\left(-\dfrac{3}{5}\right)\left(-\dfrac{4}{5}\right) = \dfrac{24}{25}$

b $\cos 2x = \cos^2 x - \sin^2 x$

$= \left(-\dfrac{4}{5}\right)^2 - \left(-\dfrac{3}{5}\right)^2 = \dfrac{7}{25}$

c $\tan x = \dfrac{2\tan\frac{x}{2}}{1-\tan^2\frac{x}{2}}$ .... Use $2A = x$ in the double angle formula.

$\dfrac{3}{4} = \dfrac{2\tan\frac{x}{2}}{1-\tan^2\frac{x}{2}}$ .... Rearrange.

$3\tan^2\frac{x}{2} + 8\tan\frac{x}{2} - 3 = 0$ .... Factorise.

$\left(3\tan\frac{x}{2} - 1\right)\left(\tan\frac{x}{2} + 3\right) = 0$ .... Solve.

$\tan\frac{x}{2} = \frac{1}{3}$ or $\tan\frac{x}{2} = -3$

If $180° < x < 270°$, then $90° < \frac{x}{2} < 135°$, which means that $\frac{x}{2}$ is in the second quadrant.

$\therefore \tan\frac{x}{2} = -3$

WORKED EXAMPLE 3.7

Solve the equation $\sin 2x = \sin x$ for $0° \leqslant x \leqslant 360°$.

Answer

$\sin 2x = \sin x$ .... Use $\sin 2x = 2\sin x \cos x$.

$2\sin x \cos x = \sin x$ .... Rearrange.

$2\sin x \cos x - \sin x = 0$ .... Factorise.

$\sin x(2\cos x - 1) = 0$

$\sin x = 0$ or $\cos x = \dfrac{1}{2}$

$\sin x = 0 \Rightarrow x = 0°, 180°, 360°$

$\cos x = \dfrac{1}{2} \Rightarrow x = 60°$

or $x = 360° - 60° = 300°$

The values of x are $0°, 60°, 180°, 300°, 360°$.

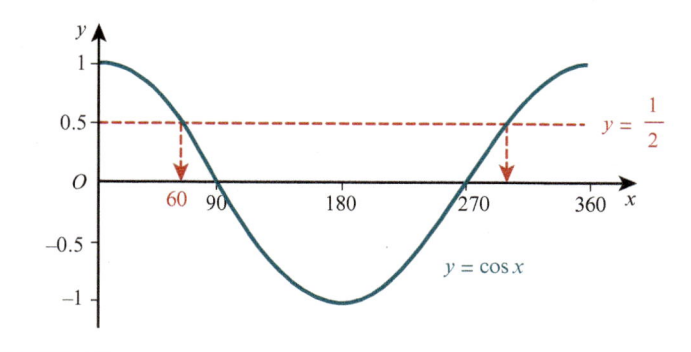

When solving equations involving $\cos 2x$ it is important that we choose the most suitable expansion of $\cos 2x$ for solving that particular equation.

WORKED EXAMPLE 3.8

Solve the equation $\cos 2x + 3\sin x = 2$ for $0 \leqslant x \leqslant 2\pi$.

Answer

$\cos 2x + 3\sin x = 2$ .... Use $\cos 2x = 1 - 2\sin^2 x$.

$1 - 2\sin^2 x + 3\sin x = 2$ .... Rearrange.

$2\sin^2 x - 3\sin x + 1 = 0$ .... Factorise.

$(2\sin x - 1)(\sin x - 1) = 0$

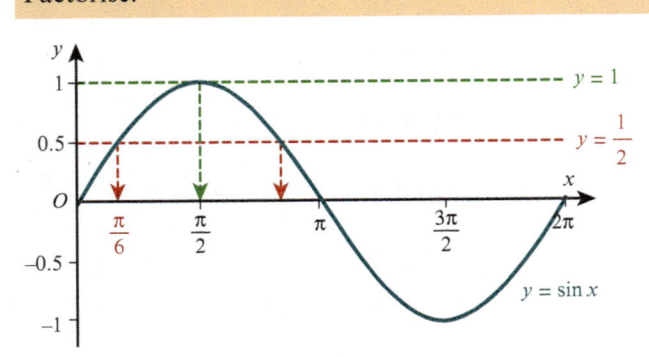

$\sin x = \dfrac{1}{2}$ or $\sin x = 1$

$\sin x = \dfrac{1}{2} \Rightarrow x = \dfrac{\pi}{6}$

or $x = \pi - \dfrac{\pi}{6} = \dfrac{5\pi}{6}$

$\sin x = 1 \Rightarrow x = \dfrac{\pi}{2}$

The values of x are $\dfrac{\pi}{6}, \dfrac{\pi}{2}, \dfrac{5\pi}{6}$.

EXERCISE 3C

1 Express each of the following as a single trigonometric ratio.
   a $2\sin 28° \cos 28°$
   b $2\cos^2 34° - 1$
   c $\dfrac{2\tan 17°}{1 - \tan^2 17°}$

2 Given that $\tan x = \dfrac{4}{3}$, where $0° < x < 90°$, find the exact value of:
   a $\sin 2x$
   b $\cos 2x$
   c $\tan 2x$
   d $\tan 3x$

3 Given that $\cos 2x = -\dfrac{527}{625}$ and that $0° < x < 90°$, find the exact value of:
   a $\sin 2x$
   b $\tan 2x$
   c $\cos x$
   d $\tan x$

4 Given that $\cos x = -\dfrac{3}{5}$ and that $90° < x < 180°$, find the exact value of:
   a $\sin 2x$
   b $\sin 4x$
   c $\tan 2x$
   d $\tan \dfrac{1}{2}x$

5 Given that $3\cos 2x + 17\sin x = 8$, find the exact value of $\sin x$.

6 Solve each equation for $0° \leqslant \theta \leqslant 360°$.
   a $2\sin 2\theta = \cos \theta$
   b $2\cos 2\theta + 3 = 4\cos \theta$
   c $2\cos 2\theta + 1 = \sin \theta$

7 Solve each equation for $0° \leq \theta \leq 180°$.

    a $2\sin 2\theta \tan \theta = 1$ **b** $3\cos 2\theta + \cos \theta = 2$

    c $2\operatorname{cosec} 2\theta + 2\tan \theta = 3\sec \theta$ **d** $2\tan 2\theta = 3\cot \theta$

    e $\tan 2\theta = 4\cot \theta$ **f** $\cot 2\theta + \cot \theta = 3$

8 Express $\cos^2 2x$ in terms of $\cos 4x$.

9 **a** Use the expansions of $\cos(2x + x)$ to show that $\cos 3x \equiv 4\cos^3 x - 3\cos x$.

   b Express $\sin 3x$ in terms of $\sin x$.

10 Solve $\tan 2\theta + 2\tan \theta = 3\cot \theta$ for $0° \leq \theta \leq 180°$.

11 **a** Prove that $\tan \theta + \cot \theta \equiv \dfrac{2}{\sin 2\theta}$.

   b Hence find the exact value of $\tan \dfrac{\pi}{12} + \cot \dfrac{\pi}{12}$.

12 **a** Prove that $2\operatorname{cosec} 2\theta \tan \theta \equiv \sec^2 \theta$.

   b Hence solve the equation $\operatorname{cosec} 2\theta \tan \theta = 2$ for $-\pi < \theta < \pi$.

13 **a** Prove that $\cos 4x + 4\cos 2x \equiv 8\cos^4 x - 3$.

   b Hence solve the equation $2\cos 4x + 8\cos 2x = 3$ for $-\pi \leq x \leq \pi$.

14 **a** Show that $\theta = 18°$ is a root of the equation $\sin 3\theta = \cos 2\theta$.

   b Express $\sin 3\theta$ and $\cos 2\theta$ in terms of $\sin \theta$.

   c Hence show that $\sin 18°$ is a root of the equation $4x^3 - 2x^2 - 3x + 1 = 0$.

   d Hence find the exact value of $\sin 18°$.

(PS) 15 Find the range of values of θ between 0 and 2π for which $\cos 2\theta > \cos \theta$.

(PS) 16 Solve the inequality $\cos 2\theta - 3\sin \theta - 2 \geq 0$ for $0° \leq \theta \leq 360°$.

(PS) 17

In triangle ABC, angle $B = x$, angle $C = 3x$, side $BC = a$, side $AC = b$.

Prove that $a = 4b\cos 2x \cos x$.

(PS) 18 **a** Use the expansions of $\cos(2x + x)$ and $\cos(2x - x)$, to prove that:

$$\cos 3x + \cos x \equiv 2\cos 2x \cos x$$

   b Solve $\cos 3x + \cos 2x + \cos x > 0$ for $0° < x < 360°$.

(PS) 19 Solve the inequality $\cos 4\theta + 3\cos 2\theta + 1 < 0$ for $0° \leq \theta \leq 360°$.

> **TIP**
>
> Solving trigonometric inequalities is not in the Cambridge syllabus. You should, however, have the skills necessary to tackle these more challenging questions.

3.4 Further trigonometric identities

This section builds on our previous work by using the compound angle formulae and the double angle formulae to prove trigonometric identities.

WORKED EXAMPLE 3.9

Prove the identity $\tan 3x \equiv \dfrac{3\tan x - \tan^3 x}{1 - 3\tan^2 x}$.

Answer

$\text{LHS} \equiv \tan(2x + x)$ — Use the compound angle formula for tan.

$\equiv \dfrac{\tan 2x + \tan x}{1 - \tan 2x \tan x}$ — Use the double angle formula.

$\equiv \dfrac{\dfrac{2\tan x}{1 - \tan^2 x} + \tan x}{1 - \tan x \left(\dfrac{2\tan x}{1 - \tan^2 x}\right)}$ — Multiply numerator and denominator by $(1 - \tan^2 x)$.

$\equiv \dfrac{2\tan x + \tan x(1 - \tan^2 x)}{1 - \tan^2 x - 2\tan^2 x}$ — Expand the brackets and simplify.

$\equiv \dfrac{3\tan x - \tan^3 x}{1 - 3\tan^2 x}$

$\equiv \text{RHS}$

WORKED EXAMPLE 3.10

Prove the identity $\dfrac{2\sin(x - y)}{\cos(x - y) - \cos(x + y)} \equiv \cot y - \cot x$.

Answer

$\text{LHS} \equiv \dfrac{2\sin(x - y)}{\cos(x - y) - \cos(x + y)}$ — Use compound angle formulae.

$\equiv \dfrac{2\sin x \cos y - 2\cos x \sin y}{(\cos x \cos y + \sin x \sin y) - (\cos x \cos y - \sin x \sin y)}$ — Simplify denominator.

$\equiv \dfrac{2\sin x \cos y - 2\cos x \sin y}{2\sin x \sin y}$ — Separate fractions.

$\equiv \dfrac{2\sin x \cos y}{2\sin x \sin y} - \dfrac{2\cos x \sin y}{2\sin x \sin y}$ — Simplify fractions.

$\equiv \dfrac{\cos y}{\sin y} - \dfrac{\cos x}{\sin x}$

$\equiv \cot y - \cot x$

$\equiv \text{RHS}$

Chapter 3: Trigonometry

EXPLORE 3.2

Odd one out

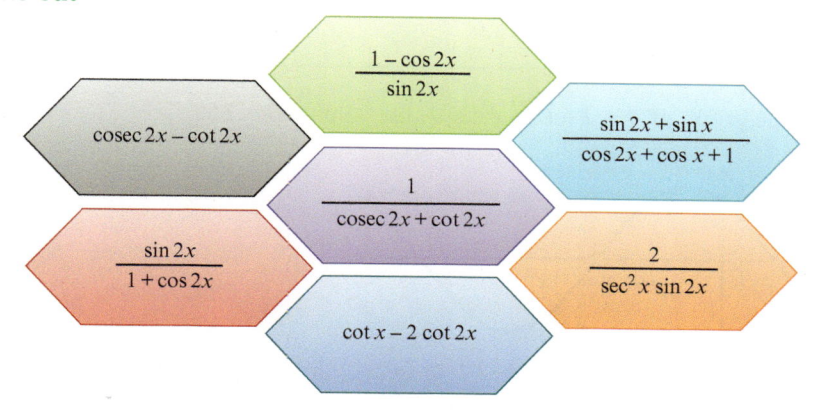

- $\dfrac{1 - \cos 2x}{\sin 2x}$
- $\csc 2x - \cot 2x$
- $\dfrac{\sin 2x + \sin x}{\cos 2x + \cos x + 1}$
- $\dfrac{1}{\csc 2x + \cot 2x}$
- $\dfrac{\sin 2x}{1 + \cos 2x}$
- $\dfrac{2}{\sec^2 x \sin 2x}$
- $\cot x - 2\cot 2x$

Find the trigonometric expression that does not match the other six expressions.

Create as many expressions of your own to match the 'odd one out'.

(Your expressions must contain at least two different trigonometric ratios.)

Compare your answers with your classmates.

EXERCISE 3D

P 1 Prove each of these identities.

    a $\tan A + \cot A \equiv 2\csc 2A$
    b $1 - \tan^2 A \equiv \cos 2A \sec^2 A$

    c $\tan 2A - \tan A \equiv \tan A \sec 2A$
    d $(\cos A + 3\sin A)^2 \equiv 5 - 4\cos 2A + 3\sin 2A$

    e $\cot 2A + \tan A \equiv \dfrac{1}{2}\csc A \sec A$
    f $\csc 2A + \cot 2A \equiv \cot A$

    g $\sec^2 A \csc^2 A \equiv 4\csc^2 2A$
    h $\sin(A+B)\sin(A-B) \equiv \sin^2 A - \sin^2 B$

P 2 Prove each of these identities.

    a $\dfrac{\sin A}{\cos B} + \dfrac{\cos A}{\sin B} \equiv \dfrac{2\cos(A-B)}{\sin 2B}$
    b $\dfrac{\cos A + \sin A}{\cos A - \sin A} \equiv \sec 2A + \tan 2A$

    c $\dfrac{1 - \tan^2 A}{1 + \tan^2 A} \equiv \cos 2A$
    d $\dfrac{\cos 2A + \sin 2A - 1}{\cos 2A - \sin 2A + 1} \equiv \tan A$

    e $\dfrac{\sin 3A + \sin A}{2\sin 2A} \equiv \cos A$
    f $\dfrac{\cos 3A - \sin 3A}{1 - 2\sin 2A} \equiv \cos A + \sin A$

    g $\dfrac{\cos 2A + 9\cos A + 5}{4 + \cos A} \equiv 2\cos A + 1$
    h $\dfrac{\cos^3 A - \sin^3 A}{\cos A - \sin A} \equiv \dfrac{2 + \sin 2A}{2}$

P 3 Use the fact that $4A = 2 \times 2A$ to show that:

    a $\dfrac{\sin 4A}{\sin A} \equiv 8\cos^3 A - 4\cos A$
    b $\cos 4A + 4\cos 2A \equiv 8\cos^4 A - 3$

P 4 Prove the identity $8\sin^2 x \cos^2 x \equiv 1 - \cos 4x$.

P 5 Prove the identity $(2\sin A + \cos A)^2 \equiv \dfrac{1}{2}(4\sin 2A - 3\cos 2A + 5)$.

P 6 Use the expansions of $\cos(3x - x)$ and $\cos(3x + x)$ to prove the identity:

    $\cos 2x - \cos 4x \equiv 2\sin 3x \sin x$

WEB LINK

Try the *Equation or identity? (II)* resource on the Underground Mathematics website.

3.5 Expressing $a\sin\theta + b\cos\theta$ in the form $R\sin(\theta \pm \alpha)$ or $R\cos(\theta \pm \alpha)$

In this section you will learn how to solve equations of the form $a\sin\theta + b\cos\theta = c$.

EXPLORE 3.3

Use graphing software to confirm that the graph of $y = 3\sin\theta + 4\cos\theta$ for $-90° \leqslant \theta \leqslant 360°$ is:

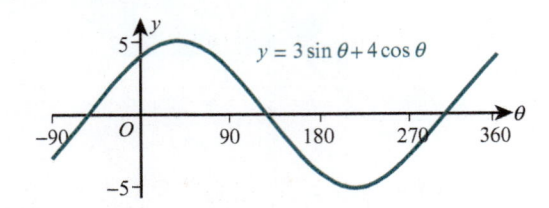

[Graph showing $y = 3\sin\theta + 4\cos\theta$ with amplitude 5, oscillating between -5 and 5 over the range $-90°$ to $360°$.]

1 What are the amplitude and period of this graph?

2 Discuss with your classmates how this graph can be obtained by transforming the graph of:

   a $y = \sin\theta$ b $y = \cos\theta$

3 Use your knowledge of transformations of functions to write the function $y = 3\sin\theta + 4\cos\theta$ in the form:

   a $y = R\sin(\theta + \alpha)$ b $R\sin(\theta + \beta)$

   (You will need to use your graph to find the approximate values of α and β.)

The formula for replacing $a\sin\theta + b\cos\theta$ by $R\sin(\theta + \alpha)$ is derived as follows.

Let $a\sin\theta + b\cos\theta = R\sin(\theta + \alpha)$ where $R > 0$ and $0° < \alpha < 90°$.

$$R\sin(\theta + \alpha) = R(\sin\theta\cos\alpha + \cos\theta\sin\alpha)$$

$$\therefore a\sin\theta + b\cos\theta = R\sin\theta\cos\alpha + R\cos\theta\sin\alpha$$

Equating coefficients of $\sin\theta$: $R\cos\alpha = a$ ----------- (1)

Equating coefficients of $\cos\theta$: $R\sin\alpha = b$ ----------- (2)

(2) ÷ (1): $\dfrac{\sin\alpha}{\cos\alpha} = \dfrac{b}{a} \Rightarrow \tan\alpha = \dfrac{b}{a}$

Squaring equations (1) and (2) and then adding gives:

$$\begin{aligned}a^2 + b^2 &= R^2\cos^2\alpha + R^2\sin^2\alpha \\ &= R^2(\cos^2\alpha + \sin^2\alpha) \quad\quad \text{Use } \cos^2\alpha + \sin^2\alpha = 1 \\ &= R^2\end{aligned}$$

Hence, $R = \sqrt{a^2 + b^2}$

$\therefore a\sin\theta + b\cos\theta$ can be written as $R\sin(\theta + \alpha)$ where $R = \sqrt{a^2 + b^2}$ and $\tan\alpha = \dfrac{b}{a}$

EXPLORE 3.4

Using a similar approach, show that:

1. $a\sin\theta - b\cos\theta \equiv R\sin(\theta - \alpha)$ where $R = \sqrt{a^2 + b^2}$ and $\tan\alpha = \dfrac{b}{a}$

2. $a\cos\theta + b\sin\theta \equiv R\cos(\theta - \alpha)$ where $R = \sqrt{a^2 + b^2}$ and $\tan\alpha = \dfrac{b}{a}$

3. $a\cos\theta - b\sin\theta \equiv R\cos(\theta + \alpha)$ where $R = \sqrt{a^2 + b^2}$ and $\tan\alpha = \dfrac{b}{a}$

Summarising the results from Explore 3.4:

KEY POINT 3.5

$a\sin\theta \pm b\cos\theta \equiv R\sin(\theta \pm \alpha)$ and $a\cos\theta \pm b\sin\theta \equiv R\cos(\theta \mp \alpha)$

where $R = \sqrt{a^2 + b^2}$, $\tan\alpha = \dfrac{b}{a}$ and $0° < \alpha < 90°$.

WORKED EXAMPLE 3.11

a Express $2\sin\theta - 3\cos\theta$ in the form $R\sin(\theta - \alpha)$, where $R > 0$ and $0° < \alpha < 90°$. Give the value of α correct to 2 decimal places.

b Hence solve the equation $2\sin\theta - 3\cos\theta = \sqrt{2}$ for $0° < \theta < 360°$.

Answer

a $2\sin\theta - 3\cos\theta = R\sin(\theta - \alpha)$

$\therefore 2\sin\theta - 3\cos\theta = R\sin\theta\cos\alpha - R\cos\theta\sin\alpha$

Equating coefficients of $\sin\theta$: $R\cos\alpha = 2$ -------- (1)

Equating coefficients of $\cos\theta$: $R\sin\alpha = 3$ ------- (2)

(2) ÷ (1): $\tan\alpha = \dfrac{3}{2} \Rightarrow \alpha = 56.31°$

Squaring equations (1) and (2) and then adding gives:

$3^2 + 2^2 = R^2 \Rightarrow R = \sqrt{13}$

$\therefore 2\sin\theta - 3\cos\theta = \sqrt{13}\sin(\theta - 56.31°)$

b $2\sin\theta - 3\cos\theta = \sqrt{2}$

$\sqrt{13}\sin(\theta - 56.31°) = \sqrt{2}$

$\sin(\theta - 56.31°) = \dfrac{\sqrt{2}}{\sqrt{13}}$

$\theta - 56.31° = 23.09°, 180° - 23.09°$

$= 23.09°, 156.91°$

$\theta = 79.4°, 213.2°$

We can find the maximum and minimum values of the expression $a\sin\theta + b\cos\theta$ by writing the expression in the form $R\sin(\theta + \alpha)$.

From the Pure Mathematics Coursebook 1, Chapter 5, Section 5.4, we know that $-1 \leq \sin(\theta + \alpha) \leq 1$, hence $-R \leq R\sin(\theta + \alpha) \leq R$.

This can be summarised as:

> **KEY POINT 3.6**
>
> The maximum value of $a\sin\theta + b\cos\theta$ is $\sqrt{a^2 + b^2}$ and occurs when $\sin(\theta + \alpha) = 1$.
>
> The minimum value of $a\sin\theta + b\cos\theta$ is $-\sqrt{a^2 + b^2}$ and occurs when $\sin(\theta + \alpha) = -1$.

WORKED EXAMPLE 3.12

Express $2\cos\theta - \sin\theta$ in the form $R\cos(\theta + \alpha)$, where $R > 0$ and $0° < \alpha < 90°$.

Hence, find the maximum and minimum values of the expression $2\cos\theta - \sin\theta$ and the values of θ in the interval $0° < \theta < 360°$ for which these occur.

Answer

$2\cos\theta - \sin\theta = R\cos(\theta + \alpha)$

$\therefore 2\cos\theta - \sin\theta = R\cos\theta\cos\alpha - R\sin\theta\sin\alpha$

Equating coefficients of $\cos\theta$: $R\cos\alpha = 2$ ---------------(1)

Equating coefficients of $\sin\theta$: $R\sin\alpha = 1$ ---------------(2)

(2) ÷ (1): $\tan\alpha = \dfrac{1}{2} \implies \alpha = 26.57°$

Squaring equations (1) and (2) and then adding gives:

$2^2 + 1^2 = R^2 \implies R = \sqrt{5}$

$\therefore 2\cos\theta - \sin\theta = \sqrt{5}\cos(\theta + 26.57°)$

Maximum value is $\sqrt{2^2 + (-1)^2}$ and occurs when $\cos(\theta + 26.57°) = 1$

$\theta + 26.57° = 360°$

$\theta = 333.4°$

Minimum value is $-\sqrt{2^2 + (-1)^2}$ and occurs when $\cos(\theta + 26.57°) = -1$

$\theta + 26.57° = 180°$

$\theta = 153.4°$

$\therefore$ Max value $= \sqrt{5}$ when $\theta = 333.4°$, min value $= -\sqrt{5}$ when $\theta = 153.4°$

> **DID YOU KNOW?**
>
> The French mathematician Joseph Fourier (1768–1830) found a way to represent any wave-like function as the sum (possibly infinite) of simple sine waves. The sum of sine functions is called a *Fourier* series. These series can be applied to vibration problems. Fourier is also credited with discovering the greenhouse effect.

EXERCISE 3E

1 **a** Express $15\sin\theta - 8\cos\theta$ in the form $R\sin(\theta - \alpha)$, where $R > 0$ and $0° < \alpha < 90°$.
Give the value of α correct to 2 decimal places.

  b Hence solve the equation $15\sin\theta - 8\cos\theta = 10$ for $0° < \theta < 360°$.

2 **a** Express $2\cos\theta - 3\sin\theta$ in the form $R\cos(\theta + \alpha)$, where $R > 0$ and $0° < \alpha < 90°$.
Give the exact value of R and the value of α correct to 2 decimal places.

  b Hence solve the equation $2\cos\theta - 3\sin\theta = 1.3$ for $0° < \theta < 360°$.

3 **a** Express $15\sin\theta - 8\cos\theta$ in the form $R\sin(\theta - \alpha)$, where $R > 0$ and $0° < \alpha < 90°$.
Give the value of α correct to 2 decimal places.

  b Hence solve the equation $15\sin\theta - 8\cos\theta = 3$ for $0° < \theta < 360°$.

  c Find the greatest value of $30\sin\theta - 16\cos\theta$ as θ varies.

4 **a** Express $4\sin\theta - 6\cos\theta$ in the form $R\sin(\theta - \alpha)$, where $R > 0$ and $0° < \alpha < 90°$.
Give the exact value of R and the value of α correct to 2 decimal places.

  b Hence solve the equation $4\sin\theta - 6\cos\theta = 3$ for $0° < \theta < 180°$.

  c Find the greatest and least values of $(4\sin\theta - 6\cos\theta)^2 - 3$ as θ varies.

5 **a** Express $3\sin\theta + 4\cos\theta$ in the form $R\sin(\theta + \alpha)$, where $R > 0$ and $0° < \alpha < 90°$.
Give the value of α correct to 2 decimal places.

  b Hence solve the equation $3\sin\theta + 4\cos\theta = 2$ for $0° < \theta < 360°$.

  c Find the least value of $3\sin\theta + 4\cos\theta + 3$ as θ varies.

6 **a** Express $\cos\theta + \sqrt{3}\sin\theta$ in the form $R\cos(\theta - \alpha)$, where $R > 0$ and $0 < \alpha < \dfrac{\pi}{2}$.
Give the exact values of R and α.

  b Hence prove that $\dfrac{1}{\left(\cos\theta + \sqrt{3}\sin\theta\right)^2} \equiv \dfrac{1}{4}\sec^2\left(\theta - \dfrac{\pi}{3}\right)$.

7 **a** Express $8\sin 2\theta + 4\cos 2\theta$ in the form $R\sin(2\theta + \alpha)$, where $R > 0$ and $0° < \alpha < 90°$.
Give the exact value of R and the value of α correct to 2 decimal places.

  b Hence solve the equation $8\sin 2\theta + 4\cos 2\theta = 3$ for $0° < \theta < 360°$.

  c Find the least value of $\dfrac{10}{(8\sin 2\theta + 4\cos 2\theta)^2}$ as θ varies.

8 **a** Express $\cos\theta - \sqrt{2}\sin\theta$ in the form $R\cos(\theta + \alpha)$, where $R > 0$ and $0° < \alpha < 90°$.
Give the exact value of R and the value of α correct to 2 decimal places.

  b Hence solve the equation $\cos\theta - \sqrt{2}\sin\theta = -1$ for $0° < \theta < 360°$.

  c Find the least value of $\dfrac{1}{\left(\sqrt{2}\cos\theta - 2\sin\theta\right)^2}$ as θ varies.

9 a Express $\cos\theta - \sin\theta$ in the form $R\cos(\theta + \alpha)$, where $R > 0$ and $0 < \alpha < \dfrac{\pi}{2}$.

   Give the exact values of R and α.

   b Show that one solution of the equation $\cos\theta - \sin\theta = \dfrac{1}{2}\sqrt{6}$ is $\theta = \dfrac{19\pi}{12}$ and find the other solution in the interval $0 < \theta < 2\pi$.

   c State the set of values of k for which the equation $\cos\theta - \sin\theta = k$ has any solutions.

10 a Express $\sin\theta - 3\cos\theta$ in the form $R\sin(\theta - \alpha)$, where $R > 0$ and $0° < \alpha < 90°$.

   Give the exact value of R and the value of α correct to 2 decimal places.

   b Hence solve the equation $\sin\theta - 3\cos\theta = -2$ for $0° < \theta < 360°$.

   c Find the greatest possible value of $1 + \sin 2\theta - 3\cos 2\theta$ as θ varies and determine the smallest positive value of θ for which this greatest value occurs.

11 a Express $\sqrt{5}\cos\theta + 2\sin\theta$ in the form $R\cos(\theta - \alpha)$, where $R > 0$ and $0° < \alpha < 90°$.

   Give the value of α correct to 2 decimal places.

   b Find the smallest positive angle θ that satisfies the equation $\sqrt{5}\cos\theta + 2\sin\theta = 3$.

   c Find the smallest positive angle θ that satisfies the equation $\sqrt{5}\cos\dfrac{1}{2}\theta + 2\sin\dfrac{1}{2}\theta = -1$.

12 a Given that $3\sec\theta + 4\csc\theta = 2\csc 2\theta$, show that $3\sin\theta + 4\cos\theta = 1$.

   b Express $3\sin\theta + 4\cos\theta$ in the form $R\sin(\theta + \alpha)$, where $R > 0$ and $0° < \alpha < 90°$.

   Give the value of α correct to 2 decimal places.

   c Hence solve the equation $3\sec\theta + 4\csc\theta = 2\csc 2\theta$ for $0° < \theta < 360°$.

PS 13 a Express $\sin(\theta + 30°) + \cos\theta$ in the form $R\sin(\theta + \alpha)$, where $R > 0$ and $0° < \alpha < 90°$.

   Give the exact values of R and α.

   b Hence solve the equation $\sin(\theta + 30°) + \cos\theta = 1$ for $0° < \theta < 360°$.

PS 14 a Find the maximum and minimum values of $7\sin^2\theta + 9\cos^2\theta + 4\sin\theta\cos\theta + 2$.

   b Hence, or otherwise, solve the equation $7\sin^2\theta + 9\cos^2\theta + 4\sin\theta\cos\theta = 10$ for $0° < \theta < 360°$.

Checklist of learning and understanding

Cosecant, secant and cotangent

$$\operatorname{cosec} \theta = \frac{1}{\sin \theta} \qquad \sec \theta = \frac{1}{\cos \theta} \qquad \cot \theta = \frac{1}{\tan \theta}$$

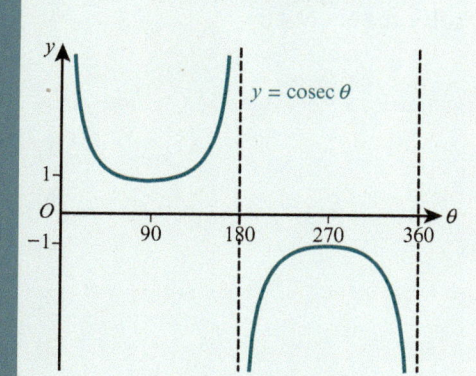

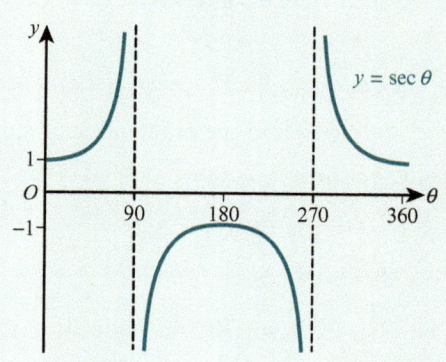

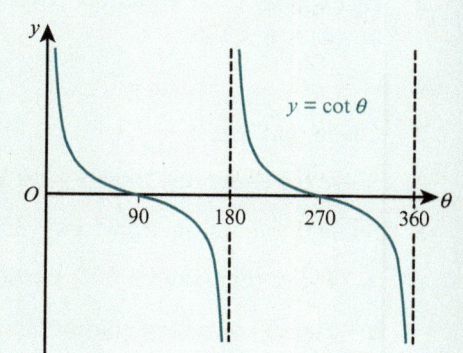

Trigonometric identities

- $1 + \tan^2 x \equiv \sec^2 x$
- $1 + \cot^2 x \equiv \operatorname{cosec}^2 x$

Compound angle formula

- $\sin(A + B) \equiv \sin A \cos B + \cos A \sin B$
- $\sin(A - B) \equiv \sin A \cos B - \cos A \sin B$
- $\cos(A + B) \equiv \cos A \cos B - \sin A \sin B$
- $\cos(A - B) \equiv \cos A \cos B + \sin A \sin B$
- $\tan(A + B) \equiv \dfrac{\tan A + \tan B}{1 - \tan A \tan B}$
- $\tan(A - B) \equiv \dfrac{\tan A - \tan B}{1 + \tan A \tan B}$

Double angle formulae

- $\sin 2A \equiv 2 \sin A \cos A$
- $\cos 2A \equiv \cos^2 A - \sin^2 A$
  $\equiv 1 - 2 \sin^2 A$
  $\equiv 2 \cos^2 A - 1$
- $\tan 2A \equiv \dfrac{2 \tan A}{1 - \tan^2 A}$

Expressing $a \sin \theta + b \cos \theta$ in the form $R \sin(\theta \pm \alpha)$ or $R \cos(\theta \pm \alpha)$

- $a \sin \theta \pm b \cos \theta = R \sin(\theta \pm \alpha)$
- $a \cos \theta \pm b \sin \theta = R \cos(\theta \mp \alpha)$

  where $R = \sqrt{a^2 + b^2}$ and $\tan \alpha = \dfrac{b}{a}$

END-OF-CHAPTER REVIEW EXERCISE 3

1. Sketch the graph of $y = 3\sec(2x - 90°)$ for $0° < x < 180°$. [3]

2. By expressing the equation $\csc\theta = 3\sin\theta + \cot\theta$ in terms of $\cos\theta$ only, solve the equation for $0° < \theta < 180°$. [5]

   Cambridge International A Level Mathematics 9709 Paper 31 Q3 June 2016

3. Given that $\cos A = \dfrac{1}{4}$, where $270° < A < 360°$, find the exact value of $\sin 2A$. [5]

4. Solve the equation $2\tan^2 x + \sec x = 1$ for $0° \leqslant x \leqslant 360°$. [6]

5. Solve the equation $2\cot^2 x + 5\csc x = 10$ for $0° < x < 360°$. [6]

6. a Prove that $\sin(x + 60°) + \cos(x + 30°) \equiv \sqrt{3}\cos x$. [3]

   b Hence solve the equation $\sin(x + 60°) + \cos(x + 30°) = \dfrac{3}{2}$ for $0° < x < 360°$. [3]

7. a Prove that $\sin(60° - x) + \cos(30° - x) \equiv \sqrt{3}\cos x$. [3]

   b Hence solve the equation $\sin(60° - x) + \cos(30° - x) = \dfrac{2}{5}\sec x$ for $0° < x < 360°$. [3]

8. i Show that the equation $\tan(x + 45°) = 6\tan x$ can be written in the form $6\tan^2 x - 5\tan x + 1 = 0$. [3]

   ii Hence solve the equation $\tan(x + 45°) = 6\tan x$, for $0° < x < 180°$. [3]

   Cambridge International AS & A Level Mathematics 9709 Paper 21 Q3 June 2010

9. a Prove that $\tan(x + 45°) - \tan(45° - x) \equiv 2\tan 2x$. [4]

   b Hence solve the equation $\tan(x + 45°) - \tan(45° - x) = 6$ for $0° < x < 180°$. [3]

10. i Express $3\cos\theta + \sin\theta$ in the form $R\cos(\theta - \alpha)$, where $R > 0$ and $0° < \alpha < 90°$, giving the exact value of R and the value of α correct to 2 decimal places. [3]

    ii Hence solve the equation $3\cos 2x + \sin 2x = 2$, giving all solutions in the interval $0° \leqslant x \leqslant 360°$. [5]

    Cambridge International AS & A Level Mathematics 9709 Paper 21 Q7 November 2013

11. a Prove that $\cos(60° - x) + \cos(300° - x) \equiv \cos x$. [3]

    b Hence

       i find the exact value of $\cos 15° + \cos 255°$ [2]

       ii solve the equation $\cos(60° - x) + \cos(300° - x) = \dfrac{1}{4}\csc x$ for $0° < x < 180°$. [3]

12. a Prove the identity $\dfrac{2\sin 2\theta - 3\cos 2\theta + 3}{\sin\theta} \equiv 4\cos\theta + 6\sin\theta$. [3]

    b Express $4\cos\theta + 6\sin\theta$ in the form $R\cos(\theta - \alpha)$, where $R > 0$ and $0 < \alpha < \dfrac{\pi}{2}$.

    Give the exact value of R and the value of α correct to 2 decimal places. [4]

    c Write down the greatest value of $\left(\dfrac{2\sin 2\theta - 3\cos 2\theta + 3}{\sin\theta}\right)^2$. [1]

13. i Express $4\sin\theta - 6\cos\theta$ in the form $R\sin(\theta - \alpha)$, where $R > 0$ and $0° < \alpha < 90°$.

    Give the exact value of R and the value of α correct to 2 decimal places. [3]

    ii Solve the equation $4\sin\theta - 6\cos\theta = 3$ for $0° \leqslant \theta \leqslant 360°$. [4]

    iii Find the greatest and least possible values of $(4\sin\theta - 6\cos\theta)^2 + 8$ as θ varies. [2]

    Cambridge International AS & A Level Mathematics 9709 Paper 21 Q8 June 2011

14 i By first expanding $\sin(2\theta + \theta)$, show that $\sin 3\theta = 3\sin\theta - 4\sin^3\theta$. [4]

ii Show that, after making the substitution $x = \dfrac{2\sin\theta}{\sqrt{3}}$, the equation $x^3 - x + \dfrac{1}{6}\sqrt{3} = 0$ can be written in the form $\sin 3\theta = \dfrac{3}{4}$. [1]

iii Hence solve the equation $x^3 - x + \dfrac{1}{6}\sqrt{3} = 0$, giving your answers correct to 3 significant figures. [4]

Cambridge International A Level Mathematics 9709 Paper 31 Q8 November 2014

15 a Prove the identity $\dfrac{1}{\sin(x + 30°) + \cos(x + 60°)} \equiv \sec x$. [3]

b Hence solve the equation $\dfrac{2}{\sin(x + 30°) + \cos(x + 60°)} = 7 - \tan^2 x$ for $0° < x < 360°$. [6]

16 a Prove the identity $\cosec^4 x - \cot^4 x \equiv \cosec^2 x + \cot^2 x$. [3]

b Hence solve the equation $\cosec^4 x - \cot^4 x = 16 - \cot x$ for $0° < x < 180°$. [6]

17 The curves C_1 and C_2 have equations $y = 1 + 4\cos 2x$ and $y = 2\cos^2 x - 4\sin 2x$.

a Show that the x-coordinates of the points where C_1 and C_2 intersect satisfy the equation $3\cos 2x + 4\sin 2x = 0$. [3]

b Express $3\cos 2x + 4\sin 2x$ in the form $R\sin(2x + \alpha)$, where $R > 0$ and $0° < \alpha < 90°$.

Give the exact value of R and the value of α correct to 2 decimal places. [3]

c Hence find all the roots of the equation $3\cos 2x + 4\sin 2x = 0$ for $0° < x < 180°$. [3]

CROSS-TOPIC REVIEW EXERCISE 1

1. Find the set of values of x satisfying the inequality $3|x-1| < |2x+1|$. [4]

   Cambridge International A Level Mathematics 9709 Paper 31 Q1 November 2012

2. Solve the equation $2|3^x - 1| = 3^x$, giving your answers correct to 3 significant figures. [4]

   Cambridge International A Level Mathematics 9709 Paper 31 Q2 November 2013

3.
   i   Solve the equation $|3x+4| = |3x-11|$. [3]

   ii Hence, using logarithms, solve the equation $|3 \times 2^y + 4| = |3 \times 2^y - 11|$, giving the answer correct to 3 significant figures. [2]

   Cambridge International AS & A Level Mathematics 9709 Paper 21 Q1 June 2015

4.
   i   Solve the equation $2|x-1| = 3|x|$. [3]

   ii Hence solve the equation $2|5^x - 1| = 3|5^x|$, giving your answer correct to 3 significant figures. [2]

   Cambridge International A Level Mathematics 9709 Paper 31 Q1 June 2016

5.

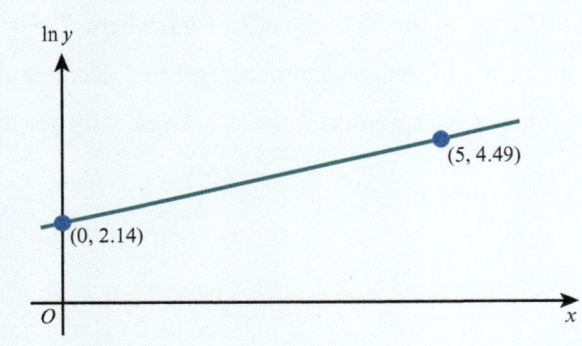

   The variables x and y satisfy the equation $y = A(b^x)$, where A and b are constants. The graph of $\ln y$ against x is a straight line passing through the points $(0, 2.14)$ and $(5, 4.49)$, as shown in the diagram. Find the values of A and b, correct to 1 decimal place. [5]

   Cambridge International AS & A Level Mathematics 9709 Paper 21 Q2 June 2012

6.
   i   It is given that x satisfies the equation $3^{2x} = 5(3^x) + 14$. Find the value of 3^x and, using logarithms, find the value of x correct to 3 significant figures. [4]

   ii Hence state the values of x satisfying the equation $3^{2|x|} = 5(3^{|x|}) + 14$. [1]

   Cambridge International AS & A Level Mathematics 9709 Paper 21 Q1 November 2016

7. The variables x and y satisfy the equation $x^n y = C$, where n and C are constants. When $x = 1.10$, $y = 5.20$, and when $x = 3.20$, $y = 1.05$.

   i   Find the values of n and C. [5]

   ii Explain why the graph of $\ln y$ against $\ln x$ is a straight line. [1]

   Cambridge International A Level Mathematics 9709 Paper 31 Q3 June 2010

8. Given that $3e^x + 8e^{-x} = 14$, find the possible values of e^x and hence solve the equation $3e^x + 8e^{-x} = 14$ correct to 3 significant figures. [6]

   Cambridge International AS & A Level Mathematics 9709 Paper 21 Q3 June 2016

9 The angles θ and ϕ lie between $0°$ and $180°$, and are such that $\tan(\theta - \phi) = 3$ and $\tan\theta + \tan\phi = 1$.
Find the possible values of θ and ϕ. [6]

Cambridge International A Level Mathematics 9709 Paper 31 Q3 November 2015

10 i Solve the equation $|4x - 1| = |x - 3|$. [3]
   ii Hence solve the equation $|4^{y+1} - 1| = |4^y - 3|$ correct to 3 significant figures. [3]

Cambridge International A Level Mathematics 9709 Paper 31 Q4 June 2013

11 The polynomial $4x^3 + ax^2 + 9x + 9$, where a is a constant, is denoted by $p(x)$. It is given that when $p(x)$ is divided by $(2x - 1)$ the remainder is 10.
   i Find the value of a and hence verify that $(x - 3)$ is a factor of $p(x)$. [3]
   ii When a has this value, solve the equation $p(x) = 0$. [4]

Cambridge International AS & A Level Mathematics 9709 Paper 21 Q5 November 2011

12 The polynomial $2x^3 - 4x^2 + ax + b$, where a and b are constants, is denoted by $p(x)$.
It is given that when $p(x)$ is divided by $(x + 1)$ the remainder is 4, and that when $p(x)$ is divided by $(x - 3)$ the remainder is 12.
   i Find the values of a and b. [5]
   ii When a and b have these values, find the quotient and remainder when $p(x)$ is divided by $(x^2 - 2)$. [3]

Cambridge International AS & A Level Mathematics 9709 Paper 21 Q7 November 2012

13 i Express $\cos x + 3 \sin x$ in the form $R \cos(x - \alpha)$, where $R > 0$ and $0° < \alpha < 90°$, giving the exact value of R and the value of α correct to 2 decimal places. [3]
   ii Hence solve the equation $\cos 2\theta + 3 \sin 2\theta = 2$, for $0° < \theta < 90°$. [5]

Cambridge International A Level Mathematics 9709 Paper 31 Q6 November 2011

14 It is given that $2\ln(4x - 5) + \ln(x + 1) = 3\ln 3$.
   i Show that $16x^3 - 24x^2 - 15x - 2 = 0$. [3]
   ii By first using the factor theorem, factorise $16x^3 - 24x^2 - 15x - 2$ completely. [4]
   iii Hence solve the equation $2\ln(4x - 5) + \ln(x + 1) = 3\ln 3$. [1]

Cambridge International A Level Mathematics 9709 Paper 31 Q6 June 2014

15 The polynomial $8x^3 + ax^2 + bx - 1$, where a and b are constants, is denoted by $p(x)$.
It is given that $(x + 1)$ is a factor of $p(x)$ and that when $p(x)$ is divided by $(2x + 1)$ the remainder is 1.
   i Find the values of a and b. [5]
   ii When a and b have these values, factorise $p(x)$ completely. [3]

Cambridge International A Level Mathematics 9709 Paper 31 Q6 November 2015

16 The polynomial $3x^3 + 2x^2 + ax + b$, where a and b are constants, is denoted by $\text{p}(x)$.
It is given that $(x - 1)$ is a factor of $\text{p}(x)$, and that when $\text{p}(x)$ is divided by $(x - 2)$ the remainder is 10.

   i   Find the values of a and b. [5]

   ii When a and b have these values, solve the equation $\text{p}(x) = 0$. [4]

Cambridge International AS and A Level Mathematics 9709 Paper 21 Q7 November 2010

17 i   The polynomial $x^3 + ax^2 + bx + 8$, where a and b are constants, is denoted by $\text{p}(x)$.
It is given that when $\text{p}(x)$ is divided by $(x - 3)$ the remainder is 14, and that when $\text{p}(x)$ is divided by $(x + 2)$ the remainder is 24. Find the values of a and b. [5]

   ii When a and b have these values, find the quotient when $\text{p}(x)$ is divided by $x^2 + 2x - 8$ and hence solve the equation $\text{p}(x) = 0$. [4]

Cambridge International AS & A Level Mathematics 9709 Paper 21 Q4 November 2013

18 i   Given that $(x + 2)$ and $(x + 3)$ are factors of $5x^3 + ax^2 + b$, find the values of the constants a and b. [4]

   ii When a and b have these values, factorise $5x^3 + ax^2 + b$ completely, and hence solve the equation $5^{3y+1} + a \times 5^{2y} + b = 0$, giving any answers correct to 3 significant figures. [5]

Cambridge International AS & A Level Mathematics 9709 Paper 21 Q5 November 2014

19 i   Find the quotient and remainder when $x^4 + x^3 + 3x^2 + 12x + 6$ is divided by $(x^2 - x + 4)$. [4]

   ii It is given that, when $x^4 + x^3 + 3x^2 + px + q$ is divided by $(x^2 - x + 4)$, the remainder is zero. Find the values of the constants p and q. [2]

   iii When p and q have these values, show that there is exactly one real value of x satisfying the equation $x^4 + x^3 + 3x^2 + px + q = 0$ and state what that value is. [3]

Cambridge International AS & A Level Mathematics 9709 Paper 21 Q6 November 2015

20 The angle α lies between $0°$ and $90°$ and is such that $2\tan^2\alpha + \sec^2\alpha = 5 - 4\tan\alpha$.

   i   Show that $3\tan^2\alpha + 4\tan\alpha - 4 = 0$ and hence find the exact value of $\tan\alpha$. [4]

   ii It is given that the angle β is such that $\cot(\alpha + \beta) = 6$. Without using a calculator, find the exact value of $\cot\beta$. [5]

Cambridge International AS & A Level Mathematics 9709 Paper 21 Q7 November 2014

21 i   Express $5\sin 2\theta + 2\cos 2\theta$ in the form $R\sin(2\theta + \alpha)$, where $R > 0$ and $0° < \alpha < 90°$, giving the exact value of R and the value of α correct to 2 decimal places.

Hence [3]

   ii solve the equation $5\sin 2\theta + 2\cos 2\theta = 4$, giving all solutions in the interval $0° \leqslant \theta \leqslant 360°$, [5]

   iii determine the least value of $\dfrac{1}{(10\sin 2\theta + 4\cos 2\theta)^2}$ as θ varies. [2]

Cambridge International AS & A Level Mathematics 9709 Paper 21 Q7 June 2013

22 The polynomial p(x) is defined by $p(x) = ax^3 + 3x^2 + bx + 12$, where a and b are constants. It is given that $(x + 3)$ is a factor of p(x). It is also given that the remainder is 18 when p(x) is divided by $(x + 2)$.

   i   Find the values of a and b. [5]

   ii When a and b have these values,

       a show that the equation p(x) = 0 has exactly one real root, [4]

       b solve the equation p(sec y) = 0 for $-180° < y < 180°$. [3]

   Cambridge International AS & A Level Mathematics 9709 Paper 21 Q7 November 2016

Chapter 4
Differentiation

In this chapter you will learn how to:

- differentiate products and quotients
- use the derivatives of e^x, $\ln x$, $\sin x$, $\cos x$, $\tan x$, together with constant multiples, sums, differences and composites
- find and use the first derivative of a function, which is defined parametrically or implicitly.

Chapter 4: Differentiation

PREREQUISITE KNOWLEDGE

Where it comes from	What you should be able to do	Check your skills
Pure Mathematics 1 Coursebook, Chapter 7	Differentiate x^n together with constant multiples, sums and differences.	1 Differentiate with respect to x. a $\quad y = 5x^3 - \dfrac{3}{x^2} + 2\sqrt{x}$ b $\quad y = \dfrac{x^8 - 4x^5 + x^2}{2x^3}$
Pure Mathematics 1 Coursebook, Chapter 7	Differentiate composite functions using the chain rule.	2 Differentiate with respect to x. a $\quad (3x - 5)^4$ b $\quad \dfrac{4}{\sqrt{1-2x}}$
Pure Mathematics 1 Coursebook, Chapter 7	Find tangents and normals to curves.	3 Find the equation of the normal to the curve $y = x^3 - 5x^2 + 2x - 1$ at the point $(1, -3)$.
Pure Mathematics 1 Coursebook, Chapter 8	Find stationary points on curves and determine their nature.	4 Find the stationary points on the curve $y = x^3 - 3x^2 + 2$ and determine their nature.

Why do we study differentiation?

In this chapter we will learn how to differentiate the product and the quotient of two simple functions. We will also learn how to find and use the derivatives of exponential, logarithmic and trigonometric functions.

All of the functions that we have differentiated so far have been of the form $y = f(x)$. In this chapter we will also learn how to differentiate functions that cannot be written in the form $y = f(x)$. For example, the function $y^2 + 2xy = 4$.

Lastly, we will learn how to find and use the derivative of a function where the variables x and y are given as a function of a third variable.

Although these topics might seem like they are just for Pure Mathematics problems, differentiation is used for calculations to do with quantum mechanics and field theories in Physics.

⏮ REWIND

This chapter builds on the work from Pure Mathematics 1 Coursebook, Chapters 7 and 8 where we learnt how to find and use the derivative of x^n.

🌐 WEB LINK

Explore the *Calculus of trigonometry and logarithms* and the *Chain rule and integration by substitution* stations on the Underground Mathematics website.

4.1 The product rule

The function $y = (x + 1)^4(3x - 2)^3$ can be considered as the product of two separate functions:

$\quad y = uv$ where $u = (x + 1)^4$ and $v = (3x - 2)^3$

To differentiate the product of two functions we can use the **product rule**.

> **KEY POINT 4.1**
>
> The product rule is:
>
> $$\frac{d}{dx}(uv) = u\frac{dv}{dx} + v\frac{du}{dx}$$

Some people find it easier to remember this rule as:

'(first function × derivative of second function) + (second function × derivative of first function)'

So for $y = (x+1)^4(3x-2)^3$,

$$\frac{dy}{dx} = \underbrace{(x+1)^4}_{\text{first}} \underbrace{\frac{d}{dx}(3x-2)^3}_{\text{differentiate second}} + \underbrace{(3x-2)^3}_{\text{second}} \underbrace{\frac{d}{dx}(x+1)^4}_{\text{differentiate first}}$$

$$= (x+1)^4 \times 9(3x-2)^2 + (3x-2)^3 \times 4(x+1)^3$$

$$= (x+1)^3(3x-2)^2\left[9(x+1) + 4(3x-2)\right]$$

$$= (x+1)^3(3x-2)^2(21x+1)$$

The product rule from first principles

Consider the function $y = uv$ where u and v are functions of x.

A small increase δx in x leads to corresponding small increases δy, δu and δv in y, u and v.

$y + \delta y = (u + \delta u)(v + \delta v)$ Expand brackets and replace y with uv.

$uv + \delta y = uv + u\delta v + v\delta u + \delta u \delta v$ Subtract uv from both sides.

$\delta y = u\delta v + v\delta u + \delta u \delta v$ Divide both sides by δx.

$$\frac{\delta y}{\delta x} = u\frac{\delta v}{\delta x} + v\frac{\delta u}{\delta x} + \delta u \frac{\delta v}{\delta x} \quad\text{----------(1)}$$

As $\delta x \to 0$, then so do δy, δu and δv and

$$\frac{\delta y}{\delta x} \to \frac{dy}{dx}, \frac{\delta u}{\delta x} \to \frac{du}{dx} \text{ and } \frac{\delta v}{\delta x} \to \frac{dv}{dx}.$$

Equation (1) becomes:

$$\frac{dy}{dx} = u\frac{dv}{dx} + v\frac{du}{dx} + 0\frac{dv}{dx}$$

$$\therefore \frac{dy}{dx} = u\frac{dv}{dx} + v\frac{du}{dx}$$

> **WORKED EXAMPLE 4.1**
>
> Find $\frac{dy}{dx}$ when $y = (2x-1)\sqrt{4x+5}$.
>
> **Answer**
>
> $y = (2x-1)(4x+5)^{\frac{1}{2}}$
>
> $$\frac{dy}{dx} = \underbrace{(2x-1)}_{\text{first}} \underbrace{\frac{d}{dx}[(4x+5)^{\frac{1}{2}}]}_{\text{differentiate second}} + \underbrace{(4x+5)^{\frac{1}{2}}}_{\text{second}} \underbrace{\frac{d}{dx}(2x-1)}_{\text{differentiate first}}$$

$$= (2x-1)\left[\frac{1}{2}(4x+5)^{-\frac{1}{2}}(4)\right] + \left(\sqrt{4x+5}\right)(2)$$
$$\underbrace{\phantom{\frac{1}{2}(4x+5)^{-\frac{1}{2}}(4)}}_{\text{Chain rule.}}$$

$$= \frac{2(2x-1)}{\sqrt{4x+5}} + 2\sqrt{4x+5} \quad \cdots\cdots\cdots\cdots \text{ Write as a single fraction.}$$

$$= \frac{2(2x-1) + 2(4x+5)}{\sqrt{4x+5}} \quad \cdots\cdots\cdots\cdots \text{ Simplify the numerator.}$$

$$= \frac{12x+8}{\sqrt{4x+5}}$$

WORKED EXAMPLE 4.2

Find the x-coordinate of the points on the curve $y = (2x-3)^2(x+5)^3$ where the gradient is 0.

Answer

$y = (2x-3)^2(x+5)^3$

$$\frac{dy}{dx} = \underbrace{(2x-3)^2}_{\text{first}} \underbrace{\frac{d}{dx}[(x+5)^3]}_{\text{differentiate second}} + \underbrace{(x+5)^3}_{\text{second}} \underbrace{\frac{d}{dx}[(2x-3)^2]}_{\text{differentiate first}}$$

$$= (2x-3)^2 \underbrace{[3(x+5)^2(1)]}_{\text{Chain rule}} + (x+5)^3 \underbrace{[2(2x-3)^1(2)]}_{\text{Chain rule}}$$

$$= 3(2x-3)^2(x+5)^2 + 4(2x-3)(x+5)^3 \quad \cdots\cdots\cdots \text{ Factorise.}$$

$$= (2x-3)(x+5)^2[3(2x-3) + 4(x+5)] \quad \cdots\cdots\cdots \text{ Simplify.}$$

$$= (2x-3)(x+5)^2(10x+11)$$

$\dfrac{dy}{dx} = 0$ when $(2x-3)(x+5)^2(10x+11) = 0$

$2x - 3 = 0 \qquad x + 5 = 0 \qquad 10x + 11 = 0$

$x = \dfrac{3}{2} \qquad\quad x = -5 \qquad\quad x = -1.1$

EXERCISE 4A

1 Use the product rule to differentiate each of the following with respect to x:

    a $x(x-2)^5$ **b** $5x(2x+1)^3$ **c** $x\sqrt{x+2}$

    d $(x-1)\sqrt{x+5}$ **e** $x^3\sqrt{2x-1}$ **f** $\sqrt{x}(x^2+2)^3$

    g $(x-3)^2(x+2)^5$ **h** $(2x-1)^5(3x+4)^4$ **i** $(2x-5)(3x^2+1)^2$

2 Find the gradient of the curve $y = x^2\sqrt{x+4}$ at the point $(-3, 9)$.

3 Find the equation of the tangent to the curve $y = (2 - x)^3(x + 1)^4$ at the point where $x = 1$.

4 Find the gradient of the tangent to the curve $y = (x + 2)(x - 1)^3$ at the point where the curve meets the y-axis.

5 Find the x-coordinate of the points on the curve $y = (3 - x)^3(x + 1)^2$ where the gradient is zero.

6 Find the x-coordinate of the point on the curve $y = (x + 2)\sqrt{1 - 2x}$ where the gradient is zero.

PS 7 a Sketch the curve $y = (x - 1)^2(5 - 2x) + 3$.

b The curve $y = (x - 1)^2(5 - 2x) + 3$ has stationary points at A and B. The straight line through A and B cuts the axes at P and Q. Find the area of the triangle POQ.

4.2 The quotient rule

We can differentiate the function $y = \dfrac{x^2 - 5}{2x + 1}$ by writing the function in the form

$y = (x^2 - 5)(2x + 1)^{-1}$ and then by applying the product rule.

Alternatively, we can consider $y = \dfrac{x^2 - 5}{2x + 1}$ as the division (quotient) of two separate functions:

$y = \dfrac{u}{v}$ where $u = x^2 - 5$ and $v = 2x + 1$.

To differentiate the quotient of two functions we can use the **quotient rule**.

KEY POINT 4.2

The quotient rule is:

$$\frac{d}{dx}\left(\frac{u}{v}\right) = \frac{v\dfrac{du}{dx} - u\dfrac{dv}{dx}}{v^2}$$

Some people find it easier to remember this rule as

$$\frac{(\text{denominator} \times \text{derivative of numerator}) - (\text{numerator} \times \text{derivative of denominator})}{(\text{denominator})^2}$$

The quotient rule from first principles

Consider the function $y = \dfrac{u}{v}$ where u and v are functions of x.

A small increase δx in x leads to corresponding small increases δy, δu and δv in y, u and v.

$$y + \delta y = \frac{u + \delta u}{v + \delta v}$$ Subtract y from both sides.

$$\delta y = \frac{u + \delta u}{v + \delta v} - y$$ Replace y with $\frac{u}{v}$.

$$\delta y = \frac{u + \delta u}{v + \delta v} - \frac{u}{v}$$ Combine fractions.

$$\delta y = \frac{v(u + \delta u) - u(v + \delta v)}{v(v + \delta v)}$$ Expand brackets and simplify.

$$\delta y = \frac{v\delta u - u\delta v}{v^2 + v\delta v}$$ Divide both sides by δx.

$$\frac{\delta y}{\delta x} = \frac{v\frac{\delta u}{\delta x} - u\frac{\delta v}{\delta x}}{v^2 + v\delta v} \quad \text{------------(1)}$$

As $\delta x \to 0$, then so do δy, δu and δv and $\frac{\delta y}{\delta x} \to \frac{dy}{dx}$, $\frac{\delta u}{\delta x} \to \frac{du}{dx}$ and $\frac{\delta v}{\delta x} \to \frac{dv}{dx}$.

Equation (1) becomes

$$\frac{d}{dx}\left(\frac{u}{v}\right) = \frac{v\frac{du}{dx} - u\frac{dv}{dx}}{v^2}$$

WORKED EXAMPLE 4.3

Find the derivative of $y = \dfrac{x^2 - 5}{2x + 1}$.

Answer

$$y = \frac{x^2 - 5}{2x + 1}$$

$$\frac{dy}{dx} = \frac{\overbrace{(2x + 1)}^{\text{denominator}} \times \overbrace{\frac{d}{dx}(x^2 - 5)}^{\text{differentiate numerator}} - \overbrace{(x^2 - 5)}^{\text{numerator}} \times \overbrace{\frac{d}{dx}(2x + 1)}^{\text{differentiate denominator}}}{\underbrace{(2x + 1)^2}_{\text{denominator squared}}}$$

$$= \frac{(2x + 1)(2x) - (x^2 - 5)(2)}{(2x + 1)^2}$$

$$= \frac{4x^2 + 2x - 2x^2 + 10}{(2x + 1)^2}$$

$$= \frac{2(x^2 + x + 5)}{(2x + 1)^2}$$

WORKED EXAMPLE 4.4

Find the derivative of $y = \dfrac{(x+2)^3}{\sqrt{x-1}}$.

Answer

$y = \dfrac{(x+2)^3}{\sqrt{x-1}}$

$\dfrac{dy}{dx} = \dfrac{\overbrace{\sqrt{x-1}}^{\text{denominator}} \times \overbrace{\dfrac{d}{dx}[(x+2)^3]}^{\text{differentiate numerator}} - \overbrace{(x+2)^3}^{\text{numerator}} \times \overbrace{\dfrac{d}{dx}\left[\sqrt{x-1}\right]}^{\text{differentiate denominator}}}{\underbrace{\left(\sqrt{x-1}\right)^2}_{\text{denominator squared}}}$

$= \dfrac{\left(\sqrt{x-1}\right)[3(x+2)^2(1)] - (x+2)^3\left[\dfrac{1}{2}(x-1)^{-\frac{1}{2}}(1)\right]}{x-1}$

$= \dfrac{3(x+2)^2\sqrt{x-1} - \dfrac{(x+2)^3}{2\sqrt{x-1}}}{x-1}$ ……… Multiply numerator and denominator by $2\sqrt{x-1}$.

$= \dfrac{6(x+2)^2(x-1) - (x+2)^3}{2(x-1)\sqrt{x-1}}$ ……… Factorise the numerator.

$= \dfrac{(x+2)^2[6(x-1) - (x+2)]}{2(x-1)^{\frac{3}{2}}}$

$= \dfrac{(x+2)^2(5x-8)}{2(x-1)^{\frac{3}{2}}}$

EXPLORE 4.1

In Worked examples 4.3 and 4.4 we differentiated the functions $y = \dfrac{x^2 - 5}{2x + 1}$ and $y = \dfrac{(x+2)^3}{\sqrt{x-1}}$ using the quotient rule. Now write each of these functions as a product, such as $y = (x^2 - 5)(2x + 1)^{-1}$, and then use the product rule to differentiate them.

Do you obtain the same answers?

Discuss with your classmates which method you prefer for functions such as these: the quotient rule or the product rule.

EXERCISE 4B

1 Use the quotient rule to differentiate each of the following with respect to x:

   a $\dfrac{2x+3}{x-4}$
   b $\dfrac{3x-5}{2-x}$
   c $\dfrac{x^2-3}{2x-1}$
   d $\dfrac{3x+1}{2-5x}$

   e $\dfrac{1-2x^2}{(x+4)^2}$
   f $\dfrac{5x^4}{(x^2-1)^2}$
   g $\dfrac{3x^2-7x}{x^2+2x+5}$
   h $\dfrac{(x+4)^2}{(x^2+1)^3}$

2 Find the gradient of the curve $y = \dfrac{x-5}{x+4}$ at the point $\left(2, -\dfrac{1}{2}\right)$.

3 Find the coordinates of the points on the curve $y = \dfrac{(x-1)^2}{2x+5}$ where the tangent is parallel to the x-axis.

4 Find the coordinates of the points on the curve $y = \dfrac{1-2x}{x-5}$ at which the gradient is 1.

5 Find the equation of the tangent of the curve $y = \dfrac{x-4}{2x+1}$ at the point where the curve crosses the y-axis.

6 Differentiate with respect to x:

   a $\dfrac{\sqrt{x}}{5x-1}$
   b $\dfrac{x-1}{\sqrt{2x+3}}$
   c $\dfrac{3-x^2}{\sqrt{x^2-1}}$
   d $\dfrac{5(x-1)^3}{\sqrt{x+2}}$

7 Find the x-coordinate of the point on the curve $y = \dfrac{x+1}{\sqrt{x-1}}$ where the gradient is 0.

8 Find the equation of the normal to the curve $y = \dfrac{x^2+1}{\sqrt{x+2}}$ at the point $(-1, 2)$.

9 The line $2x - 2y = 5$ intersects the curve $2x^2y - x^2 - 26y - 35 = 0$ at three points.

   a Find the x-coordinates of the points of intersection.
   b Find the gradient of the curve at each of the points of intersection.

4.3 Derivatives of exponential functions

The derivative of e^x

In Chapter 2, we learnt about the natural exponential function $f(x) = e^x$. This function has a very special property. If we use graphing software to draw the graph of $f(x) = e^x$ together with the graph of its gradient function we find that the two curves are identical.

> **KEY POINT 4.3**
>
> Hence, we have the rule:
>
> $$\dfrac{d}{dx}(e^x) = e^x$$

An explanation of how this rule can be obtained is as follows.

Consider the function $f(x) = e^x$ and two points whose x-coordinates are x and $x + \delta x$ where δx is a small increase in x.

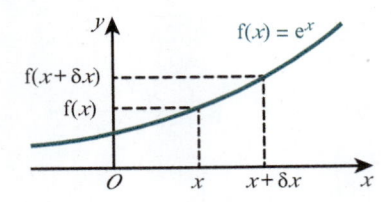

$$\frac{dy}{dx} = \lim_{\delta x \to 0} \frac{f(x + \delta x) - f(x)}{(x + \delta x) - x} = \lim_{\delta x \to 0} \frac{e^{x+\delta x} - e^x}{\delta x} = \lim_{\delta x \to 0} \frac{e^x(e^{\delta x} - 1)}{\delta x}$$

Now consider $\dfrac{e^{\delta x} - 1}{\delta x}$ for small values of δx.

δx	0.1	0.01	0.001	0.0001
$\dfrac{e^{\delta x} - 1}{\delta x}$	1.051 709	1.005 017	1.000 500	1.000 050

From the table, we can see that as $\delta x \to 0$, $\dfrac{e^{\delta x} - 1}{\delta x} \to 1$.

$\therefore \dfrac{dy}{dx} = e^x$

> **TIP**
>
> $\lim\limits_{\delta x \to 0} \dfrac{f(x + \delta x) - f(x)}{(x + \delta x) - x}$ means the limit of $\dfrac{f(x + \delta x) - f(x)}{(x + \delta x) - x}$ as $\delta x \to 0$.

The derivative of $e^{f(x)}$

Consider the function $y = e^{f(x)}$.

Let $y = e^u$ where $u = f(x)$

$\dfrac{dy}{du} = e^u \qquad \dfrac{du}{dx} = f'(x)$

Using the chain rule:
$$\frac{dy}{dx} = \frac{dy}{du} \times \frac{du}{dx}$$
$$= e^u \times f'(x)$$
$$= f'(x) \times e^{f(x)}$$
$$\frac{d}{dx}[e^{f(x)}] = f'(x) \times e^{f(x)}$$

> **KEY POINT 4.4**
>
> In particular,
> $$\frac{d}{dx}[e^{ax+b}] = a e^{ax+b}$$

WORKED EXAMPLE 4.5

Differentiate with respect to x.

a $\quad e^{2x}$
b $\quad xe^{-5x}$
c $\quad \dfrac{e^{3x}}{x^2}$

Answer

a $\quad \dfrac{d}{dx}(e^{2x}) = \underbrace{2}_{\text{differentiate index}} \times \underbrace{e^{2x}}_{\text{original function}} = 2e^{2x}$

b $\quad \dfrac{d}{dx}(xe^{-5x}) = x \times \dfrac{d}{dx}(e^{-5x}) + e^{-5x} \times \dfrac{d}{dx}(x)$ .... Product rule.

$\qquad = x \times (-5e^{-5x}) + e^{-5x} \times (1)$

$\qquad = -5xe^{-5x} + e^{-5x}$

$\qquad = e^{-5x}(1 - 5x)$

c $\quad \dfrac{d}{dx}\left(\dfrac{e^{3x}}{x^2}\right) = \dfrac{x^2 \times \dfrac{d}{dx}(e^{3x}) - e^{3x} \times \dfrac{d}{dx}(x^2)}{x^4}$ .... Quotient rule.

$\qquad = \dfrac{x^2 \times 3e^{3x} - e^{3x} \times 2x}{x^4}$

$\qquad = \dfrac{e^{3x}(3x - 2)}{x^3}$

WORKED EXAMPLE 4.6

The equation of a curve is $y = e^{2x} - 9e^x + 7x$. Find the exact value of the x-coordinate at each of the stationary points and determine the nature of each stationary point.

Answer

$y = e^{2x} - 9e^x + 7x$ .... Differentiate.

$\dfrac{dy}{dx} = 2e^{2x} - 9e^x + 7$

$\dfrac{d^2y}{dx^2} = 4e^{2x} - 9e^x$

Stationary points occur when $\dfrac{dy}{dx} = 0$.

$$2e^{2x} - 9e^x + 7 = 0 \quad \text{............................ Factorise.}$$
$$(2e^x - 7)(e^x - 1) = 0$$
$$(2e^x - 7) = 0 \text{ or } (e^x - 1) = 0$$
$$e^x = \frac{7}{2} \text{ or } e^x = 1$$
$$x = \ln\left(\frac{7}{2}\right) \text{ or } x = 0$$

When $x = 0$, $\frac{d^2y}{dx^2}$ is negative $\Rightarrow$ maximum point.

When $x = \ln\left(\frac{7}{2}\right)$, $\frac{d^2y}{dx^2}$ is positive $\Rightarrow$ minimum point.

EXERCISE 4C

1 Differentiate with respect to x.

   a e^{5x} b e^{-4x} c $2e^{6x}$

   d $3e^{-5x}$ e $4e^{\frac{x}{2}}$ f e^{2x-7}

   g e^{x^2-3} h $2x + 3e^{\sqrt{x}}$ i $5\sqrt{e^x} - \frac{1}{e^{2x}}$

   j $2(e^{3x} - 1)$ k $\frac{3e^{2x} + e^{-2x}}{2}$ l $5(e^{x^2} - 2x)$

2 a Sketch the graph of the function $y = 1 - e^{2-x}$.

   b Find the equation of the normal to the curve $y = 1 - e^{2-x}$ at the point where $y = 0$.

M 3 The mass, m grams, of a radioactive substance remaining t years after a given time, is given by the formula $m = 300e^{-0.00012t}$. Find the rate at which the mass is decreasing when $t = 2000$.

4 Differentiate with respect to x.

   a xe^x b $x^2 e^{3x}$ c $5xe^{-2x}$

   d $2\sqrt{x}\, e^x$ e $\dfrac{e^{6x}}{x}$ f $\dfrac{e^{-2x}}{\sqrt{x}}$

   g $\dfrac{e^x - 1}{e^x + 2}$ h $xe^{3x} + \dfrac{e^{6x}}{2}$ i $\dfrac{x^2 e^x - x}{e^x + 2}$

5 Find the gradient of the curve $y = \dfrac{8}{5 + e^{2x}}$ at the point where $x = 0$.

6 Find the exact coordinates of the stationary point on the curve $y = xe^x$.

7 The curve $y = 2e^{2x} + e^{-x}$ cuts the y-axis at the point P. Find the equation of the tangent to the curve at the point P and state the coordinates of the point where this tangent cuts the x-axis.

8 Find the exact coordinates of the stationary point on the curve $y = (x - 4)e^x$ and determine its nature.

9 Find the exact coordinates of the stationary point on the curve $y = \dfrac{e^{2x}}{x^2}$ and determine its nature.

10 The equation of a curve is $y = x^2 e^{-x}$.

    a Find the x-coordinates of the stationary points of the curve and determine the nature of these stationary points.

    b Show that the equation of the normal to the curve at the point where $x = 1$ is $e^2 x + ey = 1 + e^2$.

11 Find the exact value of the x-coordinates of the points on the curve $y = x^2 e^{-2x}$ at which $\dfrac{d^2 y}{dx^2} = 0$.

12 Find the coordinates of the stationary point on the curve $y = \dfrac{e^{2x-1}}{x}$.

E P 13 By writing 2 as $e^{\ln 2}$, prove that $\dfrac{d}{dx}(2^x) = 2^x \ln 2$.

PS 14 The equation of a curve is $y = x(3^x)$.

    Find the exact value of the gradient of the tangent to the curve at the point where $x = 1$.

4.4 Derivatives of natural logarithmic functions

EXPLORE 4.2

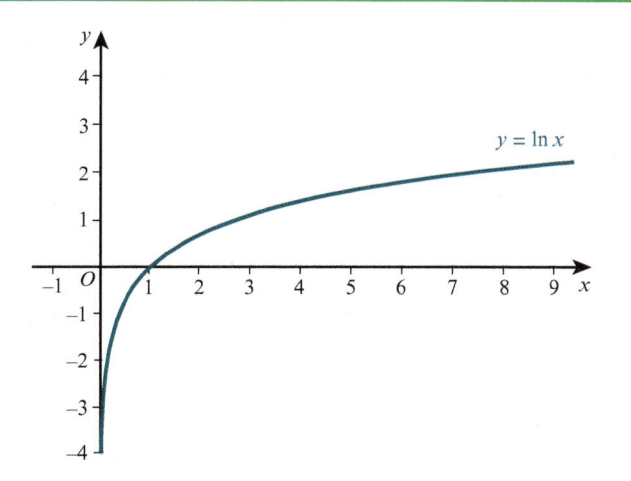

Here is a sketch of the graph of $y = \ln x$. Can you sketch a graph of its derivative (gradient function)? Discuss your sketch with your classmates.

Now sketch the graphs using graphing software. Can you suggest a formula for the derivative of $\ln x$? (It might be helpful to look at the coordinates of some of the points on the graph of the derivative.)

The derivative of $\ln x$

In Chapter 2, you learnt that if $y = \ln x$, then $e^y = x$.

In Section 4.3, you learnt that if $y = e^x$, then $\dfrac{dy}{dx} = e^x$.

Using these two results we can find the rule for differentiating $\ln x$.

$\quad y = \ln x$

$\quad e^y = x$ Differentiate both sides with respect to y.

$\quad e^y = \dfrac{dx}{dy}$ Rearrange and replace e^y by x.

$\quad \dfrac{dy}{dx} = \dfrac{1}{x}$

KEY POINT 4.5

$$\frac{d}{dx}(\ln x) = \frac{1}{x}$$

The derivative of ln f(x)

Consider the function $y = \ln f(x)$.

Let $y = \ln u$ where $u = f(x)$

$$\frac{dy}{du} = \frac{1}{u} \qquad \frac{du}{dx} = f'(x)$$

Using the chain rule:
$$\frac{dy}{dx} = \frac{dy}{du} \times \frac{du}{dx}$$
$$= \frac{1}{u} \times f'(x)$$
$$= \frac{f'(x)}{f(x)}$$

$$\frac{d}{dx}[\ln f(x)] = \frac{f'(x)}{f(x)}$$

KEY POINT 4.6

In particular,

$$\frac{d}{dx}[\ln(ax + b)] = \frac{a}{ax + b}$$

WORKED EXAMPLE 4.7

Differentiate with respect to x.

a $\ln 2x$ \qquad b $\ln(4x - 3)$ \qquad c $\ln \sqrt{5 - x}$

Answer

a $\quad \dfrac{d}{dx}(\ln 2x) = \dfrac{2}{2x}$

$\qquad\qquad\quad = \dfrac{1}{x}$

b $\quad \dfrac{d}{dx}[\ln(4x - 3)] = \dfrac{4}{4x - 3}$ \qquad 'Inside' differentiated.

\qquad\qquad\qquad\qquad\qquad\qquad\qquad\qquad 'Inside'

c **Method 1:**

$$\frac{d}{dx}\left[\ln\sqrt{5-x}\right] = \frac{\frac{1}{2}(5-x)^{-\frac{1}{2}}(-1)}{\sqrt{5-x}}$$ .... 'Inside' differentiated.

$$= -\frac{1}{2(5-x)}$$ .... 'Inside'

Method 2: using the rules of logarithms before differentiating.

$$\frac{d}{dx}\left[\ln\sqrt{5-x}\right] = \frac{d}{dx}\left[\ln(5-x)^{\frac{1}{2}}\right]$$ .... Use $\ln a^m = m \ln a$

$$= \frac{d}{dx}\left[\frac{1}{2}\ln(5-x)\right]$$

$$= \frac{1}{2} \times \frac{d}{dx}[\ln(5-x)]$$

$$= \frac{1}{2} \times \frac{-1}{5-x}$$ .... 'Inside' differentiated.

$$= -\frac{1}{2(5-x)}$$ .... 'Inside'

WORKED EXAMPLE 4.8

Differentiate with respect to x.

a $2x^4 \ln 5x$ **b** $\dfrac{\ln 2x}{x^3}$

Answer

a $\dfrac{d}{dx}(2x^4 \ln 5x) = 2x^4 \times \dfrac{d}{dx}(\ln 5x) + \ln 5x \times \dfrac{d}{dx}(2x^4)$ .... Product rule.

$$= 2x^4 \times \frac{5}{5x} + \ln 5x \times 8x^3$$

$$= 2x^3 + 8x^3 \ln 5x$$

b $\dfrac{d}{dx}\left(\dfrac{\ln 2x}{x^3}\right) = \dfrac{x^3 \times \dfrac{d}{dx}(\ln 2x) - \ln 2x \times \dfrac{d}{dx}(x^3)}{(x^3)^2}$ .... Quotient rule.

$$= \frac{x^3 \times \dfrac{2}{2x} - \ln 2x \times 3x^2}{x^6}$$

$$= \frac{x^2 - 3x^2 \ln 2x}{x^6}$$

$$= \frac{1 - 3\ln 2x}{x^4}$$

EXERCISE 4D

1 Differentiate with respect to x.

   a $\ln 3x$
   b $\ln 7x$
   c $\ln(2x+1)$
   d $5 + \ln(x^2 + 1)$
   e $\ln(2x-1)^2$
   f $\ln\sqrt{x-3}$
   g $\ln(x+3)^5$
   h $3x + \ln\left(\dfrac{2}{x}\right)$
   i $5x + \ln\left(\dfrac{2}{1-2x}\right)$
   j $\ln(\ln x)$
   k $\ln\left(2 - \sqrt{x}\right)^2$
   l $\ln(5x + \ln x)$

PS 2 The answers to **question 1 parts a** and **b** are the same. Why is this the case? How many different ways can you justify this?

3 Differentiate with respect to x.

   a $x \ln x$
   b $2x^3 \ln x$
   c $x \ln(2x + 1)$
   d $3x \ln 2x$
   e $x \ln(\ln x)$
   f $\dfrac{\ln 5x}{x}$
   g $\dfrac{2}{\ln x}$
   h $\dfrac{\ln(3x-2)}{x}$
   i $\dfrac{\ln(2x+1)}{4x-1}$

4 a Sketch the graph of the function $y = \ln(2x - 3)$.

   b Find the gradient of the curve $y = \ln(2x - 3)$ at the point where $x = 5$.

5 Find the gradient of the curve $y = e^{2x} - 5\ln(2x+1)$ at the point where $x = 0$.

6 A curve has equation $y = x^2 \ln 5x$.

   Find the value of $\dfrac{dy}{dx}$ and $\dfrac{d^2y}{dx^2}$ at the point where $x = 2$.

7 The equation of a curve is $y = x^2 \ln x$. Find the exact coordinates of the stationary point on this curve and determine whether it is a maximum or a minimum point.

8 The equation of a curve is $y = \dfrac{\ln x}{x}$. Find the exact coordinates of the stationary point on this curve and determine whether it is a maximum or a minimum point.

9 Find the equation of the tangent to the curve $y = \ln(5x - 4)$ at the point where $x = 1$.

10 Use the laws of logarithms to help differentiate these expressions with respect to x.

   a $\ln\sqrt{5x-1}$
   b $\ln\left(\dfrac{1}{3x+2}\right)$
   c $\ln[x(x+1)^5]$
   d $\ln\left(\dfrac{2x+3}{x-1}\right)$
   e $\ln\left(\dfrac{1-3x}{x^2}\right)$
   f $\ln\left[\dfrac{x(x-2)}{x+4}\right]$
   g $\ln\left[\dfrac{3-x}{(x+4)(x-1)}\right]$
   h $\ln\left[\dfrac{8}{(x+1)^2(x-2)}\right]$
   i $\ln\left[\dfrac{(x+2)(2x-1)}{x(x+5)}\right]$

11 Find $\dfrac{dy}{dx}$, in terms of x, for each of the following.

   a $e^y = 2x^2 - 1$
   b $e^y = 3x^3 + 2x$
   c $e^y = (x+1)(x-5)$

> **TIP**
>
> Take the natural logarithm of both sides of the equation before differentiating.

PS **12** A curve has equation $x = \dfrac{1}{5}\left[e^{y(2x-3)} + 4\right]$.

Find the value of $\dfrac{dy}{dx}$ when $x = 1$.

4.5 Derivatives of trigonometric functions

EXPLORE 4.3

Graphing software has been used to draw the graphs of $y = \sin x$ and $y = \cos x$ for $0 \leq x \leq 2\pi$ together with their gradient (derived) functions.

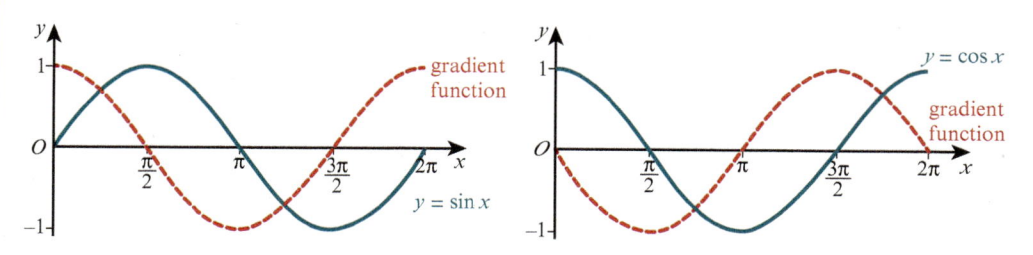

Discuss with your classmates why, for the function $y = \sin x$, you would expect the graph of the gradient function to have this shape. Do the same for the graph of $y = \cos x$.

Can you suggest how to complete the following formulae?

$\dfrac{d}{dx}(\sin x) = \qquad \dfrac{d}{dx}(\cos x) =$

The derivative of sin x

Consider the function $f(x) = \sin x$, where x is measured in radians, and two points whose x-coordinates are x and $x + \delta x$ where δx is a small increase in x.

$$\dfrac{dy}{dx} = \lim_{\delta x \to 0} \dfrac{f(x + \delta x) - f(x)}{(x + \delta x) - x}$$

$$= \lim_{\delta x \to 0} \dfrac{\sin(x + \delta x) - \sin x}{\delta x} \qquad \text{Expand } \sin(x + \delta x)$$

$$= \lim_{\delta x \to 0} \dfrac{\sin x \cos \delta x + \cos x \sin \delta x - \sin x}{\delta x} \qquad \text{As } \delta x \to 0,\ \cos \delta x \to 1$$
$$\text{and } \sin \delta x \to \delta x.$$

$$= \cos x$$

KEY POINT 4.7

$$\frac{d}{dx}(\sin x) = \cos x$$

Similarly, we can show that

$$\frac{d}{dx}(\cos x) = -\sin x$$

We can find the derivative of $\tan x$ using these two results together with the quotient rule.

$$\frac{d}{dx}(\tan x) = \frac{d}{dx}\left(\frac{\sin x}{\cos x}\right)$$ Use the quotient rule.

$$= \frac{\cos x \times \frac{d}{dx}(\sin x) - \sin x \times \frac{d}{dx}(\cos x)}{(\cos x)^2}$$

$$= \frac{\cos x \times \cos x - \sin x \times (-\sin x)}{\cos^2 x}$$

$$= \frac{\cos^2 x + \sin^2 x}{\cos^2 x}$$ Use $\cos^2 x + \sin^2 x = 1$.

$$= \frac{1}{\cos^2 x}$$ Use $\frac{1}{\cos x} = \sec x$.

$$= \sec^2 x$$

KEY POINT 4.8

$$\frac{d}{dx}(\tan x) = \sec^2 x$$

WORKED EXAMPLE 4.9

Differentiate with respect to x.

    a $2\sin x$ **b** $x\cos x$ **c** $\dfrac{\tan x}{x^2}$ **d** $(3 + 2\sin x)^5$

Answer

    a $\dfrac{d}{dx}(2\sin x) = 2\dfrac{d}{dx}(\sin x)$

                        $= 2\cos x$

    b $\dfrac{d}{dx}(x\cos x) = x \times \dfrac{d}{dx}(\cos x) + \cos x \times \dfrac{d}{dx}(x)$ Product rule.

                             $= -x\sin x + \cos x$

c $\quad \dfrac{d}{dx}\left(\dfrac{\tan x}{x^2}\right) = \dfrac{x^2 \times \dfrac{d}{dx}(\tan x) - \tan x \times \dfrac{d}{dx}(x^2)}{x^4}$ ..... Quotient rule.

$\quad = \dfrac{x^2 \times \sec^2 x - \tan x \times 2x}{x^4}$

$\quad = \dfrac{x\sec^2 x - 2\tan x}{x^3}$

d $\quad \dfrac{d}{dx}[(3 + 2\sin x)^5] = 5(3 + 2\sin x)^4 \times 2\cos x$ ..... Chain rule.

$\quad = 10\cos x(3 + 2\sin x)^4$

Derivatives of $\sin(ax + b)$, $\cos(ax + b)$ and $\tan(ax + b)$

Consider the function $y = \sin(ax + b)$ where x is measured in radians.

Let $\quad y = \sin u \quad$ where $\quad u = ax + b$

$\dfrac{dy}{du} = \cos u \qquad\qquad \dfrac{du}{dx} = a$

Using the chain rule: $\dfrac{dy}{dx} = \dfrac{dy}{du} \times \dfrac{du}{dx}$

$\qquad\qquad\qquad\quad = \cos u \times a$

$\qquad\qquad\qquad\quad = a\cos(ax + b)$

KEY POINT 4.9

$\dfrac{d}{dx}[\sin(ax + b)] = a\cos(ax + b)$

Similarly, it can be shown that:

$\dfrac{d}{dx}[\cos(ax + b)] = -a\sin(ax + b)$

$\dfrac{d}{dx}[\tan(ax + b)] = a\sec^2(ax + b)$

TIP

It is important to remember that, in calculus, all angles are measured in radians unless a question tells you otherwise.

WORKED EXAMPLE 4.10

Differentiate with respect to x.

a $\quad 2\sin 3x \qquad\qquad$ b $\quad 4x\tan 2x \qquad\qquad$ c $\quad \dfrac{\cos\left(2x - \dfrac{\pi}{4}\right)}{x^2} \qquad\qquad$ d $\quad (3 - 2\cos 5x)^4$

Answer

a $\quad \dfrac{d}{dx}(2\sin 3x) = 2\dfrac{d}{dx}(\sin 3x)$

$\qquad\qquad\qquad\quad = 2 \times \cos 3x \times (3)$

$\qquad\qquad\qquad\quad = 6\cos 3x$

b $\dfrac{d}{dx}(4x\tan 2x) = 4x \times \dfrac{d}{dx}(\tan 2x) + \tan 2x \times \dfrac{d}{dx}(4x)$ .... Product rule.

$\quad = 4x \times \sec^2 2x\,(2) + \tan 2x \times (4)$

$\quad = 8x\sec^2 2x + 4\tan 2x$

c $\dfrac{d}{dx}\left[\dfrac{\cos\left(2x - \dfrac{\pi}{4}\right)}{x^2}\right] = \dfrac{x^2 \times \dfrac{d}{dx}\left[\cos\left(2x - \dfrac{\pi}{4}\right)\right] - \cos\left(2x - \dfrac{\pi}{4}\right) \times \dfrac{d}{dx}[x^2]}{(x^2)^2}$ ... Quotient rule.

$\quad = \dfrac{-2x^2 \sin\left(2x - \dfrac{\pi}{4}\right) - 2x\cos\left(2x - \dfrac{\pi}{4}\right)}{x^4}$

$\quad = \dfrac{-2x\sin\left(2x - \dfrac{\pi}{4}\right) - 2\cos\left(2x - \dfrac{\pi}{4}\right)}{x^3}$

d $\dfrac{d}{dx}\left[(3 - 2\cos 5x)^4\right] = 4(3 - 2\cos 5x)^3 \times 10\sin 5x$ .... Chain rule.

$\quad = 40\sin 5x(3 - 2\cos 5x)^3$

EXERCISE 4E

1 Differentiate with respect to x.
- **a** $2 + \sin x$
- **b** $2\sin x + 3\cos x$
- **c** $2\cos x - \tan x$
- **d** $3\sin 2x$
- **e** $4\tan 5x$
- **f** $2\cos 3x - \sin 2x$
- **g** $\tan(3x + 2)$
- **h** $\sin\left(2x + \dfrac{\pi}{3}\right)$
- **i** $2\cos\left(3x - \dfrac{\pi}{6}\right)$

2 Differentiate with respect to x.
- **a** $\sin^3 x$
- **b** $5\cos^2 3x$
- **c** $\sin^2 x - 2\cos x$
- **d** $(3 - \cos x)^4$
- **e** $2\sin^3\left(2x + \dfrac{\pi}{6}\right)$
- **f** $3\cos^4 x + 2\tan^2\left(2x - \dfrac{\pi}{4}\right)$

3 Differentiate with respect to x.
- **a** $x\sin x$
- **b** $5x\cos 3x$
- **c** $x^2 \tan x$
- **d** $x\cos^3 2x$
- **e** $\dfrac{5}{\cos 3x}$
- **f** $\dfrac{x}{\cos x}$
- **g** $\dfrac{\tan x}{x}$
- **h** $\dfrac{\sin x}{2 + \cos x}$
- **i** $\dfrac{\sin x}{3x - 1}$
- **j** $\dfrac{1}{\sin^3 2x}$
- **k** $\dfrac{3x}{\sin 2x}$
- **l** $\dfrac{\sin x + \cos x}{\sin x - \cos x}$

4 Differentiate with respect to x.
- **a** $e^{\sin x}$
- **b** $e^{\cos 2x}$
- **c** $e^{\tan 3x}$
- **d** $e^{(\sin x - \cos x)}$
- **e** $e^x \cos x$
- **f** $e^x \sin 2x$
- **g** $e^x(2\cos x - \sin x)$
- **h** $x^3 e^{\cos x}$
- **i** $\ln(\cos x)$
- **j** $x\ln(\sin x)$
- **k** $\dfrac{\cos 2x}{e^{2x+1}}$
- **l** $\dfrac{x\sin 2x}{e^{2x}}$

5 Find the gradient of the curve $y = 3\sin 2x - 5\tan x$ at the point where $x = 0$.

6 Find the exact value of the gradient of the curve $y = 2\sin 3x - 4\cos x$ at the point $\left(\dfrac{\pi}{3}, -2\right)$.

7 Given that $y = \sin^2 x$ for $0 \leqslant x \leqslant \pi$, find the exact values of the x-coordinates of the points on the curve where the gradient is $\dfrac{\sqrt{3}}{2}$.

P 8 Prove that the gradient of the curve $y = \dfrac{5}{2 - \tan x}$ is always positive.

9 a By writing $\sec x$ as $\dfrac{1}{\cos x}$, find $\dfrac{d}{dx}(\sec x)$.

   b By writing $\operatorname{cosec} x$ as $\dfrac{1}{\sin x}$, find $\dfrac{d}{dx}(\operatorname{cosec} x)$.

   c By writing $\cot x$ as $\dfrac{\cos x}{\sin x}$, find $\dfrac{d}{dx}(\cot x)$.

P 10 Prove that the normal to the curve $y = x\sin x$ at the point $P\left(\dfrac{\pi}{2}, \dfrac{\pi}{2}\right)$ intersects the x-axis at the point $(\pi, 0)$.

11 The equation of a curve is $y = 5\sin 3x - 2\cos x$. Find the equation of the tangent to the curve at the point $\left(\dfrac{\pi}{3}, -1\right)$. Give the answer in the form $y = mx + c$, where the values of m and c are correct to 3 significant figures.

12 A curve has equation $y = 3\cos 2x + 4\sin 2x + 1$ for $0 \leqslant x \leqslant \pi$. Find the x-coordinates of the stationary points of the curve, giving your answer correct to 3 significant figures.

13 A curve has equation $y = e^x \cos x$ for $0 \leqslant x \leqslant \dfrac{\pi}{2}$. Find the exact value of the x-coordinate of the stationary point of the curve and determine the nature of this stationary point.

14 A curve has equation $y = \dfrac{\sin 2x}{e^{2x}}$ for $0 \leqslant x \leqslant \dfrac{\pi}{2}$. Find the exact value for the x-coordinate of the stationary point of this curve.

15 A curve has equation $y = \dfrac{e^{3x}}{\sin 3x}$ for $0 < x < \dfrac{\pi}{2}$. Find the exact value for the x-coordinate of the stationary point of this curve and determine the nature of this point.

PS 16 A curve has equation $y = \sin 2x - x$ for $0 \leqslant x \leqslant 2\pi$. Find the x-coordinates of the stationary points of the curve, and determine the nature of these stationary points.

PS 17 A curve has equation $y = \tan x \cos 2x$ for $0 \leqslant x < \dfrac{\pi}{2}$. Find the x-coordinate of the stationary point on the curve giving your answer correct to 3 significant figures.

4.6 Implicit differentiation

All the functions that we have differentiated so far have been of the form $y = f(x)$. These are called **explicit functions** as y is given explicitly in terms of x. However, many functions cannot be expressed in this form, for example $x^3 + 5xy + y^3 = 16$.

When a function is given as an equation connecting x and y, where y is not the subject, it is called an **implicit function**.

The chain rule $\left(\dfrac{dy}{dx} = \dfrac{dy}{du} \times \dfrac{du}{dx}\right)$ and the product rule $\left(\dfrac{d}{dx}(uv) = u\dfrac{dv}{dx} + v\dfrac{du}{dx}\right)$ are used extensively to differentiate implicit functions.

This is illustrated in the following Worked examples.

WORKED EXAMPLE 4.11

Differentiate each of these expressions with respect to x.

a $\quad y^3$
b $\quad 4x^2 y^5$
c $\quad x^3 + 5xy$

Answer

a $\quad \dfrac{d}{dx}(y^3) = \dfrac{d}{dy}(y^3) \times \dfrac{dy}{dx}$ .... Chain rule.

$\quad = 3y^2 \dfrac{dy}{dx}$

b $\quad \dfrac{d}{dx}(4x^2 y^5) = 4x^2 \dfrac{d}{dx}(y^5) + y^5 \dfrac{d}{dx}(4x^2)$ .... Product rule.

$\quad = 4x^2 \dfrac{d}{dx}(y^5) + 8xy^5$ .... Chain rule: $\dfrac{d}{dx}(y^5) = 5y^4 \dfrac{dy}{dx}$.

$\quad = 20x^2 y^4 \dfrac{dy}{dx} + 8xy^5$

c $\quad \dfrac{d}{dx}(x^3 + 5xy) = \dfrac{d}{dx}(x^3) + \dfrac{d}{dx}(5xy)$ .... Use the product rule for $\dfrac{d}{dx}(5xy)$.

$\quad = 3x^2 + 5x \dfrac{d}{dx}(y) + y \dfrac{d}{dx}(5x)$

$\quad = 3x^2 + 5x \dfrac{dy}{dx} + 5y$

WORKED EXAMPLE 4.12

Find $\dfrac{dy}{dx}$ for the curve $x^3 + y^3 = 4xy$. Hence find the gradient of the curve at the point (2, 2).

Answer

$\quad x^3 + y^3 = 4xy$ .... Differentiate with respect to x.

$\dfrac{d}{dx}(x^3) + \dfrac{d}{dx}(y^3) = \dfrac{d}{dx}(4xy)$

$3x^2 + \dfrac{d}{dy}(y^3) \times \dfrac{dy}{dx} = 4x \dfrac{d}{dx}(y) + y \dfrac{d}{dx}(4x)$

$3x^2 + 3y^2 \dfrac{dy}{dx} = 4x \dfrac{dy}{dx} + 4y$ .... Rearrange terms.

$3y^2 \dfrac{dy}{dx} - 4x \dfrac{dy}{dx} = 4y - 3x^2$ .... Factorise.

$(3y^2 - 4x) \dfrac{dy}{dx} = 4y - 3x^2$ .... Rearrange.

$\dfrac{dy}{dx} = \dfrac{4y - 3x^2}{3y^2 - 4x}$

When $x = 2$ and $y = 2$,

$$\frac{dy}{dx} = \frac{4(2) - 3(2)^2}{3(2)^2 - 4(2)} = -1$$

Gradient of the curve at the point (2, 2) is −1.

WORKED EXAMPLE 4.13

The equation of a curve is $3x^2 + 2xy + y^2 = 6$. Find the coordinates of the two stationary points on the curve.

Answer

$3x^2 + 2xy + y^2 = 6$ .... Differentiate with respect to x.

$\frac{d}{dx}(3x^2) + \frac{d}{dx}(2xy) + \frac{d}{dx}(y^2) = \frac{d}{dx}(6)$ .... Use product rule for $\frac{d}{dx}(2xy)$.

$6x + 2x\frac{d}{dx}(y) + y\frac{d}{dx}(2x) + \frac{d}{dx}(y^2) = 0$ .... Use chain rule for $\frac{d}{dx}(y^2)$.

$6x + 2x\frac{dy}{dx} + 2y + 2y\frac{dy}{dx} = 0$ .... Divide by 2 and rearrange.

$x\frac{dy}{dx} + y\frac{dy}{dx} = -y - 3x$ .... Factorise.

$(x + y)\frac{dy}{dx} = -y - 3x$ .... Rearrange.

$\frac{dy}{dx} = \frac{-y - 3x}{x + y}$

Stationary points occur when $\frac{dy}{dx} = 0$.

$-y - 3x = 0$

$y = -3x$

Substituting $y = -3x$ into the equation of the curve gives:

$3x^2 + 2x(-3x) + (-3x)^2 = 6$

$6x^2 = 6$

$x = \pm 1$

The stationary points are (−1, 3) and (1, −3).

EXERCISE 4F

1 Differentiate each expression with respect to x.

- **a** y^5
- **b** $x^3 + 2y^2$
- **c** $5x^2 + \ln y$
- **d** $2 + \sin y$
- **e** $6x^2 y^3$
- **f** $y^2 + xy$
- **g** $x^3 - 7xy + y^3$
- **h** $x \sin y + y \cos x$
- **i** $x^3 \ln y$
- **j** $x \cos 2y$
- **k** $5y + e^x \sin y$
- **l** $2xe^{\cos y}$

2 Find $\dfrac{dy}{dx}$ for each of these functions.

    a $x^3 + 2xy + y^3 = 10$ **b** $x^2y + y^2 = 5x$ **c** $2x^2 + 5xy + y^2 = 8$

    d $x \ln y = 2x + 5$ **e** $2e^x y + e^{2x} y^3 = 10$ **f** $\ln(xy) = 4 - y^2$

    g $xy^3 = 2 \ln y$ **h** $\ln x - 2 \ln y + 5xy = 3$

3 Find the gradient of the curve $x^2 + 3xy - 5y + y^3 = 22$ at the point $(1, 3)$.

4 Find the gradient of the curve $2x^3 - 4xy + y^3 = 16$ at the point where the curve crosses the x-axis.

5 A curve has equation $2x^2 + 3y^2 - 2x + 4y = 4$. Find the equation of the tangent to the curve at the point $(1, -2)$.

6 The equation of a curve is $4x^2 y + 8 \ln x + 2 \ln y = 4$. Find the equation of the normal to the curve at the point $(1, 1)$.

7 The equation of a curve is $5x^2 + 2xy + 2y^2 = 45$.

    a   Given that there are two points on the curve where the tangent is parallel to the x-axis, show by differentiation that, at these points, $y = -5x$.

    b   Hence find the coordinates of the two points.

8 The equation of a curve is $y^2 - 4xy - x^2 = 20$.

    a   Find the coordinates of the two points on the curve where $x = 4$.

    b   Show that at one of these points the tangent to the curve is parallel to the x-axis.

    c   Find the equation of the tangent to the curve at the other point.

9 The equation of a curve is $y^3 - 12xy + 16 = 0$.

    a   Show that the curve has no stationary points.

    b   Find the coordinates of the point on the curve where the tangent is parallel to the y-axis.

10 Find the gradient of the curve $5e^x y^2 + 2e^x y = 88$ at the point $(0, 4)$.

11 The equation of a curve is $x^2 - 4x + 6y + 2y^2 = 12$. Find the coordinates of the two points on the curve at which the gradient is $\dfrac{4}{3}$.

12 The equation of a curve is $2x + y \ln x = 4y$. Find the equation of the tangent to the curve at the point with coordinates $\left(1, \dfrac{1}{2}\right)$.

E PS   13 The equation of a curve is $y = x^x$. Find the exact value of the x-coordinate of the stationary point on this curve.

PS   14 Find the stationary points on the curve $x^2 - xy + y^2 = 48$. By finding $\dfrac{d^2y}{dx^2}$ determine the nature of each of these stationary points.

4.7 Parametric differentiation

Sometimes variables x and y are given as a function of a third variable t.

For example, $x = 1 + 4\sin t$ and $y = 2\cos t$.

The variable t is called a **parameter** and the two equations are called the parametric equations of the curve.

We can find the curve given by the parametric equations $x = 1 + 4\sin t$, $y = 2\cos t$ for $0 \leqslant t \leqslant 2\pi$ by finding the values of x and y for particular values of t.

t	0	$\frac{1}{4}\pi$	$\frac{1}{2}\pi$	$\frac{3}{4}\pi$	π	$\frac{5}{4}\pi$	$\frac{3}{2}\pi$	$\frac{7}{4}\pi$	2π
x	1	3.83	5	3.83	1	−1.83	−3	−1.83	1
y	2	1.41	0	−1.41	−2	−1.41	0	1.41	2

Plotting the coordinates on a grid gives the following graph.

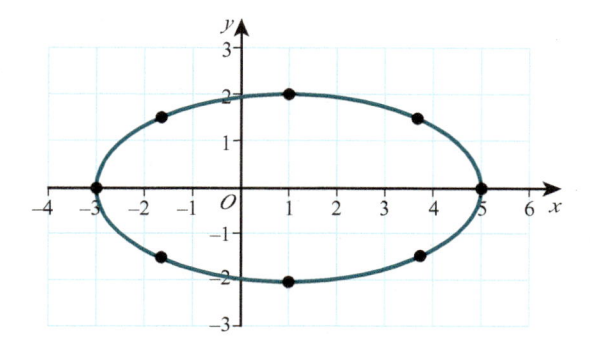

EXPLORE 4.4

1 Use tables of values for t, x and y to sketch each of the following curves. Use graphing software to check your answers.

   a $x = t^3$, $y = t^2$

   b $x = t^2$, $y = t^3 - t$

   c $x = t + \dfrac{1}{t}$, $y = t - \dfrac{1}{t}$

2 Use graphing software to draw the curves given by these parametric equations.

   a $x = \sin t$, $y = \sin 2t$

   b $x = \sin 2t$, $y = \sin 3t$

   c $x = \sin 4t$, $y = \sin 3t$

3 The curves in **question 2** are called Lissajous curves. Use the internet to find out more about these curves. (A Lissajous curve is shown at the start of this chapter.)

When a curve is given in parametric form, in terms of the parameter t, we can use the chain rule to find $\dfrac{dy}{dx}$ in terms of t.

WORKED EXAMPLE 4.14

The parametric equations of a curve are $x = t^2$, $y = 5 - 2t$.

a Find $\dfrac{dy}{dx}$ in terms of the parameter t.

b Find the equation of the tangent to the curve at the point where $t = 2$.

Answer

a $x = t^2 \implies \dfrac{dx}{dt} = 2t$

$y = 5 - 2t \implies \dfrac{dy}{dt} = -2$

$\dfrac{dy}{dx} = \dfrac{dy}{dt} \times \dfrac{dt}{dx}$ .... Chain rule.

$= -2 \times \dfrac{1}{2t}$

$= -\dfrac{1}{t}$

b When $t = 2$, gradient $= -\dfrac{1}{2}$, $x = 4$ and $y = 1$.

Using $y - y_1 = m(x - x_1)$

$y - 1 = -\dfrac{1}{2}(x - 4)$

$y = -\dfrac{1}{2}x + 3$

WORKED EXAMPLE 4.15

The parametric equations of a curve are $x = 1 + 2\sin^2\theta$, $y = 4\tan\theta$ for $\dfrac{\pi}{2} < \theta < \dfrac{3\pi}{2}$.

a Find $\dfrac{dy}{dx}$ in terms of the parameter θ.

b Find the coordinates of the point on the curve where the tangent is parallel to the y-axis.

Answer

a $x = 1 + 2\sin^2\theta \implies \dfrac{dx}{d\theta} = 4\sin\theta\cos\theta$

$y = 4\tan\theta \implies \dfrac{dy}{d\theta} = 4\sec^2\theta = \dfrac{4}{\cos^2\theta}$

$\dfrac{dy}{dx} = \dfrac{dy}{d\theta} \times \dfrac{d\theta}{dx}$ .... Chain rule.

$= \dfrac{4}{\cos^2\theta} \times \dfrac{1}{4\sin\theta\cos\theta}$

$= \dfrac{1}{\sin\theta\cos^3\theta}$

b Tangent is parallel to the y-axis when:

$\sin\theta \cos^3\theta = 0$

$\sin\theta = 0$ or $\cos\theta = 0$

$\sin\theta = 0 \Rightarrow \theta = \pi \Rightarrow (1, 0)$

There are no solutions to $\cos\theta = 0$ in the range $\frac{\pi}{2} < \theta < \frac{3\pi}{2}$.

∴ The point where the tangent is parallel to the y-axis is $(1, 0)$.

EXERCISE 4G

1 For each of the following parametric equations, find $\frac{dy}{dx}$ in terms of the given parameter.

a $x = 2t^3$, $y = t^2 - 5$

b $x = 2 + \sin 2\theta$, $y = 4\theta + 2\cos 2\theta$

c $x = 2\theta - \sin 2\theta$, $y = 2 - \cos 2\theta$

d $x = 3\tan\theta$, $y = 2\sin 2\theta$

e $x = 1 + \tan\theta$, $y = \cos\theta$

f $x = \cos 2\theta - \cos\theta$, $y = \sin^2\theta$

g $x = 1 + 2\sin^2\theta$, $y = 4\tan\theta$

h $x = 2 + e^{-t}$, $y = e^t - e^{-t}$

i $x = e^{2t}$, $y = t^3 e^t + 1$

j $x = 2\ln(t + 3)$, $y = 4e^t$

k $x = \ln(1 - t)$, $y = \frac{5}{t}$

l $x = 1 + \sqrt{t}$, $y = 2\ln t$

2 The parametric equations of a curve are $x = 3t$, $y = t^3 + 4t^2 - 3t$. Find the two values of t for which the curve has gradient 0.

3 The parametric equations of a curve are $x = 2\sin\theta$, $y = 1 - 3\cos 2\theta$. Find the exact gradient of the curve at the point where $\theta = \frac{\pi}{3}$.

4 The parametric equations of a curve are $x = 2 + \ln(t - 1)$, $y = t + \frac{4}{t}$ for $t > 1$. Find the coordinates of the only point on the curve at which the gradient is equal to 0.

5 The parametric equations of a curve are $x = e^{2t}$, $y = 1 + 2te^t$. Find the equation of the normal to the curve at the point where $t = 0$.

6 The parametric equations of a curve are $x = e^{2t}$, $y = t^2 e^{-t} - 1$.

a Show that $\frac{dy}{dx} = \frac{t(2-t)}{2e^{3t}}$.

b Show that the tangent to the curve at the point $(1, -1)$ is parallel to the x-axis and find the exact coordinates of the other point on the curve at which the tangent is parallel to the x-axis.

7 A curve is defined by the parametric equations $x = \tan\theta$, $y = 2\sin 2\theta$, for $0 < \theta < \dfrac{\pi}{2}$.

   a Show that $\dfrac{dy}{dx} = 4\cos^2\theta(2\cos^2\theta - 1)$.

   b Hence, find the coordinates of the stationary point.

8 The parametric equations of a curve are $x = t + 4\ln t$, $y = t + \dfrac{9}{t}$, for $t > 0$.

   a Show that $\dfrac{dy}{dx} = \dfrac{t^2 - 9}{t^2 + 4t}$.

   b The curve has one stationary point. Find the y-coordinate of this point and determine whether it is a maximum or a minimum point.

9 The parametric equations of a curve are $x = 1 + 2\sin^2\theta$, $y = 1 + 2\tan\theta$. Find the equation of the normal to the curve at the point where $\theta = \dfrac{\pi}{4}$.

10 The parametric equations of a curve are $x = 2\sin\theta + \cos 2\theta$, $y = 1 + \cos 2\theta$, for $0 \leqslant \theta \leqslant \dfrac{\pi}{2}$.

   a Show that $\dfrac{dy}{dx} = \dfrac{2\sin\theta}{2\sin\theta - 1}$.

   b Find the coordinates of the point on the curve where the tangent is parallel to the x-axis.

   c Show that the tangent to the curve at the point $\left(\dfrac{3}{2}, \dfrac{3}{2}\right)$ is parallel to the y-axis.

PS 11 The parametric equations of a curve are $x = \ln(\tan t)$, $y = 2\sin 2t$, for $0 < t < \dfrac{\pi}{2}$.

   a Show that $\dfrac{dy}{dx} = \sin 4t$.

   b Hence show that at the point where $x = 0$ the tangent is parallel to the x-axis.

WEB LINK

Try the *Parametric points* resource on the Underground Mathematics website.

Checklist of learning and understanding

Product rule

- $\dfrac{d}{dx}(uv) = u\dfrac{dv}{dx} + v\dfrac{du}{dx}$

Quotient rule

- $\dfrac{d}{dx}\left(\dfrac{u}{v}\right) = \dfrac{v\dfrac{du}{dx} - u\dfrac{dv}{dx}}{v^2}$

Exponential functions

- $\dfrac{d}{dx}(e^x) = e^x \qquad \dfrac{d}{dx}\left[e^{ax+b}\right] = ae^{ax+b} \qquad \dfrac{d}{dx}\left[e^{f(x)}\right] = f'(x) \times e^{f(x)}$

Logarithmic functions

- $\dfrac{d}{dx}(\ln x) = \dfrac{1}{x} \qquad \dfrac{d}{dx}[\ln(ax+b)] = \dfrac{a}{ax+b} \qquad \dfrac{d}{dx}[\ln(f(x))] = \dfrac{f'(x)}{f(x)}$

Trigonometric functions

- $\dfrac{d}{dx}(\sin x) = \cos x \qquad \dfrac{d}{dx}[\sin(ax+b)] = a\cos(ax+b)$

- $\dfrac{d}{dx}(\cos x) = -\sin x \qquad \dfrac{d}{dx}[\cos(ax+b)] = -a\sin(ax+b)$

- $\dfrac{d}{dx}(\tan x) = \sec^2 x \qquad \dfrac{d}{dx}[\tan(ax+b)] = a\sec^2(ax+b)$

END-OF-CHAPTER REVIEW EXERCISE 4

1. The parametric equations of a curve are
   $$x = 1 + \ln(t-2), \; y = t + \frac{9}{t}, \text{ for } t > 2.$$

   i Show that $\dfrac{dy}{dx} = \dfrac{(t^2 - 9)(t-2)}{t^2}$. [3]

   ii Find the coordinates of the only point on the curve at which the gradient is equal to 0. [3]

   Cambridge International AS & A Level Mathematics 9709 Paper 21 Q4 November 2010

2. Find the value of $\dfrac{dy}{dx}$ when $x = 4$ in each of the following cases:

   i $y = x\ln(x-3)$, [4]

   ii $y = \dfrac{x-1}{x+1}$. [3]

   Cambridge International AS & A Level Mathematics 9709 Paper 21 Q5 June 2011

3. The parametric equations of a curve are $x = e^{3t}, \; y = t^2 e^t + 3$.

   i Show that $\dfrac{dy}{dx} = \dfrac{t(t+2)}{3e^{2t}}$. [4]

   ii Show that the tangent to the curve at the point $(1, 3)$ is parallel to the x-axis. [2]

   iii Find the exact coordinates of the other point on the curve at which the tangent is parallel to the x-axis. [2]

   Cambridge International AS & A Level Mathematics 9709 Paper 21 Q7 November 2011

4. Find the gradient of each of the following curves at the point for which $x = 0$.

   i $y = 3\sin x + \tan 2x$ [3]

   ii $y = \dfrac{6}{1 + e^{2x}}$ [3]

   Cambridge International AS & A Level Mathematics 9709 Paper 21 Q2 June 2014

5. The equation of a curve is $2x^2 + 3xy + y^2 = 3$.

   i Find the equation of the tangent to the curve at the point $(2, -1)$, giving your answer in the form $ax + by + c = 0$, where a, b and c are integers. [6]

   ii Show that the curve has no stationary points. [4]

   Cambridge International AS & A Level Mathematics 9709 Paper 21 Q7 June 2014

6. The equation of a curve is $y^3 + 4xy = 16$.

   i Show that $\dfrac{dy}{dx} = -\dfrac{4y}{3y^2 + 4x}$. [4]

   ii Show that the curve has no stationary points. [2]

   iii Find the coordinates of the point on the curve where the tangent is parallel to the y-axis. [4]

   Cambridge International AS & A Level Mathematics 9709 Paper 21 Q7 June 2015

7. The equation of a curve is $y = 6\sin x - 2\cos 2x$.

   Find the equation of the tangent to the curve at the point $\left(\dfrac{1}{6}\pi, 2\right)$. Give the answer in the form $y = mx + c$, where the values of m and c are correct to 3 significant figures. [5]

   Cambridge International AS & A Level Mathematics 9709 Paper 21 Q3 June 2015

8

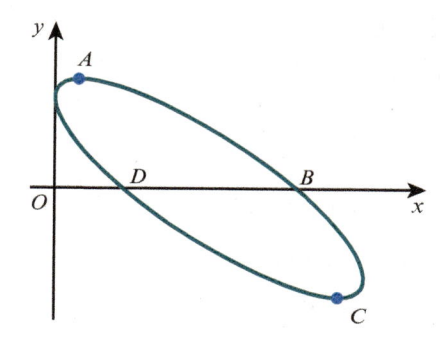

The parametric equations of a curve are $x = 6\sin^2 t$, $y = 2\sin 2t + 3\cos 2t$, for $0 \leq t \leq \pi$. The curve crosses the x-axis at the points B and D and the stationary points are A and C, as shown in the diagram.

i Show that $\dfrac{dy}{dx} = \dfrac{2}{3}\cot 2t - 1$. [5]

ii Find the values of t at A and C, giving each answer correct to 3 decimal places. [3]

iii Find the value of the gradient of the curve at B. [3]

Cambridge International AS & A Level Mathematics 9709 Paper 21 Q7 November 2015

9 The equation of a curve is $3x^2 + 4xy + y^2 = 24$. Find the equation of the normal to the curve at the point $(1, 3)$, giving your answer in the form $ax + by + c = 0$ where a, b and c are integers. [8]

Cambridge International AS & A Level Mathematics 9709 Paper 21 Q6 November 2016

10

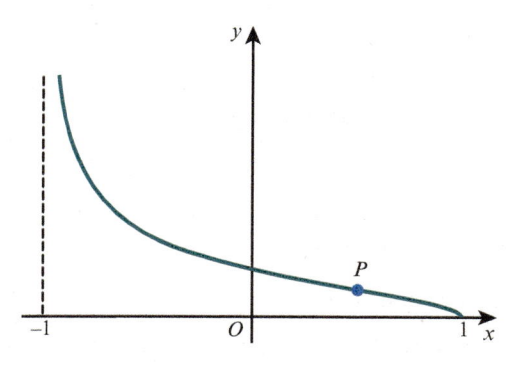

The diagram shows the curve $y = \sqrt{\left(\dfrac{1-x}{1+x}\right)}$.

i By first differentiating $\dfrac{1-x}{1+x}$, obtain an expression for $\dfrac{dy}{dx}$ in terms of x. Hence show that the gradient of the normal to the curve at the point (x, y) is $(1+x)\sqrt{(1-x^2)}$. [5]

ii The gradient of the normal to the curve has its maximum value at the point P shown in the diagram. Find, by differentiation, the x-coordinate of P. [4]

Cambridge International A Level Mathematics 9709 Paper 31 Q9 June 2010

11 The parametric equations of a curve are $x = 3(1 + \sin^2 t)$, $y = 2\cos^3 t$.

   Find $\dfrac{dy}{dx}$ in terms of t, simplifying your answer as far as possible. [5]

   Cambridge International A Level Mathematics 9709 Paper 31 Q2 November 2011

12 The equation of a curve is $\ln(xy) - y^3 = 1$.

   i Show that $\dfrac{dy}{dx} = \dfrac{y}{x(3y^3 - 1)}$. [4]

   ii Find the coordinates of the point where the tangent to the curve is parallel to the y-axis, giving each coordinate correct to 3 significant figures. [4]

   Cambridge International A Level Mathematics 9709 Paper 31 Q7 November 2012

13 The curve with equation $6e^{2x} + ke^y + e^{2y} = c$, where k and c are constants, passes through the point P with coordinates $(\ln 3, \ln 2)$.

   i Show that $58 + 2k = c$. [2]

   ii Given also that the gradient of the curve at P is -6, find the values of k and c. [5]

   Cambridge International A Level Mathematics 9709 Paper 31 Q5 June 2011

Chapter 5
Integration

In this chapter you will learn how to:

- extend the idea of 'reverse differentiation' to include the integration of e^{ax+b}, $\dfrac{1}{ax+b}$, $\sin(ax+b)$, $\cos(ax+b)$ and $\sec^2(ax+b)$
- use trigonometrical relationships in carrying out integration
- understand and use the trapezium rule to estimate the value of a definite integral.

Cambridge International AS & A Level Mathematics: Pure Mathematics 2 & 3

PREREQUISITE KNOWLEDGE

Where it comes from	What you should be able to do	Check your skills
Chapter 4	Differentiate standard functions.	1 Find $\dfrac{dy}{dx}$ when: a $y = 3\sin 2x - 5\cos x$ b $y = e^{5x-2}$ c $y = \ln(2x + 1)$
Chapter 4	Use standard trigonometric relationships to prove trigonometric identities.	2 Prove that $\cos 4x + 4\cos 2x \equiv 8\cos^4 x - 3$.

Why do we study integration?

In this chapter we will learn how to integrate exponential, logarithmic and trigonometrical functions and how to apply integration to solve problems.

It is important that you have a good grasp of the work covered in Chapter 3 on trigonometry, as you will be expected to know and use the trigonometrical relationships when solving integration problems.

P2 Lastly, we will learn how to find an estimate for $\int_a^b f(x)\,dx$ using a numerical method. This is used when we are not able to find the value of $\int_a^b f(x)\,dx$ using an algebraic method.

5.1 Integration of exponential functions

Since integration is the reverse process of differentiation, the rules for integrating exponential functions are:

KEY POINT 5.1

$$\int e^x\,dx = e^x + c \qquad \int e^{ax+b}\,dx = \frac{1}{a}e^{ax+b} + c$$

WORKED EXAMPLE 5.1

Find:

a $\displaystyle\int e^{3x}\,dx$ b $\displaystyle\int e^{-2x}\,dx$ c $\displaystyle\int e^{5x-4}\,dx$

Answer

a $\displaystyle\int e^{3x}\,dx = \frac{1}{3}e^{3x} + c$ b $\displaystyle\int e^{-2x}\,dx = \frac{1}{-2}e^{-2x} + c$ c $\displaystyle\int e^{5x-4}\,dx = \frac{1}{5}e^{5x-4} + c$

$\qquad\qquad\qquad\qquad\qquad\quad = -\dfrac{1}{2}e^{-2x} + c$

REWIND

This chapter builds on the work we did in Pure Mathematics 1 Coursebook, Chapter 9, where we learnt how to integrate $(ax + b)^n$, where $n \neq -1$.

WEB LINK

Explore the *Calculus of trigonometry and logarithms* station on the Underground Mathematics website.

REWIND

In Chapter 4, we learnt the following rules for differentiating exponential functions:

$$\frac{d}{dx}(e^x) = e^x$$

$$\frac{d}{dx}(e^{ax+b}) = ae^{ax+b}$$

WORKED EXAMPLE 5.2

Evaluate $\int_1^2 9e^{3x-1} \, dx$.

Answer

$\int_1^2 9e^{3x-1} \, dx = \left[\dfrac{9}{3} e^{3x-1} \right]_1^2$ .... Substitute limits.

$\qquad = (3e^5) - (3e^2)$

$\qquad = 3e^2(e^3 - 1)$

WORKED EXAMPLE 5.3

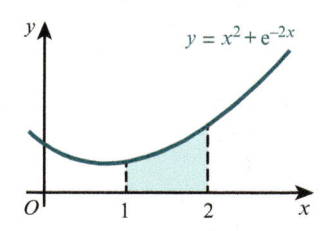

[Graph showing $y = x^2 + e^{-2x}$ with shaded region between $x=1$ and $x=2$]

Find the area of the shaded region.

Answer

Area $= \int_1^2 (x^2 + e^{-2x}) \, dx$

$\qquad = \left[\dfrac{x^3}{3} - \dfrac{1}{2} e^{-2x} \right]_1^2$ .... Substitute limits.

$\qquad = \left(\dfrac{8}{3} - \dfrac{1}{2} e^{-4} \right) - \left(\dfrac{1}{3} - \dfrac{1}{2} e^{-2} \right)$ .... Simplify.

$\qquad = \dfrac{7}{3} + \dfrac{1}{2} e^{-2} - \dfrac{1}{2} e^{-4}$

$\qquad \approx 2.39 \text{ units}^2$

EXERCISE 5A

1 Find:

a $\int e^{2x} \, dx$
b $\int e^{-4x} \, dx$
c $\int 6e^{3x} \, dx$
d $\int 4e^{\frac{1}{2}x} \, dx$
e $\int 2e^{-x} \, dx$
f $\int e^{2x+4} \, dx$
g $\int e^{3x-1} \, dx$
h $\int 6e^{2-3x} \, dx$
i $\int 2e^{8x-3} \, dx$

2 Find:

a $\displaystyle\int e^{-x}(1+e^x)\,dx$

b $\displaystyle\int 5e^x(2+e^{3x})\,dx$

c $\displaystyle\int (e^{2x}-1)^2\,dx$

d $\displaystyle\int \frac{4+e^{2x}}{e^{2x}}\,dx$

e $\displaystyle\int \frac{8-e^x}{2e^{2x}}\,dx$

f $\displaystyle\int \frac{(e^x-2)^2}{e^{2x}}\,dx$

3 Evaluate:

a $\displaystyle\int_0^2 e^{3x}\,dx$

b $\displaystyle\int_0^{\frac{1}{2}} e^{4x}\,dx$

c $\displaystyle\int_0^{\ln 2} 5e^{-2x}\,dx$

d $\displaystyle\int_0^3 e^{1+2x}\,dx$

e $\displaystyle\int_0^1 \frac{8}{e^{2x-1}}\,dx$

f $\displaystyle\int_0^1 (e^x+1)^2\,dx$

g $\displaystyle\int_0^1 (e^x+e^{2x})^2\,dx$

h $\displaystyle\int_0^1 \left(2e^x - \frac{5}{e^x}\right)^2 dx$

i $\displaystyle\int_0^2 \frac{(e^x+1)^2}{e^{2x}}\,dx$

4 A curve is such that $\dfrac{dy}{dx} = 6e^{2x} + 2e^{-x}$. Given that the curve passes through the point $(0, 2)$, find the equation of the curve.

5 A curve is such that $\dfrac{d^2y}{dx^2} = 20e^{-2x}$. Given that $\dfrac{dy}{dx} = -8$ when $x = 0$ and that the curve passes through the point $\left(1, \dfrac{5}{e^2}\right)$, find the equation of the curve.

6 Find the exact area of the region bounded by the curve $y = 1 + e^{2x-5}$, the x-axis and the lines $x = 1$ and $x = 3$.

7

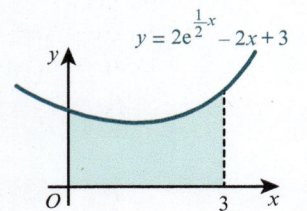

$y = 2e^{\frac{1}{2}x} - 2x + 3$

Find the exact area of the shaded region.

8 a Show that $\dfrac{d}{dx}(xe^x - e^x) = xe^x$.

b

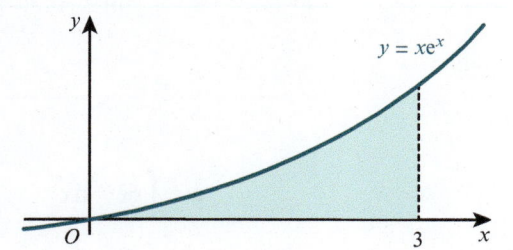

$y = xe^x$

Use your result from **part a** to evaluate the area of the shaded region.

9 a Find $\displaystyle\int_0^a (4e^{-2x} + 5e^{-x})\,dx$, where a is a positive constant.

b Hence find the value of $\displaystyle\int_0^\infty (4e^{-2x} + 5e^{-x})\,dx$.

10

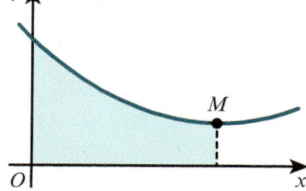

The diagram shows the curve $y = 2e^x + 8e^{-x} - 7$ and its minimum point M.

Find the area of the shaded region.

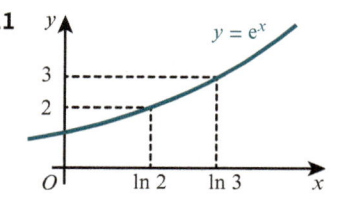

11

The diagram shows the graph of $y = e^x$. The points $(\ln 2, 2)$ and $(\ln 3, 3)$ lie on the curve.

   a Find the value of $\int_{\ln 2}^{\ln 3} e^x \, dx$.

   b Hence show that $\int_2^3 \ln y \, dy = \ln\left(\dfrac{27}{4e}\right)$.

5.2 Integration of $\dfrac{1}{ax+b}$

Since integration is the reverse process of differentiation, the rules for integration are:

KEY POINT 5.2

$$\int \frac{1}{x} \, dx = \ln x + c, \quad x > 0 \qquad \int \frac{1}{ax+b} \, dx = \frac{1}{a} \ln(ax+b) + c, \quad ax+b > 0$$

REWIND

In Chapter 4, we learnt the following rules for differentiating logarithmic functions:

$\dfrac{d}{dx}(\ln x) = \dfrac{1}{x}, \, x > 0$

$\dfrac{d}{dx}[\ln(ax+b)] = \dfrac{a}{ax+b}, \, ax+b > 0$

It is important to remember that $\ln x$ is only defined for $x > 0$.

WORKED EXAMPLE 5.4

Find each of these integrals and state the values of x for which the integral is valid.

  a $\int \dfrac{2}{x} \, dx$ **b** $\int \dfrac{4}{2x+1} \, dx$ **c** $\int \dfrac{6}{2-3x} \, dx$

Answer

a $\int \dfrac{2}{x} \, dx = 2 \int \dfrac{1}{x} \, dx$

            $= 2 \ln x + c, \quad x > 0$

b $\int \dfrac{4}{2x+1} \, dx = 4 \int \dfrac{1}{2x+1} \, dx$

                 $= 4\left(\dfrac{1}{2}\right) \ln(2x+1) + c$ Valid for $2x+1 > 0$.

                 $= 2 \ln(2x+1) + c, \quad x > -\dfrac{1}{2}$

c $\quad\int \dfrac{6}{2-3x}\,dx = 6\int \dfrac{1}{2-3x}\,dx$

$\qquad\qquad\qquad = 6\left(\dfrac{1}{-3}\right)\ln(2-3x) + c \quad\cdots\cdots\cdots$ Valid for $2-3x > 0$.

$\qquad\qquad\qquad = -2\ln(2-3x) + c,\ x < \dfrac{2}{3}$

EXPLORE 5.1

Fiona is asked to find $\int x^{-1}\,dx$.

She tries to use the formula $\int x^n\,dx = \dfrac{1}{n+1}x^{n+1} + c$ to obtain her answer.

Fiona is also asked to find $\int (2x+3)^{-1}\,dx$.

She tries to use the formula $\int (ax+b)^n\,dx = \dfrac{1}{a(n+1)}(ax+b)^{n+1} + c$ to obtain her answer.

Discuss with your classmates why Fiona's methods do not work.

EXPLORE 5.2

Rafiu and Fausat are asked to find $\int \dfrac{1}{2(3x+1)}\,dx$.

Rafiu writes: $\int \dfrac{1}{2(3x+1)}\,dx = \int \dfrac{1}{6x+2}\,dx$

$\qquad\qquad\qquad = \dfrac{1}{6}\ln(6x+2) + c$

Fausat writes: $\int \dfrac{1}{2(3x+1)}\,dx = \dfrac{1}{2}\int \dfrac{1}{3x+1}\,dx$

$\qquad\qquad\qquad = \left(\dfrac{1}{2}\right)\left(\dfrac{1}{3}\right)\ln(3x+1) + c$

$\qquad\qquad\qquad = \dfrac{1}{6}\ln(3x+1) + c$

Decide who is correct and discuss the reasons for your decision with your classmates.

Consider the integrals $\int_{2}^{3}\dfrac{1}{x}\,dx$ and $\int_{-3}^{-2}\dfrac{1}{x}\,dx$.

From the symmetry properties of the graph of $y = \dfrac{1}{x}$ we can see that the shaded areas that represent the two integrals are equal in magnitude. However, one of the areas is below the x-axis, which suggests that $\int_{-3}^{-2}\dfrac{1}{x}\,dx = -\int_{2}^{3}\dfrac{1}{x}\,dx$.

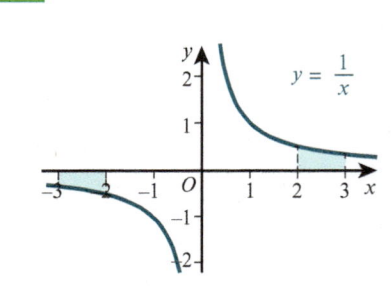

Evaluating $\int_2^3 \frac{1}{x} dx$ gives:

$$\int_2^3 \frac{1}{x} dx = \left[\ln x \right]_2^3 = \ln 3 - \ln 2 = \ln \frac{3}{2}$$

This implies that $\int_{-3}^{-2} \frac{1}{x} dx = -\ln \frac{3}{2} = \ln \frac{2}{3}$.

If we try using integration to find the value of $\int_{-3}^{-2} \frac{1}{x} dx$ we obtain:

$$\int_{-3}^{-2} \frac{1}{x} dx = \left[\ln x \right]_{-3}^{-2} = \ln(-2) - \ln(-3) = \ln \frac{2}{3}$$

There is, however, a problem with this calculation in that $\ln x$ is only defined for $x > 0$ so $\ln(-2)$ and $\ln(-3)$ do not actually exist.

Hence for $x < 0$, we say that $\int \frac{1}{x} dx = \ln|x| + c$.

In conclusion, the integration formulae are written as:

KEY POINT 5.3

$$\int \frac{1}{x} dx = \ln|x| + c \qquad \int \frac{1}{ax+b} dx = \frac{1}{a} \ln|ax+b| + c$$

WEB LINK

A more in-depth explanation for this formula can be found in the *Two for one* resource on the Underground Mathematics website.

TIP

It is normal practice to only include the modulus sign when finding definite integrals.

WORKED EXAMPLE 5.5

Find the value of $\int_2^3 \frac{6}{2-3x} dx$.

Answer

$\int_2^3 \frac{6}{2-3x} dx = \left[\left(\frac{6}{-3} \right) \ln|2-3x| \right]_2^3$ ····· Substitute limits.

$= (-2 \ln 7) - (-2 \ln 4)$

$= -2 \ln 7 + 2 \ln 4$ ····· Simplify.

$= 2 \ln \frac{4}{7}$

EXERCISE 5B

1 Find:

    a $\int \frac{6}{x} dx$ b $\int \frac{1}{2x} dx$ c $\int \frac{1}{3x+1} dx$

    d $\int \frac{6}{2x-5} dx$ e $\int \frac{5}{2-3x} dx$ f $\int \frac{3}{2(5x-1)} dx$

2 Evaluate:

    a $\int_0^5 \frac{1}{x+2} dx$ b $\int_1^4 \frac{1}{2x+1} dx$ c $\int_{-1}^4 \frac{3}{2x+5} dx$

    d $\int_2^4 \frac{2}{4x+1} dx$ e $\int_0^2 \frac{3}{2x-7} dx$ f $\int_2^3 \frac{4}{3-2x} dx$

3 Evaluate:

a $\displaystyle\int_4^{10}\left(2+\frac{5}{3x-2}\right)dx$

b $\displaystyle\int_1^3\left(\frac{3}{x}-\frac{4}{2x-1}\right)dx$

c $\displaystyle\int_{-1}^3\left(2x-1+\frac{4}{2x+3}\right)dx$

4 a Given that $\dfrac{4x}{2x-1}\equiv 2+\dfrac{A}{2x-1}$, find the value of the constant A.

b Hence show that $\displaystyle\int_1^5\frac{4x}{2x-1}dx=8+\ln 9$.

5 a Find the quotient and remainder when $6x^2+5x$ is divided by $2x-5$.

b Hence show that $\displaystyle\int_0^1\frac{6x^2+5x}{2x-5}dx=\frac{23}{2}-25\ln\left(\frac{5}{3}\right)$.

6 A curve is such that $\dfrac{dy}{dx}=2x+\dfrac{3}{x+e}$.

Given that the curve passes through the point (e, e^2), find the equation of the curve.

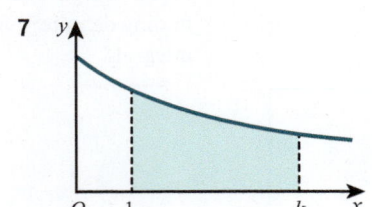

PS 7

The diagram shows part of the curve $y=\dfrac{2}{x+3}$. Given that the shaded region has area 4, find the value of k.

PS 8 The points $P(1, -2)$ and $Q(2, k)$ lie on the curve for which $\dfrac{dy}{dx}=3-\dfrac{2}{x}$. The tangents to the curve at the points P and Q intersect at the point R. Find the coordinates of R.

5.3 Integration of $\sin(ax+b)$, $\cos(ax+b)$ and $\sec^2(ax+b)$

In Chapter 4, we learnt how to differentiate some trigonometric functions:

$\dfrac{d}{dx}(\sin x)=\cos x$ $\qquad\qquad$ $\dfrac{d}{dx}[\sin(ax+b)]=a\cos(ax+b)$

$\dfrac{d}{dx}(\cos x)=-\sin x$ $\qquad\qquad$ $\dfrac{d}{dx}[\cos(ax+b)]=-a\sin(ax+b)$

$\dfrac{d}{dx}(\tan x)=\sec^2 x$ $\qquad\qquad$ $\dfrac{d}{dx}[\tan(ax+b)]=a\sec^2(ax+b)$

KEY POINT 5.4

Since integration is the reverse process of differentiation, the rules for integrating are:

$$\int \cos x \, dx = \sin x + c \qquad \int \cos(ax + b) \, dx = \frac{1}{a} \sin(ax + b) + c$$

$$\int \sin x \, dx = -\cos x + c \qquad \int \sin(ax + b) \, dx = -\frac{1}{a} \cos(ax + b) + c$$

$$\int \sec^2 x \, dx = \tan x + c \qquad \int \sec^2(ax + b) \, dx = \frac{1}{a} \tan(ax + b) + c$$

TIP

It is important to remember that the formulae for differentiating and integrating these trigonometric functions only apply when x is measured in radians.

WORKED EXAMPLE 5.6

Find:

a $\int \sin 2x \, dx$ **b** $\int \cos 3x \, dx$ **c** $\int 4 \sec^2 \frac{x}{2} \, dx$

Answer

a $\int \sin 2x \, dx = -\frac{1}{2} \cos 2x + c$ **b** $\int \cos 3x \, dx = \frac{1}{3} \sin 3x + c$ **c** $\int 4 \sec^2 \frac{x}{2} \, dx = 4 \int \sec^2 \frac{x}{2} \, dx$

$$= 4 \times \left(2 \tan \frac{x}{2}\right) + c$$

$$= 8 \tan \frac{x}{2} + c$$

WORKED EXAMPLE 5.7

Find the exact value of $\int_0^{\frac{\pi}{4}} (3 - 2 \sin 2x) \, dx$.

Answer

$\int_0^{\frac{\pi}{4}} (3 - 2 \sin 2x) \, dx = \left[3x + \frac{2}{2} \cos 2x \right]_0^{\frac{\pi}{4}}$ Substitute limits.

$= \left(\frac{3\pi}{4} + \cos \frac{\pi}{2} \right) - (0 + \cos 0)$

$= \left(\frac{3\pi}{4} + 0 \right) - (0 + 1)$

$= \frac{3\pi}{4} - 1$

EXERCISE 5C

1 Find:

a $\int \sin 3x \, dx$
b $\int \cos 4x \, dx$
c $\int \sin \frac{x}{2} \, dx$

d $\int 3\sin 2x \, dx$
e $\int 5\cos 3x \, dx$
f $\int \sec^2 2x \, dx$

g $\int 2\cos(1 - 5x) \, dx$
h $\int 3\sin(2x + 1) \, dx$
i $\int 2\sec^2(5x - 2) \, dx$

2 Evaluate:

a $\int_0^{\frac{1}{6}\pi} \cos 4x \, dx$
b $\int_0^{\frac{2}{3}\pi} \sin\left(\frac{1}{2}x\right) dx$
c $\int_0^{\frac{1}{6}\pi} \sec^2 2x \, dx$

d $\int_0^{\frac{1}{2}\pi} (\sin 2x + \cos x) \, dx$
e $\int_0^{\frac{1}{6}\pi} (\cos 2x + \sin x) \, dx$
f $\int_{-\frac{1}{4}\pi}^{\frac{1}{4}\pi} (5 - 2\sec^2 x) \, dx$

3 a Find $\dfrac{d}{dx}(x\sin x + \cos x)$.

b Hence find the exact value of $\displaystyle\int_0^{\frac{1}{3}\pi} x\cos x \, dx$.

4 A curve is such that $\dfrac{dy}{dx} = 1 - 3\sin 2x$.

Given that the curve passes through the point $\left(\dfrac{\pi}{4}, 0\right)$, find the equation of the curve.

5 A curve is such that $\dfrac{d^2y}{dx^2} = -12\sin 2x - 2\cos x$.

Given that $\dfrac{dy}{dx} = 4$ when $x = 0$ and that the curve passes through the point $\left(\dfrac{\pi}{2}, -3\right)$, find the equation of the curve.

6 The point $\left(\dfrac{\pi}{2}, 5\right)$ lies on the curve for which $\dfrac{dy}{dx} = 4\sin\left(2x - \dfrac{\pi}{2}\right)$.

a Find the equation of the curve.

b Find the equation of the normal to the curve at the point where $x = \dfrac{\pi}{3}$.

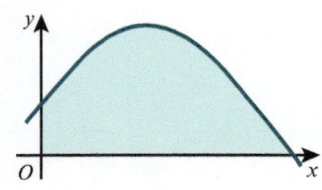

PS 7

The diagram shows part of the curve $y = 1 + \sqrt{3}\sin 2x + \cos 2x$. Find the exact value of the area of the shaded region.

8

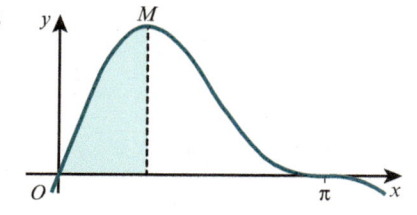

The diagram shows part of the curve $y = 3\sin 2x + 6\sin x$ and its maximum point M.

Find the exact area of the shaded region.

9

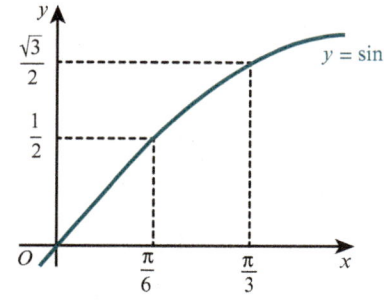

The diagram shows part of the graph of $y = \sin x$. The points $\left(\dfrac{\pi}{6}, \dfrac{1}{2}\right)$ and $\left(\dfrac{\pi}{3}, \dfrac{\sqrt{3}}{2}\right)$ lie on the curve.

a Find the exact value of $\displaystyle\int_{\frac{\pi}{6}}^{\frac{\pi}{3}} \sin x \, dx$.

b Hence show that $\displaystyle\int_{\frac{1}{2}}^{\frac{\sqrt{3}}{2}} (\sin^{-1} y) \, dy = \dfrac{\pi}{12}\left(2\sqrt{3} - 1\right) - \dfrac{\sqrt{3} - 1}{2}$.

10

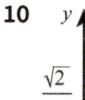

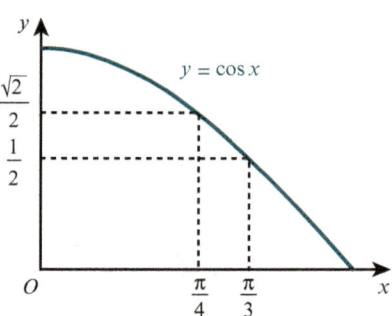

The diagram shows part of the graph of $y = \cos x$. The points $\left(\dfrac{\pi}{4}, \dfrac{\sqrt{2}}{2}\right)$ and $\left(\dfrac{\pi}{3}, \dfrac{1}{2}\right)$ lie on the curve.

a Find the exact value of $\displaystyle\int_{\frac{\pi}{4}}^{\frac{\pi}{3}} \cos x \, dx$.

b Hence show that $\displaystyle\int_{\frac{1}{2}}^{\frac{\sqrt{2}}{2}} (\cos^{-1} y) \, dy = \dfrac{\pi}{24}\left(3\sqrt{2} - 4\right) + \dfrac{\sqrt{3} - \sqrt{2}}{2}$.

5.4 Further integration of trigonometric functions

Sometimes, when it is not immediately obvious how to integrate a trigonometric function, we can use trigonometric identities to rewrite the function in a form that we can then integrate.

To integrate $\sin^2 x$ and $\cos^2 x$ we must use the double angle identities:

$$\cos 2x \equiv 1 - 2\sin^2 x \quad \text{and} \quad \cos 2x \equiv 2\cos^2 x - 1$$

Rearranging these two identities gives:

$$\sin^2 x \equiv \frac{1}{2}(1 - \cos 2x) \quad \text{and} \quad \cos^2 x \equiv \frac{1}{2}(1 + \cos 2x)$$

WORKED EXAMPLE 5.8

Find:

a $\displaystyle\int \sin^2 x \, dx$

b $\displaystyle\int 3\cos^2 2x \, dx$

Answer

a $\displaystyle\int \sin^2 x \, dx = \frac{1}{2}\int (1 - \cos 2x) \, dx$ .... Use $\sin^2 x \equiv \frac{1}{2}(1 - \cos 2x)$.

$\quad = \dfrac{1}{2}\left(x - \dfrac{1}{2}\sin 2x\right) + c$

$\quad = \dfrac{1}{2}x - \dfrac{1}{4}\sin 2x + c$

b $\displaystyle\int 3\cos^2 2x \, dx = \frac{3}{2}\int (1 + \cos 4x) \, dx$ .... Use $\cos^2 2x \equiv \frac{1}{2}(1 + \cos 4x)$.

$\quad = \dfrac{3}{2}\left(x + \dfrac{1}{4}\sin 4x\right) + c$

$\quad = \dfrac{3}{2}x + \dfrac{3}{8}\sin 4x + c$

WORKED EXAMPLE 5.9

Find $\displaystyle\int \sin^4 x \, dx$.

Answer

$\sin^4 x = (\sin^2 x)^2$ .... Use $\sin^2 x \equiv \frac{1}{2}(1 - \cos 2x)$.

$\quad = \left[\dfrac{1}{2}(1 - \cos 2x)\right]^2$

$\quad = \dfrac{1}{4}(1 - 2\cos 2x + \cos^2 2x)$ .... Use $\cos^2 2x \equiv \frac{1}{2}(1 + \cos 4x)$.

$\quad = \dfrac{1}{4}\left(1 - 2\cos 2x + \dfrac{1}{2} + \dfrac{1}{2}\cos 4x\right)$

$\quad = \dfrac{3}{8} - \dfrac{1}{2}\cos 2x + \dfrac{1}{8}\cos 4x$

$\displaystyle\int \sin^4 x \, dx = \int \left(\dfrac{3}{8} - \dfrac{1}{2}\cos 2x + \dfrac{1}{8}\cos 4x\right) dx$

$\quad = \dfrac{3}{8}x - \dfrac{1}{4}\sin 2x + \dfrac{1}{32}\sin 4x + c$

To integrate $\tan^2 x$ we use the identity $1 + \tan^2 x \equiv \sec^2 x$.

Rearranging this identity gives $\tan^2 x \equiv \sec^2 x - 1$.

WORKED EXAMPLE 5.10

Find the exact value of $\int_0^{\frac{\pi}{4}} 5\tan^2 x \, dx$.

Answer

$\int_0^{\frac{\pi}{4}} 5\tan^2 x \, dx = \int_0^{\frac{\pi}{4}} (5\sec^2 x - 5) \, dx$ — Use $\tan^2 x \equiv \sec^2 x - 1$.

$\phantom{\int_0^{\frac{\pi}{4}} 5\tan^2 x \, dx} = \left[5\tan x - 5x \right]_0^{\frac{\pi}{4}}$ — Substitute limits.

$\phantom{\int_0^{\frac{\pi}{4}} 5\tan^2 x \, dx} = \left(5 - \frac{5\pi}{4} \right) - (0 - 0)$

$\phantom{\int_0^{\frac{\pi}{4}} 5\tan^2 x \, dx} = 5 - \frac{5\pi}{4}$

Worked example 5.11 uses the principle that integration is the reverse process of differentiation.

WORKED EXAMPLE 5.11

a Show that if $y = \sec x$ then $\dfrac{dy}{dx} = \tan x \sec x$.

b Hence find the exact value of $\int_0^{\frac{1}{4}\pi} (\cos 2x + 5\tan x \sec x) \, dx$.

Answer

a $y = \dfrac{1}{\cos x}$ — Use the quotient rule.

$\dfrac{dy}{dx} = \dfrac{(\cos x)(0) - (1)(-\sin x)}{\cos^2 x}$

$\phantom{\dfrac{dy}{dx}} = \dfrac{\sin x}{\cos^2 x}$

$\phantom{\dfrac{dy}{dx}} = \tan x \sec x$

b $\int_0^{\frac{1}{4}\pi} (\cos 2x + 5\tan x \sec x) \, dx$ — Use $\int \tan x \sec x \, dx = \sec x$.

$\phantom{\int_0^{\frac{1}{4}\pi}} = \left[\dfrac{1}{2}\sin 2x + 5\sec x \right]_0^{\frac{1}{4}\pi}$ — Substitute limits.

$\phantom{\int_0^{\frac{1}{4}\pi}} = \left(\dfrac{1}{2} + 5\sqrt{2} \right) - (0 + 5)$

$\phantom{\int_0^{\frac{1}{4}\pi}} = 5\sqrt{2} - \dfrac{9}{2}$

WORKED EXAMPLE 5.12

a Prove that $2\operatorname{cosec} 2x \tan x \equiv \sec^2 x$.

b Hence find the exact value of $\displaystyle\int_{-\frac{1}{3}\pi}^{\frac{1}{3}\pi} (5 + \operatorname{cosec} 2x \tan x)\,dx$.

Answer

a $2\operatorname{cosec} 2x \tan x \equiv \dfrac{2\tan x}{\sin 2x}$ .... Use $\tan x = \dfrac{\sin x}{\cos x}$ and $\sin 2x = 2\sin x \cos x$.

$\equiv \dfrac{2\sin x}{(2\sin x \cos x)\cos x}$ .... Simplify.

$\equiv \dfrac{1}{\cos^2 x}$

$\equiv \sec^2 x$

b $\displaystyle\int_{-\frac{1}{3}\pi}^{\frac{1}{3}\pi}(5+\operatorname{cosec} 2x \tan x)\,dx = \int_{-\frac{1}{3}\pi}^{\frac{1}{3}\pi}\left(5+\dfrac{1}{2}\sec^2 x\right)dx$ .... Use **part a**.

$= \left[5x + \dfrac{1}{2}\tan x\right]_{-\frac{1}{3}\pi}^{\frac{1}{3}\pi}$ .... Simplify.

$= \left(\dfrac{5\pi}{3} + \dfrac{\sqrt{3}}{2}\right) - \left(-\dfrac{5\pi}{3} - \dfrac{\sqrt{3}}{2}\right)$

$= \dfrac{10\pi}{3} + \sqrt{3}$

WORKED EXAMPLE 5.13

a Use the expansions of $\cos(3x - x)$ and $\cos(3x + x)$ to show that:
$\cos 2x - \cos 4x \equiv 2\sin 3x \sin x$

b Hence show that $\displaystyle\int_0^{\frac{1}{4}\pi} \sin 3x \sin x\,dx = \dfrac{1}{4}$.

Answer

a $\cos 2x - \cos 4x \equiv \cos(3x - x) - \cos(3x + x)$

$\equiv (\cos 3x \cos x + \sin 3x \sin x) - (\cos 3x \cos x - \sin 3x \sin x)$

$\equiv 2\sin 3x \sin x$

b $\displaystyle\int_0^{\frac{1}{4}\pi}\sin 3x \sin x\,dx = \dfrac{1}{2}\int_0^{\frac{1}{4}\pi}(\cos 2x - \cos 4x)\,dx$ .... Use **part a**.

$= \dfrac{1}{2}\left[\dfrac{1}{2}\sin 2x - \dfrac{1}{4}\sin 4x\right]_0^{\frac{1}{4}\pi}$ .... Simplify.

$= \dfrac{1}{2}\left[\left(\dfrac{1}{2} - 0\right) - (0 - 0)\right]$

$= \dfrac{1}{4}$

EXERCISE 5D

1 Find:

   a $\displaystyle\int 3\cos^2 x \, dx$ **b** $\displaystyle\int 4\cos^2\left(\dfrac{x}{2}\right) dx$ **c** $\displaystyle\int \sin^2 3x \, dx$

   d $\displaystyle\int 2\tan^2 x \, dx$ **e** $\displaystyle\int 6\tan^2(3x) \, dx$ **f** $\displaystyle\int \cos^4 x \, dx$

2 Find the value of:

   a $\displaystyle\int_0^{\frac{1}{3}\pi} \sin^2 x \, dx$ **b** $\displaystyle\int_{\frac{1}{6}\pi}^{\frac{1}{3}\pi} 3\tan^2 x \, dx$ **c** $\displaystyle\int_0^{\frac{1}{3}\pi} \cos^2 x \, dx$

   d $\displaystyle\int_0^{\frac{1}{8}\pi} \cos^2 2x \, dx$ **e** $\displaystyle\int_0^{\frac{1}{6}\pi} 4\sin^2 x \, dx$ **f** $\displaystyle\int_0^{\frac{1}{6}\pi} \tan^2 2x \, dx$

3 Find the value of:

   a $\displaystyle\int_0^{\frac{1}{6}\pi} \left(\cos^2 x - \dfrac{1}{\cos^2 x}\right) dx$ **b** $\displaystyle\int_0^{\frac{1}{6}\pi} (\cos^2 x + \sin 2x) \, dx$ **c** $\displaystyle\int_0^{\frac{1}{4}\pi} (2\sin x + \cos x)^2 \, dx$

   d $\displaystyle\int_0^{\frac{1}{3}\pi} (\cos x - 3\sin x)^2 \, dx$ **e** $\displaystyle\int_0^{\frac{1}{6}\pi} \dfrac{1 + \cos^4 x}{\cos^2 x} \, dx$ **f** $\displaystyle\int_0^{\frac{1}{6}\pi} \dfrac{5}{1 + \cos 4x} \, dx$

4 a By differentiating $\dfrac{\cos x}{\sin x}$, show that if $y = \cot x$ then $\dfrac{dy}{dx} = -\csc^2 x$.

   b Hence, show that $\displaystyle\int_{\frac{1}{6}\pi}^{\frac{1}{4}\pi} \csc^2 2x \, dx = \dfrac{\sqrt{3}}{6}$.

5 a Show that $\tan x + \cot x \equiv \dfrac{2}{\sin 2x}$.

   b Hence show that $\displaystyle\int_{\frac{1}{3}\pi}^{\frac{1}{2}\pi} \dfrac{8}{\tan x + \cot x} \, dx = 1$.

6 a Use the expansions of $\sin(5x + 2x)$ and $\sin(5x - 2x)$ to show that:

     $\sin 7x + \sin 3x \equiv 2\sin 5x \cos 2x$

   b Hence show that $\displaystyle\int_{\frac{1}{6}\pi}^{\frac{1}{3}\pi} 2\sin 5x \cos 2x \, dx = \dfrac{11 - 3\sqrt{3}}{42}$.

7 a Show that $\sin 3x \equiv 3\sin x - 4\sin^3 x$.

   b Hence show that $\displaystyle\int_0^{\frac{1}{3}\pi} 2\sin^3 x \, dx = \dfrac{5}{12}$.

8 a Show that $\cos 3x \equiv 4\cos^3 x - 3\cos x$.

   b Hence show that $\displaystyle\int_0^{\frac{1}{6}\pi} (4\cos^3 x + 2\cos x) \, dx = \dfrac{17}{6}$.

9 The diagram shows part of the curve $y = \cos x + 2\sin x$. Find the exact volume of the solid formed when the shaded region is rotated through 360° about the x-axis.

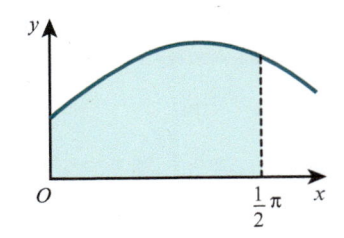

10 a Show that $\sec^2 x + \sec x \tan x \equiv \dfrac{1}{1 - \sin x}$.

b Show that if $y = \sec x$ then $\dfrac{dy}{dx} = \sec x \tan x$.

c Hence show that $\displaystyle\int_0^{\frac{1}{6}\pi} \dfrac{4}{1 - \sin 2x}\, dx = 2(1 + \sqrt{3})$.

5.5 The trapezium rule

We already know that the area, A, under the curve $y = f(x)$ between $x = a$ and $x = b$ can be found by evaluating $\displaystyle\int_a^b f(x)\, dx$. Sometimes we might not be able to find the value of $\displaystyle\int_a^b f(x)\, dx$ by the integration methods that we have covered so far or maybe the function cannot be integrated algebraically.

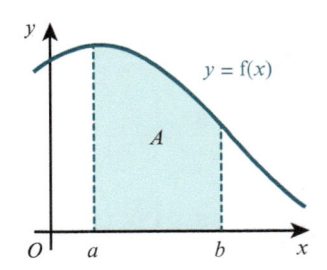

In these situations an approximate answer can be found using a numerical method called the **trapezium rule**.

This numerical method involves splitting the area under the curve $y = f(x)$ between $x = a$ and $x = b$ into equal width strips.

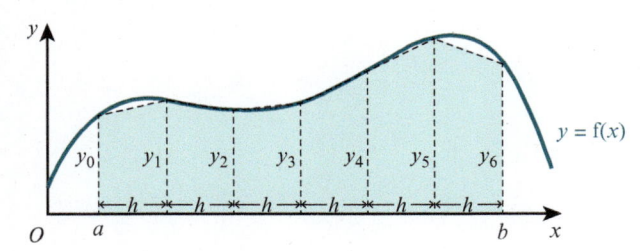

The area of each trapezium-shaped strip can be found using the formula:

$$\text{area} = \frac{1}{2}(\text{sum of parallel sides}) \times \text{width}$$

The sum of the areas of all the strips gives an approximate value for the area under the curve and hence an approximate value for $\displaystyle\int_a^b f(x)\, dx$.

For the previous diagram, there are 6 strips, each of width h, where $h = \dfrac{b - a}{6}$ and the length of the vertical edges (ordinates) of the strips are y_0, y_1, y_2, y_3, y_4, y_5 and y_6.

The sum of the areas of the trapezium-shaped strips is:

$$\frac{h}{2}(y_0 + y_1) + \frac{h}{2}(y_1 + y_2) + \frac{h}{2}(y_2 + y_3) + \frac{h}{2}(y_3 + y_4) + \frac{h}{2}(y_4 + y_5) + \frac{h}{2}(y_5 + y_6)$$

$$= \frac{h}{2}[y_0 + y_6 + 2(y_1 + y_2 + y_3 + y_4 + y_5)]$$

When there are n strips and the ordinates are $y_0, y_1, y_2, y_3, \ldots, y_{n-1}, y_n$ then the sum of the areas of the strips is:

$$\frac{h}{2}(y_0 + y_1) + \frac{h}{2}(y_1 + y_2) + \frac{h}{2}(y_2 + y_3) + \ldots + \frac{h}{2}(y_{n-2} + y_{n-1}) + \frac{h}{2}(y_{n-1} + y_n)$$

Hence we can find an approximate value for $\int_a^b f(x)\,dx$ using:

KEY POINT 5.5

$$\int_a^b f(x)\,dx \approx \frac{h}{2}[y_0 + y_n + 2(y_1 + y_2 + y_3 + \ldots + y_{n-1})] \text{ where } h = \frac{b-a}{n}.$$

An easy way to remember this rule in terms of the ordinates is:

$$\int_a^b f(x)\,dx \approx \text{half width of strip} \times (\text{first} + \text{last} + \text{twice the sum of all the others}).$$

TIP

- The width of the strip is the interval along the x-axis.
- The number of ordinates is one more than the number of strips.
- You should not use a numerical method when an algebraic method is available to you, unless you are specifically asked to do so.

WORKED EXAMPLE 5.14

The diagram shows the curve $y = x - 2\ln x$.

Use the trapezium rule with 2 intervals to estimate the shaded area, giving your answer correct to 2 decimal places. State, with a reason, whether the trapezium rule gives an under-estimate or an over-estimate of the true value.

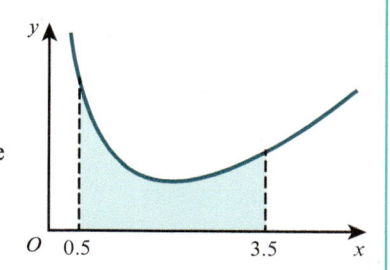

Answer

$a = 0.5$, $b = 3.5$ and $h = 1.5$

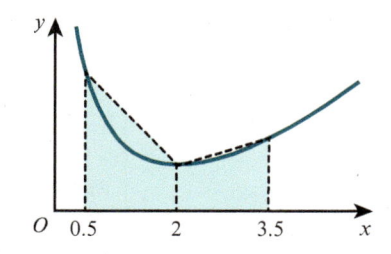

x	0.5	2	3.5
y	1.8863	0.6137	0.9945
	y_0	y_1	y_2

$$\text{Area} \approx \frac{h}{2}[y_0 + y_2 + 2y_1]$$

$$\approx \frac{1.5}{2}[1.8863 + 0.9945 + 2 \times 0.6137]$$

$$\approx 3.08115\ldots$$

$$\approx 3.08 \text{ (to 2 decimal places)}$$

It can be seen from the diagram that this is an over-estimate since the top edges of the strips all lie above the curve.

WORKED EXAMPLE 5.15

The diagram shows the curve $y = 2\sqrt{4x - x^2}$.

Use the trapezium rule with 4 intervals to estimate the value of $\int_0^4 2\sqrt{4x - x^2}\,dx$ giving your answer correct to 2 decimal places. State, with a reason, whether the trapezium rule gives an under-estimate or an over-estimate of the true value.

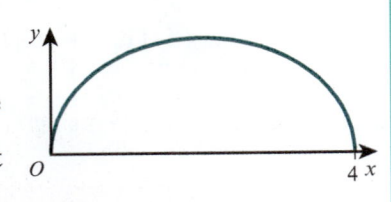

Answer

$a = 0, b = 4$ and $h = 1$

x	0	1	2	3	4
y	0	$2\sqrt{3}$	4	$2\sqrt{3}$	0
	y_0	y_1	y_2	y_3	y_4

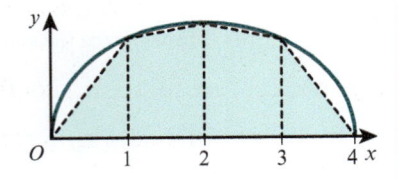

$$\int_0^4 2\sqrt{4x - x^2}\,dx \approx \frac{h}{2}[y_0 + y_4 + 2(y_1 + y_2 + y_3)]$$

$$\approx \frac{1}{2}\left[0 + 0 + 2\left(2\sqrt{3} + 4 + 2\sqrt{3}\right)\right]$$

$$\approx 4\sqrt{3} + 4$$

$$\approx 10.93 \text{ (to 2 decimal places)}$$

It can be seen from the diagram that this is an under-estimate since the top edges of the strips all lie below the curve.

Worked example 5.14 involves a 'concave' curve and the diagram shows that the trapezium rule gives an over-estimate for the area.

Worked example 5.15 involves a 'convex' curve and the diagram shows that the trapezium rule gives an under-estimate for the area.

If a curve is partially convex and partially concave over the required interval it is not so easy to predict whether the trapezium rule will give an under-estimate or an over-estimate for the area.

The more strips that are used, the more accurate the estimate will be.

> **DID YOU KNOW?**
>
> If you double the number of strips in a trapezium rule approximation, the error reduces to approximately one quarter the previous error.

EXERCISE 5E

1 Use the trapezium rule with 2 intervals to estimate the value of each of these definite integrals. Give your answers correct to 2 decimal places.

a $\int_2^4 \sqrt{x^2 - 2}\,dx$

b $\int_0^2 |e^x - 12|\,dx$

c $\int_0^4 \frac{2}{3 + e^x}\,dx$

d $\int_1^7 \frac{2}{\ln(x + 5)}\,dx$

e $\int_0^{\frac{1}{3}\pi} \frac{1}{2 + \tan x}\,dx$

f $\int_2^{12} \log_{10} x\,dx$

2

The diagram shows part of the curve $y = x^2 e^{-x}$. Use the trapezium rule with 4 intervals to estimate the value of $\int_1^5 x^2 e^{-x} \, dx$, giving your answer correct to 2 decimal places.

3 a Use the trapezium rule with 6 intervals to estimate the value of
$$\int_{\frac{1}{4}\pi}^{\frac{7}{4}\pi} \operatorname{cosec}\left(\frac{1}{2}x\right) dx,$$
giving your answer correct to 2 decimal places.

b Use a sketch of the graph of $y = \operatorname{cosec}\left(\frac{1}{2}x\right)$ for $0 \leqslant x \leqslant 2\pi$, to explain whether the trapezium rule gives an under-estimate or an over-estimate of the true value of $\int_{\frac{1}{4}\pi}^{\frac{7}{4}\pi} \operatorname{cosec}\left(\frac{1}{2}x\right) dx$.

4

The diagram shows the curve $y = e^x \cos x$ for $0 \leqslant x \leqslant \frac{1}{2}\pi$. Use the trapezium rule with 3 intervals to estimate the value of $\int_0^{\frac{1}{2}\pi} e^x \cos x \, dx$, giving your answer correct to 2 decimal places. State, with a reason, whether the trapezium rule gives an under-estimate or an over-estimate of the true value of $\int_0^{\frac{1}{2}\pi} e^x \cos x \, dx$.

5

The diagram shows part of the curve $y = \dfrac{e^x}{2x}$. Use the trapezium rule with 4 intervals to estimate the area of the shaded region, giving your answer correct to 2 decimal places.

State, with a reason, whether the trapezium rule gives an under-estimate or an over-estimate of the true value of the shaded area.

> **WEB LINK**
>
> Try the *When does the trapezium rule give the exact answer here?* resource on the Underground Mathematics website.

Checklist of learning and understanding

Integration formulae

- $\int e^x \, dx = e^x + c$
- $\int e^{ax+b} \, dx = \dfrac{1}{a} e^{ax+b} + c$
- $\int \dfrac{1}{x} \, dx = \ln|x| + c$
- $\int \dfrac{1}{ax+b} \, dx = \dfrac{1}{a} \ln|ax+b| + c$
- $\int \cos x \, dx = \sin x + c$
- $\int \cos(ax+b) \, dx = \dfrac{1}{a} \sin(ax+b) + c$
- $\int \sin x \, dx = -\cos x + c$
- $\int \sin(ax+b) \, dx = -\dfrac{1}{a} \cos(ax+b) + c$
- $\int \sec^2 x \, dx = \tan x + c$
- $\int \sec^2(ax+b) \, dx = \dfrac{1}{a} \tan(ax+b) + c$

The trapezium rule

- The trapezium rule can be used to find an approximate value for $\int_a^b f(x) \, dx$. If the region under the curve is divided into n strips each of width h, then:

$$\int_a^b f(x) \, dx \approx \dfrac{h}{2} [y_0 + y_n + 2(y_1 + y_2 + y_3 + \ldots + y_{n-1})] \text{ where } h = \dfrac{b-a}{n}.$$

END-OF-CHAPTER REVIEW EXERCISE 5

1. Show that $\int_{\frac{1}{6}\pi}^{\frac{1}{3}\pi} (1 + 2\tan^2 x)\,dx = \frac{4\sqrt{3}}{3} - \frac{\pi}{6}$. [4]

2. Find the exact value of $\int_{-4}^{-2} \left(\frac{4}{1-2x} - 2x\right)dx$, giving your answer in the form $a + \ln\left(\frac{b}{c}\right)$, where a, b and c are integers. [4]

3.

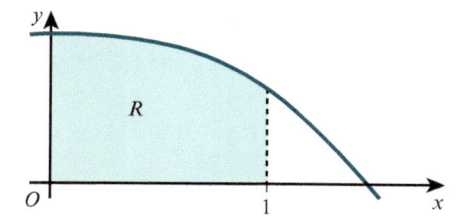

   The diagram shows part of the curve $y = 2 - x^2 \ln(x+1)$. The shaded region R is bounded by the curve and by the lines $x = 0$, $x = 1$ and $y = 0$. Use the trapezium rule with 4 intervals to estimate the area of R giving your answer correct to 2 decimal places. State, with a reason, whether the trapezium rule gives an under-estimate or an over-estimate of the true value of the area of R. [4]

4. Find the exact value of $\int_0^{\frac{1}{6}\pi} (\cos 3x - \sin 2x)\,dx$. [5]

5. Show that $\int_0^1 (3e^x - 2)^2\,dx = \frac{9}{2}e^2 - 12e + \frac{23}{2}$. [5]

6. a Find $\int_0^k (5e^{-2x} + 2e^{-3x})\,dx$, where k is a positive constant. [4]

   b Hence find the exact value of $\int_0^\infty (5e^{-2x} + 2e^{-3x})\,dx$. [1]

7. Show that $\int_2^4 \frac{3}{5x+1}\,dx = \frac{3}{5}\ln\frac{21}{11}$. [5]

8. a Find the value of the constant A such that $\frac{8x}{2x+5} \equiv 4 + \frac{A}{2x+5}$. [2]

   b Hence show that $\int_1^3 \frac{8x}{2x+5}\,dx = 8 - 10\ln\frac{11}{7}$. [5]

9. i Show that $12\sin^2 x \cos^2 x \equiv \frac{3}{2}(1 - \cos 4x)$. [3]

   ii Hence show that $\int_{\frac{1}{4}\pi}^{\frac{1}{3}\pi} 12\sin^2 x \cos^2 x\,dx = \frac{\pi}{8} + \frac{3\sqrt{3}}{16}$. [3]

   Cambridge International AS & A Level Mathematics 9709 Paper 21 Q3 June 2013

10. a Find $\int 4e^x(3 + e^{2x})\,dx$. [3]

    b Show that $\int_{-\frac{1}{4}\pi}^{\frac{1}{4}\pi} (3 + 2\tan^2 \theta)\,d\theta = \frac{1}{2}(8 + \pi)$. [4]

    Cambridge International AS & A Level Mathematics 9709 Paper 21 Q6 June 2011

11 i Show that $(2\sin x + \cos x)^2$ can be written in the form $\dfrac{5}{2} + 2\sin 2x - \dfrac{3}{2}\cos 2x$. [5]

ii Hence find the exact value of $\displaystyle\int_0^{\frac{1}{4}\pi}(2\sin x + \cos x)^2\,dx$. [4]

Cambridge International AS & A Level Mathematics 9709 Paper 21 Q7 June 2012

12 i By differentiating $\dfrac{\cos x}{\sin x}$, show that if $y = \cot x$ then $\dfrac{dy}{dx} = -\operatorname{cosec}^2 x$. [3]

ii By expressing $\cot^2 x$ in terms of $\operatorname{cosec}^2 x$ and using the result of **part i**, show that $\displaystyle\int_{\frac{1}{4}\pi}^{\frac{1}{2}\pi}\cot^2 x\,dx = 1 - \dfrac{1}{4}\pi$. [4]

iii Express $\cos 2x$ in terms of $\sin^2 x$ and hence show that $\dfrac{1}{1-\cos 2x}$ can be expressed as $\dfrac{1}{2}\operatorname{cosec}^2 x$. Hence, using the result of **part i**, find $\displaystyle\int \dfrac{1}{1-\cos 2x}\,dx$. [3]

Cambridge International AS & A Level Mathematics 9709 Paper 21 Q8 June 2010

13 i Find $\displaystyle\int \dfrac{1+\cos^4 2x}{\cos^2 2x}\,dx$. [3]

ii Without using a calculator, find the exact value of $\displaystyle\int_4^{14}\left(2 + \dfrac{6}{3x-2}\right)dx$, giving your answer in the form $\ln(ae^b)$, where a and b are integers. [5]

Cambridge International AS & A Level Mathematics 9709 Paper 21 Q7 June 2016

14

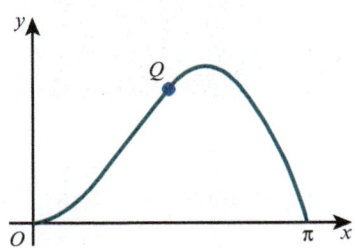

The diagram shows the curve $y = x\sin x$, for $0 \leqslant x \leqslant \pi$. The point $Q\left(\dfrac{1}{2}\pi, \dfrac{1}{2}\pi\right)$ lies on the curve.

i Show that the normal to the curve at Q passes through the point $(\pi, 0)$. [5]

ii Find $\dfrac{d}{dx}(\sin x - x\cos x)$. [2]

iii Hence evaluate $\displaystyle\int_0^{\frac{1}{2}\pi} x\sin x\,dx$. [3]

Cambridge International AS & A Level Mathematics 9709 Paper 21 Q8 November 2010

15 i Using the expansion of $\cos(3x - x)$ and $\cos(3x + x)$, prove that $\dfrac{1}{2}(\cos 2x - \cos 4x) \equiv \sin 3x \sin x$. [3]

ii Hence show that $\displaystyle\int_{\frac{1}{6}\pi}^{\frac{1}{3}\pi}\sin 3x \sin x\,dx = \dfrac{1}{8}\sqrt{3}$. [3]

Cambridge International A Level Mathematics 9709 Paper 31 Q4 June 2010

Chapter 6
Numerical solutions of equations

In this chapter you will learn how to:

- locate approximately a root of an equation, by means of graphical considerations and/or searching for a sign change
- understand the idea of, and use the notation for, a sequence of approximations that converges to a root of an equation
- understand how a given simple iterative formula of the form $x_{n+1} = F(x_n)$ relates to the equation being solved, and use a given iteration, or an iteration based on a given rearrangement of an equation, to determine a root to a prescribed degree of accuracy.

PREREQUISITE KNOWLEDGE

Where it comes from	What you should be able to do	Check your skills
IGCSE / O Level Mathematics	Substitute numbers for letters in complicated formulae and algebraic expressions, including those involving logarithmic functions, exponential functions and major and minor trigonometric functions.	1 Evaluate each of the following expressions when $x = 1.5$. a $\frac{2}{3}\left(x + \frac{1}{x^2}\right)$ b $100 \sin x - 99x$, (x in radians) c e^{-x} d $\ln x + 5$
IGCSE / O Level Mathematics	Rearrange complicated formulae and equations.	2 Rearrange each formula so that x is the subject. a $y = 4x^2 + 13$, given $x > 0$ b $0 = \frac{y - 3x + 7}{w}$ c $3 = \sqrt[3]{\frac{1+y}{x^5}}$
Pure Mathematics 1 Coursebook, Chapters 1 and 2	Understand the relationship between a graph and its associated algebraic equation, and use the relationship between points of intersection of graphs and solutions of equations.	3 Show that the x-coordinates of the solutions to the simultaneous equations: $y = x^2 - 6x + 10$ $y = x$ also satisfy the equation $x^2 - 7x + 10 = 0$.
Pure Mathematics 1 Coursebook, Chapter 1	Know that the values of x that make both sides of an equation equal are called the roots of the equation. Also, the roots of $f(x) = 0$ are the x-intercepts on the graph of $y = f(x)$.	4 Find the roots of the equation $x^2 - 7x + 10 = 0$.
IGCSE / O Level Mathematics Pure Mathematics 1 Coursebook, Chapter 5	Sketch graphs of quadratic or cubic functions or the trigonometric functions sine, cosine, tangent.	5 Sketch the graph, $y = f(x)$, of each function. a $f(x) = x^3 + 1$ b $f(x) = \cos 2x - 3$ c $f(x) = 4x^2 - 16x$
Chapters 2 and 3	Sketch graphs of exponential or logarithmic functions or the trigonometric functions cosecant, secant, cotangent.	6 Sketch the graph, $y = g(x)$, of each function. a $g(x) = \frac{e^x}{2}$ b $g(x) = \ln(x - 1)$ c $g(x) = \csc x$

Chapter 6: Numerical solutions of equations

Why solve equations numerically?

You are used to solving equations using direct, algebraic methods such as factorising or using the quadratic formula.

You might be surprised to learn that not all equations can be solved using a direct, algebraic process.

Numerical methods are ways of calculating approximate solutions to equations. They are extremely powerful problem-solving tools and many are available. They are widely used in engineering, computing, finance and many other applications.

A well-programmed computer is quickly able to solve a complicated equation using a numerical method. So why do we need to understand how the method works?

Knowing how a numerical method works helps us to:

- understand when the results are likely to be reasonable
- understand how to use available software correctly
- select an appropriate method when choices are available
- write our own programs when we need to do so, if we have the programming skills!

To be a powerful problem-solving tool, the method must be used correctly. This might mean the method needs to work quickly or work to obtain a particular level of accuracy. When used incorrectly, numerical methods might be very slow or even fail to work, as we will see later in this chapter.

In this chapter we will focus on one numerical method, the numerical solution of an equation using an **iterative formula**.

6.1 Finding a starting point

As we will see throughout this chapter, a good starting point is very important.

We need a method to find an approximate solution to an equation first.

We might find this using, for example, a graphical approach or the change of sign approach.

For suitable functions, each of these approaches will produce two values: one above and one below the solution that we are finding.

WORKED EXAMPLE 6.1

By sketching graphs of $y = x^3$ and $y = 4 - x$, show that the equation $x^3 + x - 4 = 0$ has a root α between 1 and 2.

Answer

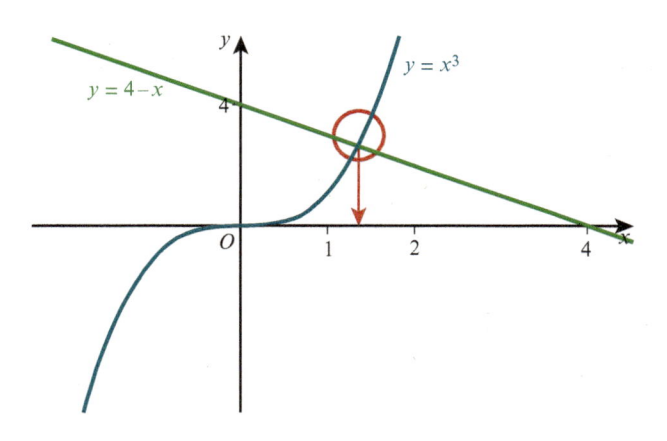

> **TIP**
>
> Make sure your sketch shows all the key features of each graph, such as the coordinates of turning points and the coordinates of any intercepts.

There is one point of intersection and its x-coordinate is between 1 and 2.

At this point $x^3 = 4 - x$.

This means that the equation $x^3 = 4 - x$ has one root between 1 and 2.

Rearranging this equation gives $x^3 + x - 4 = 0$.

Hence $x^3 + x - 4 = 0$ has a root, α, between 1 and 2.

WORKED EXAMPLE 6.2

Show, by calculation, that the equation $f(x) = x^5 + x - 1 = 0$ has a root α between 0 and 1.

Answer

Find the value of α such that $f(\alpha) = 0$. .... Why? Think about it and check the explanation that follows.

$f(0) = 0^5 + 0 - 1 = -1$
$f(1) = 1^5 + 1 - 1 = 1$ .... As α is between 0 and 1, then $f(0)$ might have the opposite sign to $f(1)$.

The change of sign indicates the presence of a root, so $0 < \alpha < 1$. .... This conclusion is important as it completes the argument.

To see why the change of sign method works in this case, look at the graph of $y = f(x)$.

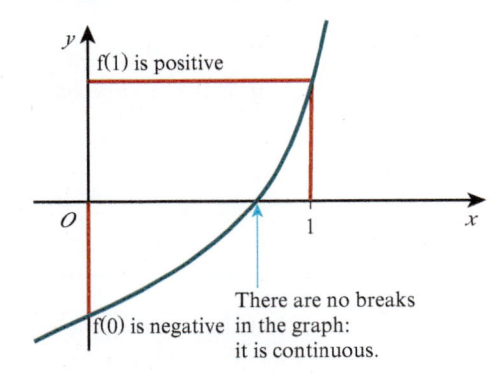

f(1) is positive

f(0) is negative There are no breaks in the graph: it is continuous.

Since it starts below the x-axis at $x = 0$ and finishes above the x-axis at $x = 1$, it must cross the x-axis somewhere in between, say at $x = \alpha$, where $0 < \alpha < 1$. So α is a root of the equation, that is, $f(\alpha) = 0$.

Chapter 6: Numerical solutions of equations

WORKED EXAMPLE 6.3

a By sketching a suitable pair of graphs, show that the equation $\cos x = 2x - 1$ (where x is in radians) has only one root for $0 \leqslant x \leqslant \frac{1}{2}\pi$.

b Verify by calculation that this root lies between $x = 0.8$ and $x = 0.9$.

Answer

a Draw the graphs of $y = \cos x$ and $y = 2x - 1$.   .... Why are these graphs suitable?

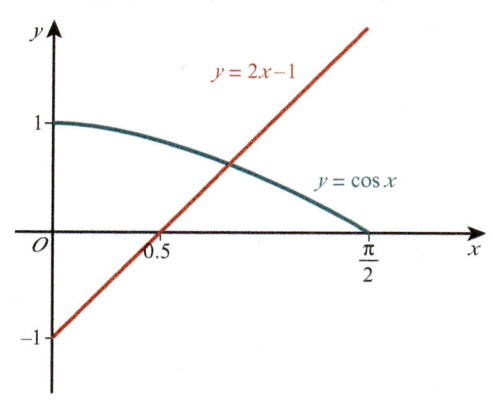

The graphs intersect once only and so the equation $\cos x = 2x - 1$ has only one root for $0 \leqslant x \leqslant \frac{1}{2}\pi$.

b $\cos x = 2x - 1$ so $\cos x - 2x + 1 = 0$

Let $f(x) = \cos x - 2x + 1$ and so $f(x) = 0$ then

$f(0.8) = \cos 0.8 - 2(0.8) + 1 = 0.0967...$

$f(0.9) = \cos 0.9 - 2(0.9) + 1 = -0.1783...$

Change of sign indicates the presence of a root.

EXERCISE 6A

1 a On the same axes, sketch the graphs of $y = \sqrt{1 + x}$ and $y = x^2$.

  b Using your answer to **part a**, explain why the equation $x^2 - \sqrt{1 + x} = 0$ has two roots.

  c Show, by calculation, that the smaller of these two roots lies between $x = -1$ and $x = 0$.

2 a By sketching graphs of $y = x^3 + 5x^2$ and $y = 5 - 2x$, determine the number of real roots of the equation $x^3 + 5x^2 + 2x - 5 = 0$.

  b Verify by calculation that the largest root of $x^3 + 5x^2 + 2x - 5 = 0$ lies between $x = 0$ and $x = 2$.

3 The curve $y = x^3 + 5x - 1$ cuts the x-axis at the point $(\alpha, 0)$.

  a By sketching the graph of $y = x^3$ and one other suitable graph, deduce that this is the only point where the curve $y = x^3 + 5x - 1$ cuts the x-axis.

  b Using calculations, show that $0.1 < \alpha < 0.5$.

4 a On the same axes, sketch the graphs of $y = \ln(x+1)$ and $y = 3x - 4$.

   b Hence deduce the number of roots of the equation $\ln(x+1) - 3x + 4 = 0$.

5 a By sketching graphs of $y = e^x$ and $y = x + 6$, determine the number of roots of the equation $e^x - x - 6 = 0$.

   b Show, by calculation, that $e^x - x - 6 = 0$ has a root that lies between $x = 2.0$ and $x = 2.1$.

6 a Show, by calculation, that $(x+2)e^{5x} = 1$ has a root between $x = 0$ and $x = -0.2$.

   b Show, by sketching the graph of $y = e^{5x}$ and one other suitable graph, that this is the only root of this equation.

7 a Show by calculation that the equation $\cos^{-1} 2x = 1 - x$ has a root α, where $0.4 \leq \alpha \leq 0.5$.

   b The diagram shows the graph of $y = \cos^{-1} 2x$ where $-0.5 \leq x \leq 0.5$.

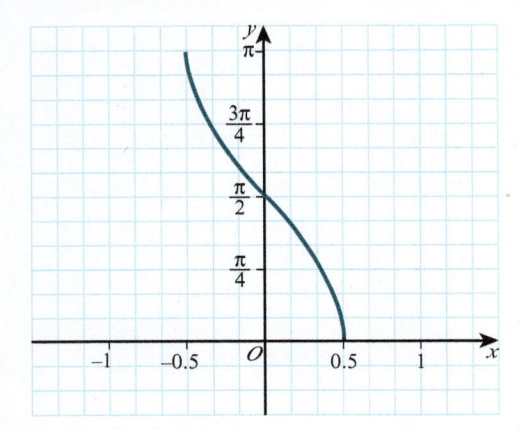

   By drawing a suitable graph on a copy of the diagram, show that α is the only root of $\cos^{-1} 2x = 1 - x$ for $-0.5 \leq x \leq 0.5$.

8 The curve with equation $y = \dfrac{\sin x}{2x + 3}$ where $x < -\dfrac{\pi}{2}$ intersects the line $y = 1$ at the point $(\alpha, 1)$.

   a By sketching the graph of $y = \sin x$ for $-2\pi < x < -\dfrac{\pi}{2}$ radians and one other suitable graph, deduce that this is the only point where the curve $y = \dfrac{\sin x}{2x + 3}$ intersects the line $y = 1$.

   b Using calculations, show that $-2 < \alpha < -1.9$.

PS 9 a By sketching a suitable pair of graphs, show that the equation $x^3 + 4x = 7x + 4$ has only one root for $0 \leq x \leq 5$.

   b Verify by calculation that this root lies between $x = 2$ and $x = 3$.

PS 10 a Show graphically that the equation $2^x = x + 4$ has exactly two roots.

   b Show, by calculation, that the larger of the two roots is between 2.7 and 2.8.

PS 11 a By sketching two suitable graphs on the same diagram, show that the equation $\cot x = x^2$ has one root between 0 and $\dfrac{\pi}{2}$ radians.

   b Show, by calculation, that this root, α, lies between 0.8 and 1.

> ⏮ **REWIND**
>
> You might need to look back at Chapter 3 before attempting Question 11.

PS **12 a** By sketching a suitable pair of graphs, deduce the number of roots of the equation $x = \tan 2x$ for $0 < x < 2\pi$.

    b Verify, by calculation, that one of these roots, α, lies between 2.1 and 2.2.

PS **13 a** Show graphically that the equation $\operatorname{cosec} x = \sin x$ has exactly two roots for $0 < x < 2\pi$.

    b Use an algebraic method to find the value of the larger root correct to 3 significant figures.

PS **14** $f(x) = 20x^3 + 8x^2 - 7x - 3$

    a The equation $20x^3 + 8x^2 - 7x - 3 = 0$ has exactly two roots.

      Without factorising the cubic equation, show, by calculation, that one of these roots is between 0.5 and 1.

    b Show, by sketching the graph of $y = f(x)$ for $-1 \leq x \leq 1$, that the smaller root is between -1 and 0.

    c Explain why it is not necessary to use a numerical method to find the two solutions of this equation.

PS **15** A guitar tuning peg is in the shape of a cylinder with a hemisphere at one end.

    The cylinder is 20 mm long and the whole peg is made from 800 mm³ of plastic.

    The base radius of the cylinder is r mm.

    a Show that $\pi r^3 + 30\pi r^2 - 1200 = 0$.

    b Show that the value of r is between 3 mm and 4 mm.

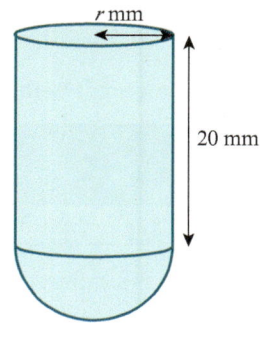

> **EXPLORE 6.1**
>
> Anil is trying to find the roots of $f(x) = |\tan x| = 0$ between 0 and 2π radians.
>
> He states:
>
> $f(3.1) > 0$ and $f(3.2) > 0$. There is no change of sign so there is no root of $|\tan x| = 0$ between 3.1 and 3.2.
>
> Mika states:
>
> $y = |\tan x|$ and it meets the x-axis where $x = \pi$ radians so Anil has made a calculation error.
>
> Discuss Anil and Mika's statements.

> **DID YOU KNOW?**
>
> Interestingly, cubic and quartic equations can be solved algebraically. This has been known since the 16th century and was shown by the Italian mathematicians Niccolò Fontana (Tartaglia) and Lodovico Ferrari. Quintic equations cannot be solved algebraically in general. This was shown by the Norwegian mathematician Niels Abel and the French mathematician Évariste Galois in the early 19th century.

6.2 Improving your solution

Once we have an approximate location for a solution, we can improve it.

One common way to do this is to use an **iterative process**, finding values that are closer and closer to α, the root we are looking to find.

For example, we have shown, by calculation, that the equation $f(x) = x^5 + x - 1 = 0$ has a root α between 0 and 1. This is a reasonable starting point but probably not accurate enough if we wanted a solution to this equation in the real world. What if we needed to know the value correct to 1 significant figure? How could we find it?

Let us consider a rearrangement of $f(x) = x^5 + x - 1 = 0$.

The root we are trying to find is a value of x, so rearranging $x^5 + x - 1 = 0$ into the form $x = F(x)$ seems reasonable (in other words, $x = $ some function of x).

One possible rearrangement of this form is $x = \sqrt[5]{1-x}$.

As we already know that $0 < \alpha < 1$, a good starting point is $x = 0.5$.

Now we can **test** our value to see how accurate it is. How?

- Work out the value of $\sqrt[5]{1-x}$ for the starting value.

    $\sqrt[5]{1 - 0.5} = \sqrt[5]{0.5} = 0.87055...$

- Compare it with x.

x	$F(x) = \sqrt[5]{1-x}$	Are x and $F(x)$ the same correct to 1 significant figure?
0.5	0.87055...	No

- Are the values the same to the required number of significant figures or decimal places?

    Yes – then use this value as the estimate for the root.
    No – then find a more accurate value.

Look at the graphs of $y = x$ and $y = \sqrt[5]{1-x}$ on the same diagram.

> **REWIND**
>
> Remember from Section 6.1 that α is just the special name given to the solution that you are looking for.

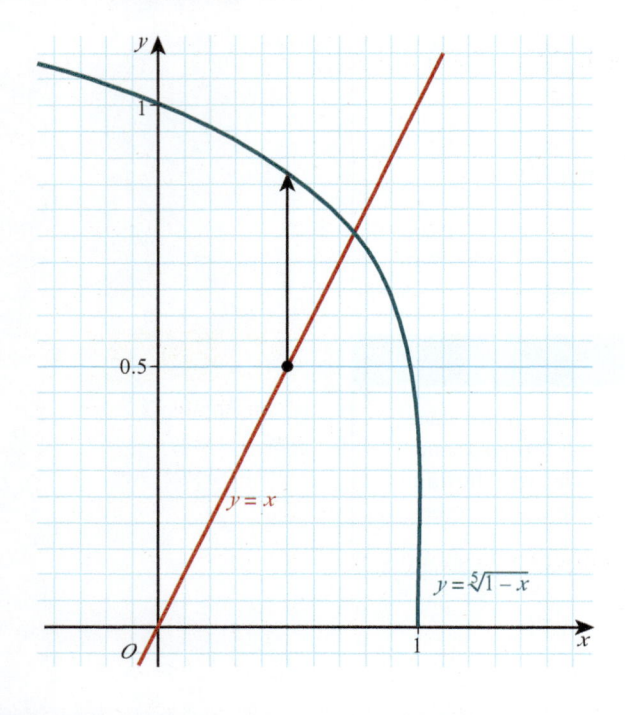

The graphs show that the values x and $F(x)$ are not very close together as the vertical distance between the line and curve is quite large.

How do we improve the accuracy of the estimate for the root?

We can see in the following diagram that 0.870 55... is a better approximation for the root α than 0.5.

Is 0.870 55... close enough?

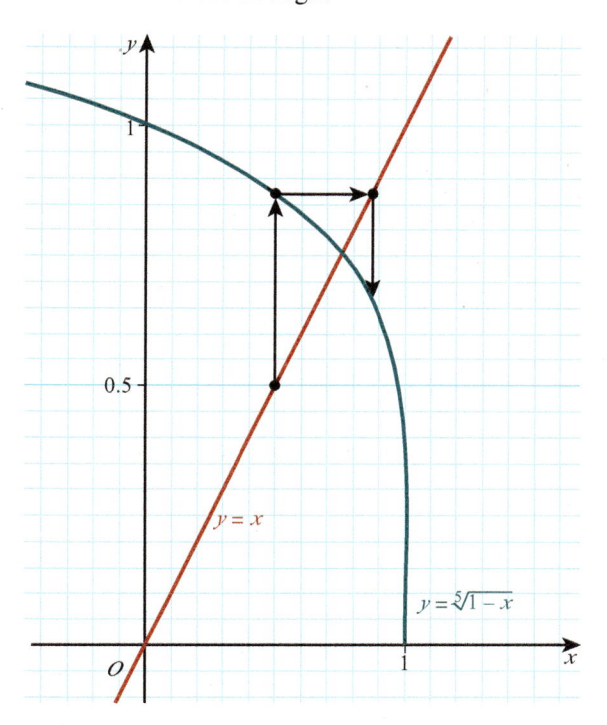

The vertical distance between the line and curve is now smaller, so the values are **converging**.

Testing $x = 0.870\,55...$ gives $\sqrt[5]{1-x} = \sqrt[5]{1-0.870\,55...} = 0.664\,38...$

x	$F(x) = \sqrt[5]{1-x}$	Are x and $F(x)$ the same correct to 1 significant figure?
0.5	0.870 55...	No
0.870 55...	0.664 38...	No

We can see that 0.664 38... is a better approximation of the root α than 0.870 55... because the values are closer together. Is 0.664 38... close enough?

When we continue testing and comparing values, we find:

x	$F(x) = \sqrt[5]{1-x}$	Are x and $F(x)$ the same correct to 1 significant figure?
0.664 38...	0.803 83...	No
0.803 83...	0.721 97...	No
0.721 97...	0.774 13...	No
0.774 13...	0.742 62...	No
0.742 62...	0.762 27...	No
0.762 27...	0.750 26...	Yes

> **TIP**
> Using a table is helpful, but still very repetitive and a lot of work!

Each test is called an **iteration**. Each iteration gives a value of x that is closer to the actual value of the root than the previous value of x.

It seems as if the value of α is 0.8 correct to 1 significant figure.

This (enlarged) diagram shows what is happening with each new value.

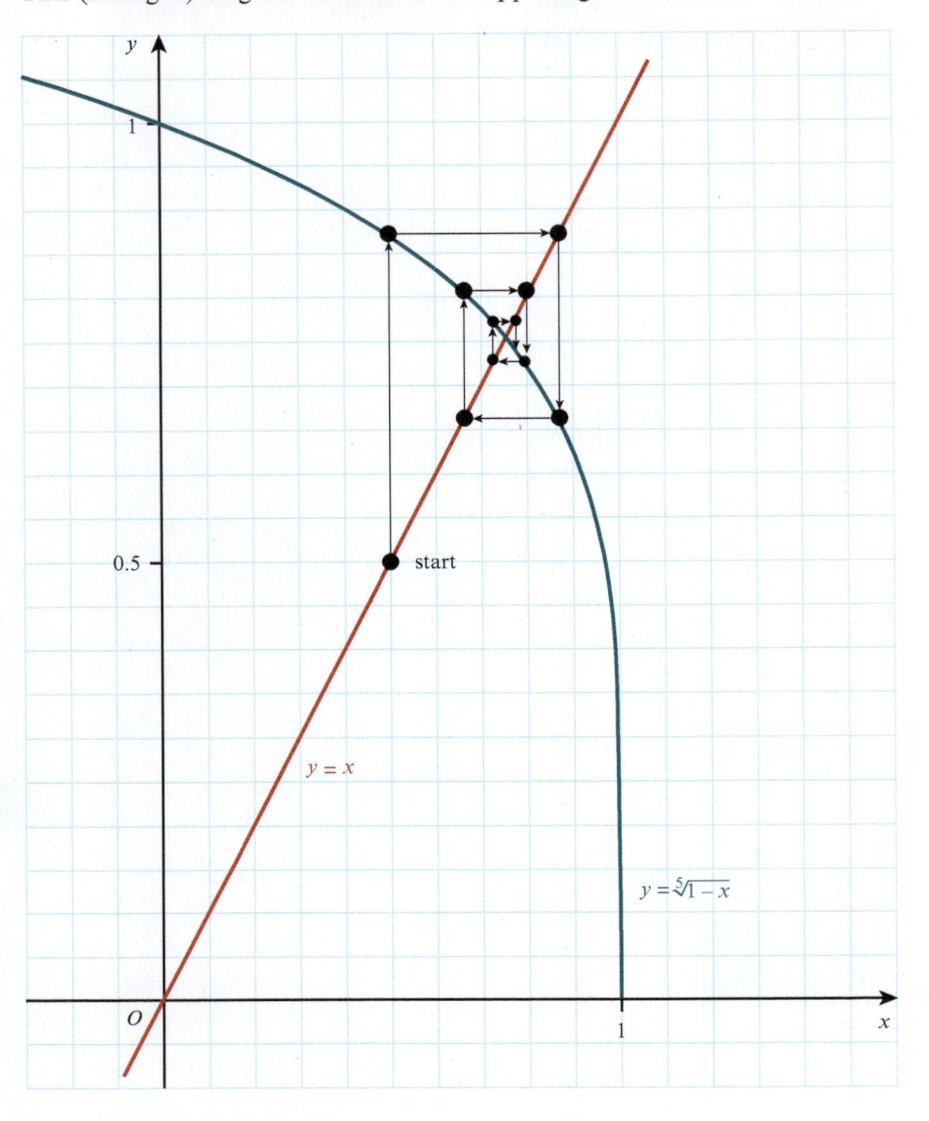

Each value that comes from $F(x)$ is used as the next value for x. The spiral is often called the cobweb pattern.

Summary

When trying to solve $f(x) = 0$, rearrange the equation to the form $x = F(x)$.

When α is the value of x such that $x = F(x)$ then α is also a solution of $f(x) = 0$.

Subscript notation and formulae

In order to avoid unnecessary repetition and to make sure the process is recorded accurately, the method outlined is usually described using subscript notation.

The first approximation of the root is called x_1 (sometimes it is called x_0).

The next approximation of the root is called x_2, and so on.

A relationship such as $x = F(x)$ where $F(x) = \sqrt[5]{1-x}$ is the basis for an iterative formula:

$$x_{n+1} = F(x_n) = \sqrt[5]{1 - x_n}$$

Looking at the values we already have we can see how much easier this is to record:

$x_1 = 0.5$	$x_{n+1} = F(x_n) = \sqrt[5]{1 - x_n}$	Are they the same correct to 1 significant figure?
	$x_2 = 0.87055...$	No
	$x_3 = 0.66438...$	No
	$x_4 = 0.80383...$	No
	$x_5 = 0.72197...$	No
	$x_6 = 0.77413...$	No
	$x_7 = 0.74262...$	No
	$x_8 = 0.76227...$	No
	$x_9 = 0.75026...$	Yes

In this format we test the accuracy of the values by comparing each value of x_n in the sequence with the previous value.

> **TIP**
>
> Does your calculator have an (Ans) key?
>
> Using the (Ans) key reduces the number of key strokes needed to use an iterative process.
>
> - Type the value for x_1 into your calculator and press ▣.
>
>   e.g. 0.5 ▣
>
> - Type the formula into your calculator using (Ans) instead of (x_n).
>
>   e.g. $\sqrt[5]{1 - \text{Ans}}$
>
> Now **each time** you press ▣ the calculator gives you the **next** value of x_n generated (so $x_2, x_3, x_4, ...$).
>
> Try this now and check that you generate the values that are given in the table.

WORKED EXAMPLE 6.4

The equation $x^2 + x - 3 = 0$ has a negative root, α.

a Show that this equation can be rearranged as $x = \dfrac{3}{x} - 1$.

b Use the iterative formula $x_{n+1} = \dfrac{3}{x_n} - 1$ with a starting value of $x_1 = -2.5$ to find α correct to 2 decimal places. Give the result of each iteration to 4 decimal places, where appropriate.

Answer

a $x^2 + x - 3 = 0$ .... Divide through by x.

$\dfrac{x^2}{x} + \dfrac{x}{x} - \dfrac{3}{x} = \dfrac{0}{x}$

$x + 1 - \dfrac{3}{x} = 0$ .... Isolate x.

$x = \dfrac{3}{x} - 1$ .... ✓

b

$x_1 = -2.5$	$x_{n+1} = \dfrac{3}{x_n} - 1$	Are these the same correct to 2 decimal places?
	$x_2 = \dfrac{3}{-2.5} - 1 = -2.2$	No
	$x_3 = \dfrac{3}{-2.2} - 1 = -2.3636$	No
	$x_4 = -2.2692$	No
	$x_5 = -2.3220$	No
	$x_6 = -2.2920$	No
	$x_7 = -2.3089$	No
	$x_8 = -2.2993$	No
	$x_9 = -2.3047$	Yes

The root **seems** to be $\alpha = -2.30$ to 2 decimal places.

We can **prove** this, using a change of sign test:

When $\alpha = -2.30$ then $-2.305 < \alpha < -2.295$ and so

$f(-2.305) = (-2.305)^2 + (-2.305) - 3 = 0.00802\ldots$

$f(-2.295) = (-2.295)^2 + (-2.295) - 3 = -0.0279\ldots$

Change of sign indicates presence of root, therefore $\alpha = -2.30$, to 2 decimal places.

> **TIP**
>
> Any value **between** -2.305 and -2.295 must round to -2.30 to 2 decimal places.
>
> The change of sign test is an important test. If the root passes the change of sign test it must be correct.

WORKED EXAMPLE 6.5

The equation $e^x - 1 = 2x$ has a root $\alpha = 0$.

  a Show by calculation that this equation also has a root, β, such that $1 < \beta < 2$.

  b Show that this equation can be rearranged as $x = \ln(2x + 1)$.

  c Use an iteration process based on the equation in **part b**, with a suitable starting value, to find β correct to 3 significant figures.

  Give the result of each step of the process to 5 significant figures.

Answer

  a Let $f(x) = e^x - 1 - 2x = 0$ and look for a change of sign.

   $f(1) = e - 1 - 2 = -0.2817\ldots$

   $f(2) = e^2 - 1 - 2(2) = 2.38905\ldots$

   Change of sign indicates presence of root.

b $e^x - 1 = 2x$ .... Isolate e.

$e^x = 2x + 1$ .... Take logs to base e.

$x = \ln(2x + 1)$ .... ✓

c A suitable iterative formula is $x_{n+1} = \ln(2x_n + 1)$ with a starting value of $x_1 = 1.5$.
(Any value between 1 and 2 would be an acceptable starting value.)

$x_1 = 1.5$	$x_{n+1} = \ln(2x_n + 1)$	Are these the same correct to 3 significant figures?
	$x_2 = \ln(2 \times 1.5 + 1) = 1.38629\ldots = 1.3863$	No
	$x_3 = \ln(2 \times 1.38629\ldots + 1) = 1.32776\ldots = 1.3278$	No
	$x_4 = 1.29623\ldots = 1.2962$	No
	$x_5 = 1.27884\ldots = 1.2788$	No
	$x_6 = 1.26910\ldots = 1.2691$	No
	$x_7 = 1.26362\ldots = 1.2636$	No
	$x_8 = 1.26051\ldots = 1.2605$	Yes

The root **seems** to be $\beta = 1.26$ correct to 3 significant figures.

We can **prove** this, using a change of sign test:

When $\beta = 1.26$ then $1.255 < \beta < 1.265$ and so:

$f(1.255) = e^{1.255} - 1 - 2(1.255) = -0.0021616\ldots$

$f(1.265) = e^{1.265} - 1 - 2(1.265) = 0.01309\ldots$

Change of sign indicates presence of root, therefore $\beta = 1.26$ correct to 3 significant figures.

You might notice that the pattern of convergence is different in this example. This is clear when you look at the graphs of $y = x$ and $y = \ln(2x + 1)$ on the same grid. The type of convergence shown in the graph on the right is often called a staircase pattern.

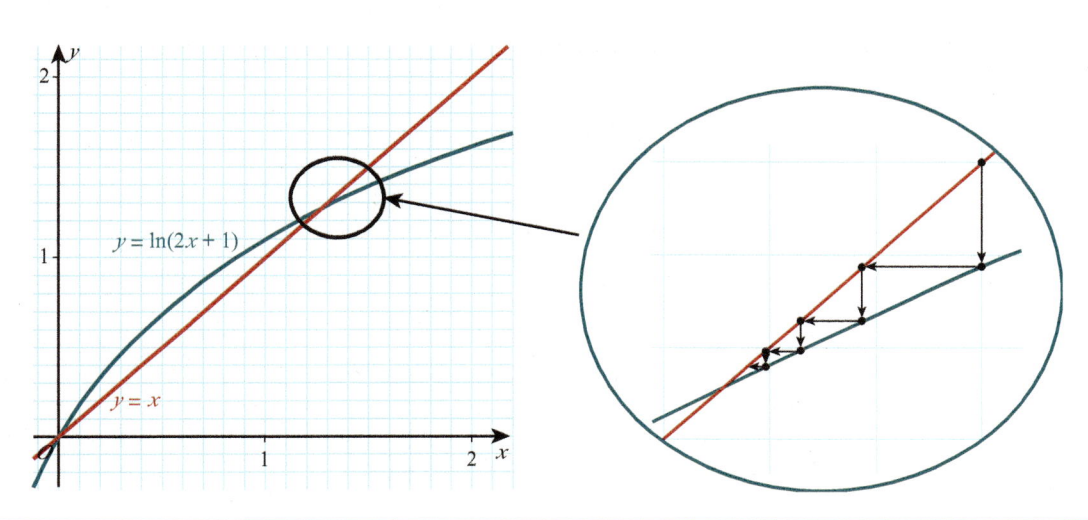

EXERCISE 6B

1. The equation $x^3 + 5x - 7 = 0$ has a root α between 1 and 1.2.

   The iterative formula $x_{n+1} = \dfrac{7 - x_n^3}{5}$ can be used to find the value of α.

   a Using a starting value of $x_1 = 1.1$, find and write down, correct to 4 decimal places, the value of each of x_2, x_3, x_4, x_5, x_6.

   b Write the value of x_6 correct to 2 decimal places and prove that this is the value of α correct to 2 decimal places.

2. a Show that the equation $\ln(x + 1) + 2x - 4 = 0$ has a root α between $x = 1$ and $x = 2$.

   b Use the iterative formula $x_{n+1} = \dfrac{4 - \ln(x_n + 1)}{2}$ with an initial value of $x_1 = 1.5$ to find the value of α correct to 3 decimal places.

3. The equation $(x - 0.5)e^{3x} = 1$ has a root α.

   a Show, by sketching the graph of $y = e^{3x}$ and one other suitable graph, that α is the only root of this equation.

   b Use the iterative formula $x_{n+1} = e^{-3x_n} + 0.5$ with $x_1 = 0.5$ to find the value of α correct to 2 decimal places. Give the value of each of your iterations to 4 decimal places.

4. a By sketching a suitable pair of graphs, show that the equation $x^3 + 10x = x + 5$ has only one root that lies between 0 and 1.

   b Use the iterative formula $x_{n+1} = \dfrac{5 - x_n^3}{9}$, with a suitable value for x_1, to find the value of this root correct to 4 decimal places. Give the result of each iteration to 6 decimal places.

5. The equation $\cos^{-1} 3x = 1 - x$ has a root α.

   a Show, by calculation, that α is between $\dfrac{\pi}{15}$ and $\dfrac{\pi}{12}$.

   b Show that the given equation can be rearranged into the form $x = \dfrac{1}{3} \cos(1 - x)$.

   c Using the iterative formula $x_{n+1} = \dfrac{1}{3} \cos(1 - x_n)$ with a suitable starting value, x_1, find the value of α correct to 3 decimal places. Give the result of each iteration to 5 decimal places.

6. The terms of a sequence with first term $x_1 = 1$ are defined by the iterative formula:

   $$x_{n+1} = \sqrt{\dfrac{5 - 2x_n - x_n^3}{5}}$$

   The terms converge to the value α.

   a Use this formula to find the value of α correct to 2 decimal places. Give the value of each term you calculate to 4 decimal places.

   b The value α is the root of an equation of the form $ax^3 + bx^2 + cx + d = 0$, where a, b, c and d are integers. Find this equation.

7. The equation $x^4 - 1 - x = 0$ has a root, α, between $x = 1$ and $x = 2$.

   a Show that α also satisfies the equation $x = \sqrt[4]{1 + x}$.

   b Write down an iterative formula based on the equation in **part a**.

   c Use your iterative formula, with a starting value of $x_1 = 1.5$, to find α correct to 2 decimal places. Give the result of each iteration to 4 decimal places.

8 The equation $\operatorname{cosec} x = x^2$ has a root, α, between 1 and 2.

   The equation can be rearranged either as $x = \sin^{-1}\left(\dfrac{1}{x^2}\right)$ or $x = \sqrt{\dfrac{1}{\sin x}}$.

   a Write down two possible iterative formulae, one based on each given rearrangement.

   Use the starting value 1.5.

   b Show that one of the formulae from **part a** fails to converge.

   c Show that the other formula from **part a** converges to α and find the value of α correct to 3 decimal places. Give the result of each iteration to 5 decimal places.

9 The terms of a sequence, defined by the iterative formula $x_{n+1} = \ln(x_n^2 + 4)$, converge to the value α. The first term of the sequence is 2.

   a Find the value of α correct to 2 decimal places. Give each term of the sequence you find to 4 decimal places.

   b The value α is a root of an equation of the form $x^2 = f(x)$. Find this equation.

10 The equation $\dfrac{\sin(x-1)}{2x-3} + 1 = 0$ has a root, α, between $x = 1$ and $x = 1.4$.

   a Show that α also satisfies the equation $x = \dfrac{3 - \sin(x-1)}{2}$.

   b Using an iterative formula based on the equation from **part a**, with a suitable starting value, find the value of α correct to 3 significant figures. Give the result of each iteration to 5 significant figures.

PS 11 The graphs of $y = \left(\dfrac{3}{2}\right)^x - 1$ and $y = x$ intersect at the points $O(0, 0)$ and A.

   a Sketch these graphs on the same diagram.

   b Using logarithms, find a suitable iterative formula that can be used to find the coordinates of the point A.

   c Calculate the length of the line OA, giving your answer correct to 2 significant figures. Give the value of any iterations you calculate to a suitable number of significant figures.

PS 12 The diagram shows a container in the shape of a cone with a cylinder on top.

   The height of the cylinder is 3 times its base radius, r.

   The volume of the container must be $5500\,\text{cm}^3$.

   The base of the cone has a radius of r cm.

   a Write down an expression for the height of the cone in terms of r.

   b Show that $2\pi r^3 + 11\pi r^2 - 5500 = 0$.

   c The equation in **part b** has a root, α.

   Show that α is also a root of the equation $r = \sqrt[3]{\dfrac{5500 - 11\pi r^2}{2\pi}}$.

   d Using an iterative formula based on the equation in **part c**, and a starting value of 8, find the value of α correct to 3 decimal places. Give the value of each of your iterations to an appropriate degree of accuracy.

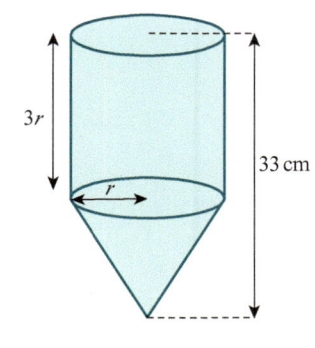

   e Explain what information is given by your answer to **part d**, in the context of the question.

PS 13 The equation $x^3 - 7x^2 + 1 = 0$ has two positive roots, α and β, which are such that α lies between 0 and 1 and β lies between 6 and 7.

   By deriving **two** suitable iterative formulae from the given equation, carry out suitable iterations to find the value of α and of β, giving each correct to 2 decimal places. Give the value of each of your iterations to 4 decimal places.

14 **a** Show graphically that the equation $\cot x = \sin x$ has a root, α, which is such that $0 < \alpha < \dfrac{\pi}{2}$.

  b Show that the equation in **part a** can be rearranged as $x = \sin^{-1}\sqrt{\cos x}$.

  c Using an iterative formula based on the equation in **part b**, with an initial value of 0.9, find the value of α correct to 2 decimal places. Give the value of each iteration to an appropriate number of decimal places.

  d The equation given in **part a** can be rearranged in other ways.

  Find an example of a formula, based on a different rearrangement of the equation given in **part a**, that also converges to the value of α, with an initial value of 0.9, in fewer iterations than the formula used in **part c**.

EXPLORE 6.2

You are going to explore why using rearrangements of equations into the form $x = F(x)$ sometimes fails.

Possible causes of failure to consider are the starting value and the formula itself.

Discuss why formulae based on:

- $x = \sqrt[5]{1 - x}$ (the cobweb pattern)
- $x = \ln(2x + 1)$ (the staircase pattern)

each converged to a root.

Look at the equation $e^x = 3x + 1$ and some related graphs.

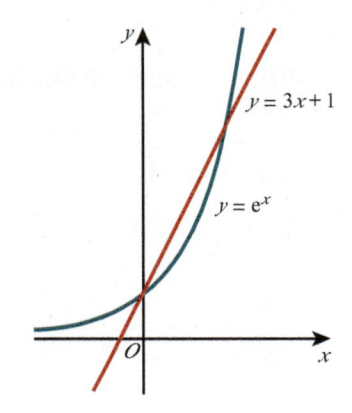

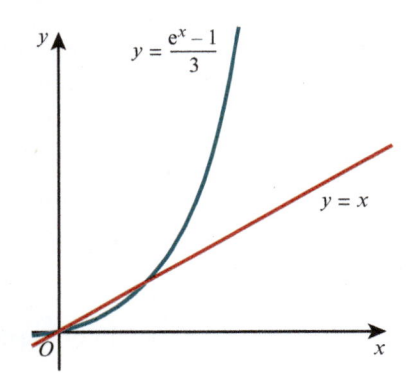

Discuss the rearrangement $x = \dfrac{e^x - 1}{3}$.

Does this give a convergent iterative formula? Explain your answer.

Look at the equations in **Exercise 6B**.

Using one of the equations in **questions 1** to **5** or **7** to **14**, draw the graph of $y = x$ and the graph of a function $F(x)$, where $x = F(x)$ is a rearrangement of the equation given and for which $x = F(x)$ will fail to converge.

Explain any similarities or differences between your example and the rearrangement $x = \dfrac{e^x - 1}{3}$.

Chapter 6: Numerical solutions of equations

6.3 Using iterative processes to solve problems involving other areas of mathematics

WORKED EXAMPLE 6.6

a Show that $\left(\dfrac{1+\cos x}{2\sin x}\right)^2 + \left(\dfrac{1-\cos x}{2\sin x}\right)^2 = \cot^2 x + \dfrac{1}{2}$.

b Hence, given that α is a root of the equation $\left(\dfrac{1+\cos x}{2\sin x}\right)^2 + \left(\dfrac{1-\cos x}{2\sin x}\right)^2 = x$, show that α is also a root of the equation $x = \tan^{-1}\sqrt{\dfrac{2}{2x-1}}$ for $0 < x < \dfrac{\pi}{2}$.

c It is given that α is the only root of the equation $x = \tan^{-1}\sqrt{\dfrac{2}{2x-1}}$ for $0 < x < \dfrac{\pi}{2}$. Verify by calculation that the value of α lies between 0.9 and 1.0.

d Using an iterative formula based on the equation in **part c**, find the value of α correct to 3 significant figures.

Answer

a $\left(\dfrac{1+\cos x}{2\sin x}\right)^2 + \left(\dfrac{1-\cos x}{2\sin x}\right)^2$ .... Square.

$= \dfrac{1+2\cos x+\cos^2 x}{4\sin^2 x} + \dfrac{1-2\cos x+\cos^2 x}{4\sin^2 x}$ .... Collect terms.

$= \dfrac{2+2\cos^2 x}{4\sin^2 x}$ .... Simplify.

$= \dfrac{1+\cos^2 x}{2\sin^2 x}$ .... Write as separate fractions.

$= \dfrac{1}{2\sin^2 x} + \dfrac{\cos^2 x}{2\sin^2 x}$ .... Simplify.

$= \dfrac{\operatorname{cosec}^2 x}{2} + \dfrac{\cot^2 x}{2}$ .... Write in terms of $\cot x$.

$= \dfrac{1+\cot^2 x}{2} + \dfrac{\cot^2 x}{2}$ .... Collect terms.

$= \cot^2 x + \dfrac{1}{2}$

b $\cot^2 x + \dfrac{1}{2} = x$ .... Rearrange.

$\cot^2 x = x - \dfrac{1}{2}$ .... Square root.

$\cot x = \sqrt{x - \dfrac{1}{2}}$ .... Left-hand side: write in terms of $\tan x$. Right-hand side: write as a single fraction.

$\dfrac{1}{\tan x} = \sqrt{\dfrac{2x-1}{2}}$ .... Take the reciprocal of each side.

$\tan x = \sqrt{\dfrac{2}{2x-1}}$ .... Make x the subject.

$x = \tan^{-1}\sqrt{\dfrac{2}{2x-1}}$

c Let $f(x) = x - \tan^{-1}\sqrt{\dfrac{2}{2x-1}}$ and so $f(x) = 0$ then:

$$f(0.9) = 0.9 - \tan^{-1}\sqrt{\dfrac{2}{2(0.9)-1}} = -0.10685\ldots$$

$$f(1) = 1 - \tan^{-1}\sqrt{\dfrac{2}{2(1)-1}} = 0.44683\ldots$$

Change of sign indicates the presence of a root.

$x_1 = 0.95$	$x_{n+1} = \tan^{-1}\sqrt{\dfrac{2}{2x_n - 1}}$	Are they the same correct to 3 significant figures?
	$x_2 = \tan^{-1}\sqrt{\dfrac{2}{2(0.95)-1}} = 0.97992\ldots$	No
	$x_3 = \tan^{-1}\sqrt{\dfrac{2}{2(0.97992\ldots)-1}} = 0.96494\ldots$	No
	$x_4 = 0.97234\ldots$	No
	$x_5 = 0.96866\ldots$	No
	$x_6 = 0.97048\ldots$	No
	$x_7 = 0.96958\ldots$	Yes

α seems to be 0.970 correct to 3 significant figures.

We can **prove** this, using a change of sign test:

When $\alpha = 0.970$ then $0.9695 < \alpha < 0.9705$ and so

$$f(0.9695) = 0.9695 - \tan^{-1}\sqrt{\dfrac{2}{2(0.9695)-1}} = -0.000571\ldots$$

$$f(0.9705) = 0.9705 - \tan^{-1}\sqrt{\dfrac{2}{2(0.9705)-1}} = 0.000924\ldots$$

Change of sign indicates presence of root.

Therefore $\alpha = 0.970$ correct to 3 significant figures.

WORKED EXAMPLE 6.7

The diagram shows the design of a company logo, ABC.

AB is an arc of a circle with centre C.

The area of the unshaded segment is to be the same as the area of the shaded triangle.

Angle ACB is θ radians.

a Show that $\theta = 2\sin\theta$.

b Showing all your working, use an iterative formula based on the equation in **part a**, with an initial value of 1.85, to find θ correct to 3 significant figures.

c Hence find the length of AB, given that AC is 8 cm.

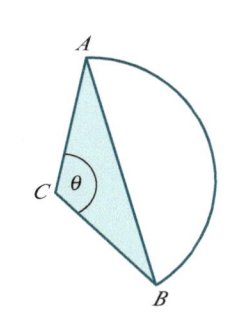

Answer

a $\frac{1}{2}r^2 \sin\theta = \frac{1}{2}r^2(\theta - \sin\theta)$ .... Divide through by $\frac{1}{2}r^2$.

$\sin\theta = \theta - \sin\theta$ .... Add $\sin\theta$.

$\theta = 2\sin\theta$ .... ✓

b

$\theta_1 = 1.85$	$\theta_{n+1} = 2\sin\theta_n$	Are they the same correct to 3 significant figures?
	$\theta_2 = 2\sin 1.85 = 1.922\,55\ldots$	No
	$\theta_3 = 2\sin 1.922\,55\ldots = 1.877\,53\ldots$	No
	$\theta_4 = 1.906\,64\ldots$	No
	$\theta_5 = 1.888\,26\ldots$	No
	$\theta_6 = 1.900\,05\ldots$	No
	$\theta_7 = 1.892\,56\ldots$	No
	$\theta_8 = 1.897\,35\ldots$	No
	$\theta_9 = 1.894\,30\ldots$	No
	$\theta_{10} = 1.896\,25\ldots$	No
	$\theta_{11} = 1.895\,00\ldots$	Yes

θ seems to be 1.90 correct to 3 significant figures.

We can **prove** this, using a change of sign test with $f(\theta) = \theta - 2\sin\theta = 0$:

When $\theta = 1.90$ then $1.895 < \theta < 1.905$ and so

$f(1.895) = 1.895 - 2\sin 1.895 = -0.000809\ldots$
$f(1.905) = 1.905 - 2\sin 1.905 = 0.01565\ldots$

Change of sign indicates presence of root, therefore $\theta = 1.90$ correct to 3 significant figures.

c Using trigonometry:

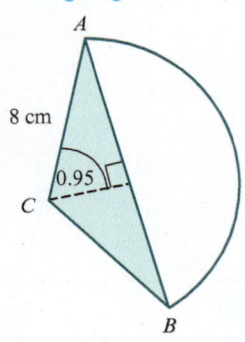

$$\sin 0.95 = \frac{0.5AB}{8}$$ ········· Multiply by 16.

$$16 \sin 0.95 = AB$$

$AB = 13.014..... = 13$ cm (correct to 2 significant figures)

EXERCISE 6C

1. The curve with equation $y = \dfrac{e^{2x} + x}{x^3}$ has a stationary point with x-coordinate lying between 1 and 2.

   a Show that the x-coordinate of this stationary point satisfies the equation $x = \dfrac{3}{2} + \dfrac{x}{e^{2x}}$.

   b Using a suitable iterative process based on the equation given in **part a**, find the value of the x-coordinate of this stationary point correct to 4 decimal places. Give each iteration to 6 decimal places.

2. The parametric equations of a curve are $x = t^2 + 6$, $y = t^4 - t^3 - 5t$.

   The curve has a stationary point for a value of t that lies between 1 and 2.

   a Show that the value of t at this stationary point satisfies the equation
   $$t = \sqrt[3]{\frac{3t^2 + 5}{4}}$$

   b Use an iterative process based on the equation in **part a** to find the value of t correct to 3 decimal places. Show the result of each iteration to 6 decimal places.

   c Hence find the coordinates of the stationary point, giving each coordinate correct to 1 significant figure.

3. In the diagram, triangle ABC is right-angled and angle BAC is θ radians. The point O is the mid point of AC and $OC = r$. Angle BOC is 2θ radians and BOC is a sector of the circle with centre O. The area of triangle ABC is 2 times the area of the shaded segment.

   a Show that θ satisfies the equation $\sin 2\theta = \theta$.

   b This equation has one root in the interval $0 < \theta < \dfrac{\pi}{2}$. Use the iterative formula $\theta_{n+1} = \sin 2\theta_n$ to determine the root correct to 2 decimal places. Give the result of each iteration to 4 decimal places.

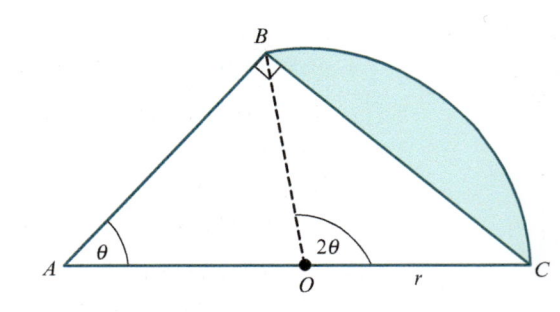

4   The diagram shows the curve $y = x^2 \cos 4x$ for $0 \leq x \leq \dfrac{\pi}{8}$. The point P is a maximum point.

   a   Show that the x-coordinate of P satisfies the equation $4x^2 \tan 4x = 2x$.

   b   Show also that the x-coordinate of P satisfies the equation $x = \dfrac{1}{4}\tan^{-1}\left(\dfrac{1}{2x}\right)$.

   c   Using an iterative formula based on the equation in **part b** with initial value $x_1 = 0.3$ find the x-coordinate of P correct to 2 decimal places. Give the result of each iteration to 4 decimal places.

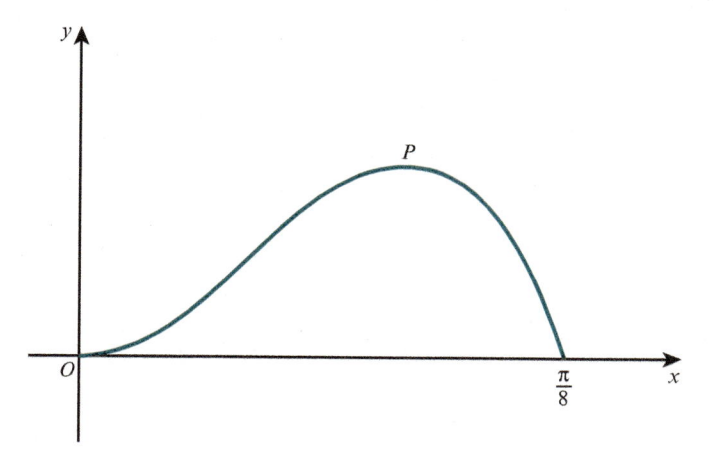

   P3   d   Use integration by parts twice to find the exact area enclosed between the curve and the x-axis from 0 to $\dfrac{\pi}{8}$.

EXPLORE 6.3

The golden ratio

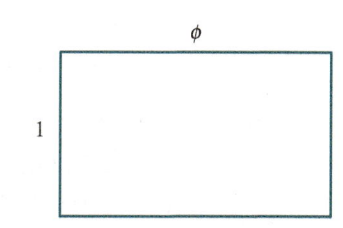

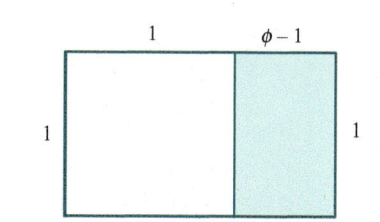

When a rectangle with sides in the ratio $1:\phi$ (phi) can be cut into a square of side 1 and a smaller rectangle, also with sides in the ratio $1:\phi$, the original rectangle is called a **golden rectangle** and the ratio $1:\phi$ is called the **golden ratio**.

This means that $\dfrac{\phi - 1}{1} = \dfrac{1}{\phi}$.

By using a suitable iterative formula based on this ratio, with a starting value of 1, find the value of ϕ correct to 5 significant figures.

Explore the link between the golden ratio and the sequence of Fibonacci numbers:

$$1, 1, 2, 3, 5, 8, ..., ...$$

DID YOU KNOW?

The relationship between the Fibonacci numbers and the golden ratio produces the golden spiral.

This can be seen in many natural settings, one of which is the spiralling pattern of the petals in a rose flower.

From a tight centre, petals flow outward in a spiral, growing wider and larger as they do so. Nature has made the best use of an incredible structure. This perfect arrangement of petals is why the small, compact rose bud grows into such a beautiful flower.

Checklist of learning and understanding

- The process of finding a sequence of values that become closer and closer to a point of intersection of $y = x$ and $y = F(x)$ is called an **iterative process**.
- Each time you generate a value of x you carry out an iteration.
- When the point of intersection is at $x = \alpha$ and your iterations are values that are getting closer and closer to α, you are converging to α.
- Start with x_1 work out $F(x_1)$. Is it accurate enough?   Yes → Stop, No →
- Let $x_2 = F(x_1)$ work out $F(x_2)$. Is it accurate enough?   Yes → Stop, No →
- Let $x_3 = F(x_2)$ work out $F(x_3)$. Is it accurate enough?   Yes → Stop, No →
  $\vdots$
- Let $x_{n+1} = F(x_n)$ work out $F(x_{n+1})$. Is it accurate enough?   Yes → Stop.
- The relationship $x_{n+1} = F(x_n)$ is called an **iterative formula**.
- The sequence of values of x given by this formula are called $x_1, x_2, x_3, \ldots, x_n$.
- When the value found is accurate enough, then it is usually given a particular name such as α.
- The success and speed of the iterative process depends on the function chosen for $F(x_n)$ and the value chosen for x_1.

Chapter 6: Numerical solutions of equations

END-OF-CHAPTER REVIEW EXERCISE 6

1. The terms of the sequence generated by the iterative formula $x_{n+1} = \dfrac{6}{7}\left(x_n + \dfrac{1}{x_n^3}\right)$ with initial value $x_1 = 1.5$, converge to α.

   a. Use this formula to find α correct to 2 decimal places. Give the result of each iteration to an appropriate number of decimal places. [3]

   b. State an equation satisfied by α, and hence find the exact value of α. [2]

2. The equation $x^3 + \dfrac{x}{2} - 8 = 0$ has one real root, denoted by α.

   a. Find, by calculation, the pair of consecutive integers between which α lies. [2]

   b. Show that, if a sequence of values given by the iterative formula $x_{n+1} = \sqrt{\dfrac{8}{x_n} - \dfrac{1}{2}}$ converges, then it converges to α. [2]

   c. Use this iterative formula to determine α correct to 1 decimal place. Give the result of each iteration to 4 decimal places. [3]

3. a. By sketching a suitable pair of graphs, show that the equation $e^{2x+1} = 14 - x^3$ has exactly one real root. [3]

   b. Show by calculation that this root lies between 0.5 and 1. [2]

   c. Show that this root also satisfies the equation $x = \dfrac{\ln(14 - x^3) - 1}{2}$. [1]

   d. Use an iteration process based on the equation in **part c**, with a suitable starting value, to find the root correct to 4 decimal places. Give the result of each step of the process to 6 decimal places. [3]

4. The sequence of values given by the iterative formula $x_{n+1} = \dfrac{x_n(1 + \sec^2 x_n) - \tan x_n}{\sec^2 x_n - 1}$ with initial value $x_1 = 1$, converges to α.

   a. Use this formula to find α correct to 2 decimal places, showing the result of each iteration to 4 decimal places. [3]

   b. Show that α also satisfies the equation $\tan x - 2x = 0$ where $1 \leqslant x < \dfrac{\pi}{2}$. [2]

5. The equation $\dfrac{\ln x^2 - 6}{2} = x - 5$ has a root α, such that $p < \alpha < q$, where p and q are consecutive integers.

   a. Show that α also satisfies the equation $x = \ln x + 2$. [1]

   b. Find the value of p and the value of q. [2]

   c. Use the iterative formula based on the equation in **part a** with starting value $x = p$, to find the value of α correct to 3 decimal places. [3]

   d. Hence find an approximate value of y that satisfies the equation $5^y - 2 = y \ln 5$. [2]

6. a. By sketching a suitable pair of graphs, show that the equation $7 - x^6 = |x^2 - 1|$ has exactly two real roots, α and β, where α is a positive constant. [3]

   b. Show that α satisfies the equation $x = \sqrt[5]{\dfrac{8}{x} - x}$. [1]

   c. Write down an iterative formula based on the equation in **part a**. Use this formula, with a suitable starting value, to find the value of α correct to 3 significant figures. Give the result of each iteration to 5 significant figures. [3]

   d. Hence write down the value of β correct to 3 significant figures. [1]

7 It is given that $\int_0^a \left(\dfrac{1}{3x+5} + 2e^{6x}\right) dx = 0.6$, where the constant $a > 0$.

   a Show that $a = \dfrac{1}{6} \ln\left(2.8 + \ln\left(\dfrac{5}{3a+5}\right)\right)$. [4]

   b Use an iterative formula based on the equation in **part a**, with a starting value of $a = 0.2$, to find the value of a correct to 3 decimal places. Give the result of each iteration to an appropriate number of decimal places. [3]

8 A curve has parametric equations
   $$x = \dfrac{1}{t}, \quad y = \sqrt{t+1}$$

   The point P on the curve has parameter p. It is given that the gradient of the curve at P is -0.4.

   a Show that $p = \dfrac{2\sqrt{5}}{5}(p+1)^{\frac{1}{4}}$. [3]

   b Use an iterative process based on the equation in **part a** to find the value of p correct to 3 decimal places. Use a starting value of 1 and show the result of each iteration to 5 decimal places. [3]

9 In the diagram, A, B and C are points on the circumference of the circle with centre O and radius r. The shaded region, ABC, is a sector of the circle with centre C. Angle OCA is equal to θ radians.

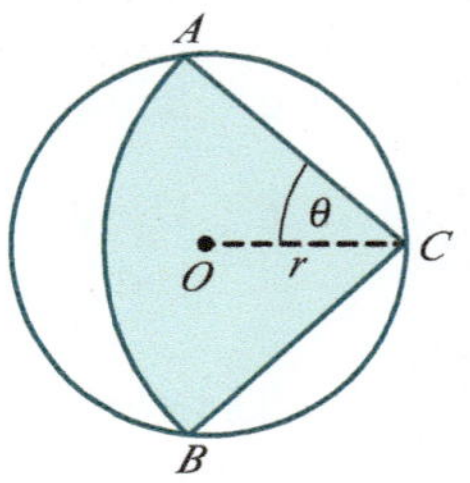

   The area of the shaded region is equal to $\dfrac{3}{8}$ of the area of the circle.

   a Show that $\theta = \cos^{-1}\sqrt{\dfrac{3\pi}{32\theta}}$. [3]

   b Verify by calculation that θ lies between 0.8 and 1.2 radians. [2]

   c Use the iterative formula $\theta_{n+1} = \cos^{-1}\sqrt{\dfrac{3\pi}{32\theta_n}}$ to find the value of θ correct to 3 decimal places. Give the result of each iteration to 5 decimal places. [3]

10 The functions f and g are defined, for $0 \leqslant x \leqslant \pi$, by $f(x) = e^{x-2}$ and $g(x) = 5 - \cos x$.

    The diagram shows the graph of $y = f(x)$ and the graph of $y = g(x)$.

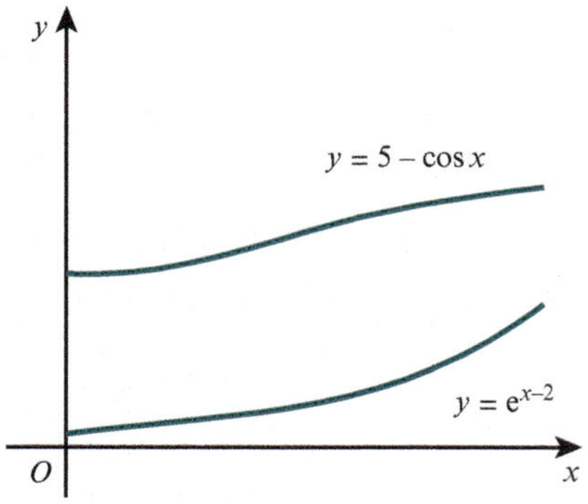

    The gradients of the curves are equal both when $x = p$ and when $x = q$.

    a Given that $p < q$, verify by calculation that p is 0.16 correct to 2 decimal places. [4]

    b Show that q satisfies the equation $q = 2 + \ln(\sin q)$. [1]

    c Given also that $1.5 < q < 2.5$, use the iterative formula $q_{n+1} = 2 + \ln(\sin q_n)$ to calculate q correct to 2 decimal places, showing the result of each iteration to 4 decimal places. [3]

11 The equation $x = \cos x + \sin x$ has a root α that lies between 1 and 1.4.

    a Show that α is also a root of the equation $x = \sqrt{2} \cos\left(x - \dfrac{\pi}{4}\right)$. [3]

    b Using the iterative formula $x_{n+1} = \sqrt{2} \cos\left(x_n - \dfrac{\pi}{4}\right)$, find the value of α, giving your answer correct to 2 decimal places. [3]

Chapter 6: Numerical solutions of equations

P3 **12 a** By sketching each of the graphs of $y = \sec x$ and $y = \left(\dfrac{\pi}{2} - x\right)\left(\dfrac{\pi}{4} + x\right)$ for $-\dfrac{\pi}{2} < x \leqslant \dfrac{\pi}{2}$ on the same diagram, show that the equation $\sec x = \left(\dfrac{\pi}{2} - x\right)\left(\dfrac{\pi}{4} + x\right)$ has exactly two real roots in the interval $-\dfrac{\pi}{2} < x \leqslant \dfrac{\pi}{2}$. [3]

b Show that the equation $\sec x = \left(\dfrac{\pi}{2} - x\right)\left(\dfrac{\pi}{4} + x\right)$ can be written in the form
$$x = \sqrt{\dfrac{2\pi x + \pi^2 - 8\sec x}{8}}.$$ [1]

c The two real roots of the equation $\sec x = \left(\dfrac{\pi}{2} - x\right)\left(\dfrac{\pi}{4} + x\right)$ in the interval $-\dfrac{\pi}{2} < x \leqslant \dfrac{\pi}{2}$ are denoted α and β.

Verify by calculation that the smaller root, α, is -0.21 correct to 2 decimal places. [2]

d Using an iterative formula based on the equation given in **part b**, with an initial value of 1, find the value of β correct to 2 decimal places. Give the result of each iteration to 4 decimal places. [3]

⏩ FAST FORWARD

You need to have studied the skills in Chapter 8 before attempting questions 13 and 14.

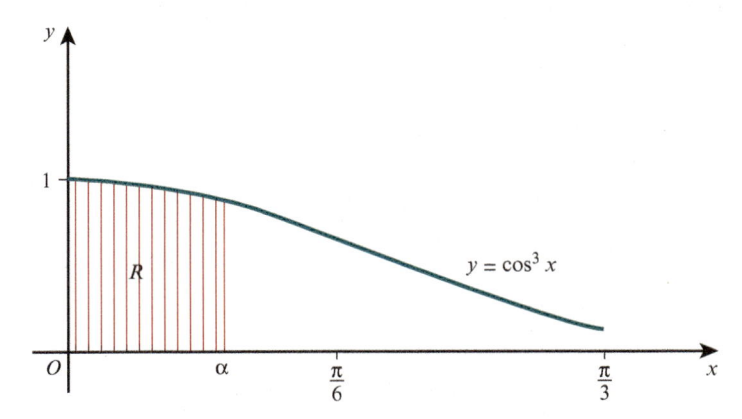

P3 **13**

The diagram shows part of the curve $y = \cos^3 x$, where x is in radians. The shaded region between the curve, the axes and the line $x = \alpha$ is denoted by R. The area of R is equal to 0.3.

a Using the substitution $u = \sin x$, find $\displaystyle\int_0^\alpha \cos^3 x \, dx$. Hence show that $\sin \alpha = \dfrac{0.9}{3 - \sin^2 \alpha}$. [6]

b Use the iterative formula $\alpha_{n+1} = \sin^{-1}\left(\dfrac{0.9}{3 - \sin^2 \alpha_n}\right)$ with $\alpha_1 = 0.2$, to find the value of α correct to 3 significant figures. Give the result of each iteration to 5 significant figures. [3]

P3 **14** It is given that $\displaystyle\int_1^a \ln x \, dx = 5$, where a is a constant greater than 1.

a Show that $a = \dfrac{4 + a}{\ln a}$. [5]

b Use an iterative formula based on the equation in **part a** to find the value of a correct to 3 decimal places. Use an initial value of 5 and give the result of each iteration to 5 decimal places. [3]

CROSS-TOPIC REVIEW EXERCISE 2

1 The equation of a curve is $y = \dfrac{1+x}{1+2x}$ for $x > -\dfrac{1}{2}$. Show that the gradient of the curve is always negative. [3]

Cambridge International A Level Mathematics 9709 Paper 31 Q1 November 2013

2

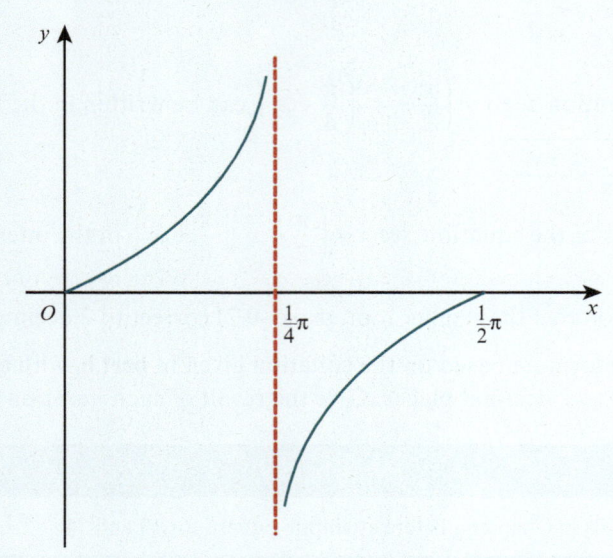

The diagram shows the part of the curve $y = \dfrac{1}{2}\tan 2x$ for $0 \leqslant x \leqslant \dfrac{1}{2}\pi$. Find the x-coordinates of the points on this part of the curve at which the gradient is 4. [5]

Cambridge International AS & A Level Mathematics 9709 Paper 21 Q3 November 2011

3 The parametric equations of a curve are $x = \ln(1 - 2t)$, $y = \dfrac{2}{t}$, for $t < 0$.

  i Show that $\dfrac{dy}{dx} = \dfrac{1-2t}{t^2}$. [3]

  ii Find the exact coordinates of the only point on the curve at which the gradient is 3. [3]

Cambridge International AS & A Level Mathematics 9709 Paper 21 Q4 November 2012

4 The equation of a curve is $x^2 y + y^2 = 6x$.

  i Show that $\dfrac{dy}{dx} = \dfrac{6-2xy}{x^2+2y}$. [4]

  ii Find the equation of the tangent to the curve at the point with coordinates (1, 2), giving your answer in the form $ax + by + c = 0$. [3]

Cambridge International AS & A Level Mathematics 9709 Paper 21 Q6 June 2010

5 **i** Use the trapezium rule with two intervals to estimate the value of $\displaystyle\int_0^1 \dfrac{1}{6+2e^x}\,dx$, giving your answer correct to 2 decimal places. [3]

  ii Find $\displaystyle\int \dfrac{(e^x - 2)^2}{e^{2x}}\,dx$. [4]

Cambridge International AS & A Level Mathematics 9709 Paper 21 Q6 November 2012

6 i By sketching a suitable pair of graphs, show that the equation $\ln x = 4 - \frac{1}{2}x$ has exactly one real root, α. [2]

   ii Verify by calculation that $4.5 < \alpha < 5.0$. [2]

   iii Use an iterative formula $x_{n+1} = 8 - 2\ln x_n$ to find α correct to 2 decimal places. Give the result of each iteration to 4 decimal places. [3]

   Cambridge International AS & A Level Mathematics 9709 Paper 21 Q4 November 2015

7 The parametric equations of a curve are $x = \dfrac{1}{\cos^3 t}$, $y = \tan^3 t$, where $0 \leqslant t < \frac{1}{2}\pi$.

   i Show that $\dfrac{dy}{dx} = \sin t$. [4]

   ii Hence show that the equation of the tangent to the curve at the point with parameter t is
   $y = x \sin t - \tan t$. [3]

   Cambridge International A Level Mathematics 9709 Paper 31 Q4 November 2014

8 It is given that the positive constant a is such that $\displaystyle\int_{-a}^{a} (4e^{2x} + 5)\, dx = 100$.

   i Show that $a = \dfrac{1}{2} \ln(50 + e^{-2a} - 5a)$. [4]

   ii Use the iterative formula $a_{n+1} = \dfrac{1}{2} \ln(50 + e^{-2a_n} - 5a_n)$ to find α correct to 3 decimal places. Give the result of each iteration to 5 decimal places. [3]

   Cambridge International AS & A Level Mathematics 9709 Paper 21 Q4 November 2016

9 The equation of a curve is $y = e^{-2x} \tan x$, for $0 \leqslant x < \frac{1}{2}\pi$.

   i Obtain an expression for $\dfrac{dy}{dx}$ and show that it can be written in the form $e^{-2x}(a + b \tan x)^2$, where a and b are constants. [5]

   ii Explain why the gradient of the curve is never negative. [1]

   iii Find the value of x for which the gradient is least. [1]

   Cambridge International A Level Mathematics 9709 Paper 31 Q5 November 2015

10 i Prove that $\tan \theta + \cot \theta \equiv \dfrac{2}{\sin 2\theta}$. [3]

   ii Hence

   a find the exact value of $\tan \dfrac{1}{8}\pi + \cot \dfrac{1}{8}\pi$, [2]

   b evaluate $\displaystyle\int_{0}^{\frac{1}{2}\pi} \dfrac{6}{\tan \theta + \cot \theta}\, d\theta$. [3]

   Cambridge International AS & A Level Mathematics 9709 Paper 21 Q5 June 2014

11 The equation of a curve is $xy(x - 6y) = 9a^3$, where a is a non-zero constant. Show that there is only one point on the curve at which the tangent is parallel to the x-axis, and find the coordinates of this point. [7]

   Cambridge International A Level Mathematics 9709 Paper 31 Q4 November 2016

12 The curve with equation $y = \dfrac{6}{x^2}$ intersects the line $y = x + 1$ at the point P.

  i   Verify by calculation that the x-coordinate of P lies between 1.4 and 1.6. [2]

  ii Show that the x-coordinate of P satisfies the equation $x = \sqrt{\left(\dfrac{6}{x+1}\right)}$. [2]

  iii Use the iterative formula $x_{n+1} = \sqrt{\left(\dfrac{6}{x_n + 1}\right)}$, with initial value $x_1 = 1.5$, to determine the x-coordinate of P correct to 2 decimal places. Give the result of each iteration to 4 decimal places. [3]

  Cambridge International AS & A Level Mathematics 9709 Paper 21 Q6 November 2010

13 The parametric equations of a curve are $x = e^{2t}$, $y = 4te^t$.

  i   Show that $\dfrac{dy}{dx} = \dfrac{2(t+1)}{e^t}$. [4]

  ii Find the equation of the normal to the curve at the point where $t = 0$. [4]

  Cambridge International AS & A Level Mathematics 9709 Paper 21 Q5 June 2013

14 For each of the following curves, find the exact gradient at the point indicated:

  i $y = 3 \cos 2x - 5 \sin x$ at $\left(\dfrac{1}{6}\pi, -1\right)$, [3]

  ii $x^3 + 6xy + y^3 = 21$ at $(1, 2)$. [5]

  Cambridge International AS & A Level Mathematics 9709 Paper 21 Q4 November 2014

15 i   By differentiating $\dfrac{1}{\cos x}$, show that if $y = \sec x$ then $\dfrac{dy}{dx} = \sec x \tan x$. [2]

  ii Show that $\dfrac{1}{\sec x - \tan x} \equiv \sec x + \tan x$. [1]

  iii Deduce that $\dfrac{1}{(\sec x - \tan x)^2} \equiv 2\sec^2 x - 1 + 2\sec x \tan x$. [2]

  iv Hence show that $\displaystyle\int_0^{\frac{1}{4}\pi} \dfrac{1}{(\sec x - \tan x)^2} \, dx = \dfrac{1}{4}\left(8\sqrt{2} - \pi\right)$. [3]

  Cambridge International A Level Mathematics 9709 Paper 31 Q5 November 2012

16 A curve is defined by the parametric equations $x = 2\tan\theta$, $y = 3\sin 2\theta$, for $0 \leq \theta < \dfrac{1}{2}\pi$.

  i   Show that $\dfrac{dy}{dx} = 6\cos^4\theta - 3\cos^2\theta$. [4]

  ii Find the coordinates of the stationary point. [3]

  iii Find the gradient of the curve at the point $\left(2\sqrt{3}, \dfrac{3}{2}\sqrt{3}\right)$. [2]

  Cambridge International AS & A Level Mathematics 9709 Paper 21 Q5 June 2016

17

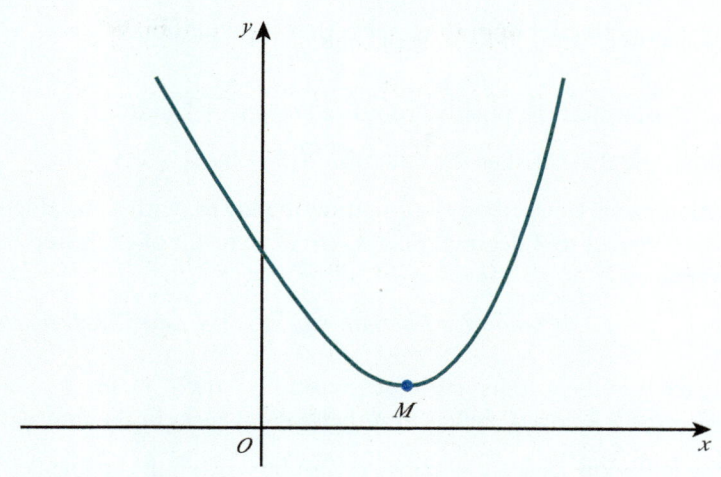

The diagram shows the curve $y = 4e^{\frac{1}{2}x} - 6x + 3$ and its minimum point M.

i Show that the x-coordinate of M can be written in the form $\ln a$, where the value of a is to be stated. [5]

ii Find the exact value of the area of the region enclosed by the curve and the lines $x = 0$, $x = 2$ and $y = 0$. [4]

Cambridge International AS & A Level Mathematics 9709 Paper 21 Q5 June 2012

18 The equation of a curve is $x^3 - 3x^2y + y^3 = 3$.

i Show that $\dfrac{dy}{dx} = \dfrac{x^2 - 2xy}{x^2 - y^2}$. [4]

ii Find the coordinates of the points on the curve where the tangent is parallel to the x-axis. [5]

Cambridge International A Level Mathematics 9709 Paper 31 Q7 June 2016

19 A curve has parametric equations $x = \dfrac{1}{(2t+1)^2}$, $y = \sqrt{(t+2)}$

The point P on the curve has parameter p and it is given that the gradient of the curve at P is -1.

i Show that $p = (p+2)^{\frac{1}{6}} - \dfrac{1}{2}$. [6]

ii Use an iterative process based on the equation in **part i** to find the value of p correct to 3 decimal places. Use a starting value of 0.7 and show the result of each iteration to 5 decimal places. [3]

Cambridge International AS & A Level Mathematics 9709 Paper 21 Q6 June 2012

20 i Find

a $\displaystyle\int \dfrac{e^{2x} + 6}{e^{2x}} \, dx$, [3]

b $\displaystyle\int 3\cos^2 x \, dx$. [3]

ii Use the trapezium rule with 2 intervals to estimate the value of $\displaystyle\int_1^2 \dfrac{6}{\ln(x+2)} \, dx$, giving your answer correct to 2 decimal places. [3]

Cambridge International AS & A Level Mathematics 9709 Paper 21 Q6 November 2013

21 i   By sketching a suitable pair of graphs, show that the equation $e^{2x} = 14 - x^2$ has exactly two real roots. [3]

   ii   Show by calculation that the positive root lies between 1.2 and 1.3. [2]

   iii Show that this root also satisfies the equation $x = \frac{1}{2}\ln(14 - x^2)$. [1]

   iv   Use an iteration process based on the equation in **part iii**, with a suitable starting value, to find the root correct to 2 decimal places. Give the result of each step of the process to 4 decimal places. [3]

   Cambridge International AS & A Level Mathematics 9709 Paper 21 Q7 June 2011

22 i   By sketching each of the graphs $y = \operatorname{cosec} x$ and $y = x(\pi - x)$ for $0 < x < \pi$, show that the equation $\operatorname{cosec} x = x(\pi - x)$ has exactly two real roots in the interval for $0 < x < \pi$. [3]

   ii   Show that the equation $\operatorname{cosec} x = x(\pi - x)$ can be written in the form $x = \dfrac{1 + x^2 \sin x}{\pi \sin x}$. [2]

   iii The two real roots of the equation $\operatorname{cosec} x = x(\pi - x)$ in the interval $0 < x < \pi$ are denoted by α and β, where $\alpha < \beta$.

   a Use the iterative formula $x_{n+1} = \dfrac{1 + x_n^2 \sin x_n}{\pi \sin x_n}$ to find α correct to 2 decimal places. Give the result of each iteration to 4 decimal places. [3]

   b Deduce the value of β correct to 2 decimal places. [1]

   Cambridge International A Level Mathematics 9709 Paper 31 Q8 June 2014

23

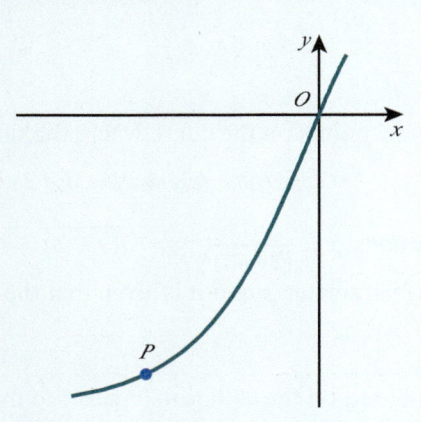

The diagram shows part of the curve with parametric equations

$$x = 2\ln(t + 2), \quad y = t^3 + 2t + 3.$$

   i   Find the gradient of the curve at the origin. [5]

   ii At the point P on the curve, the value of the parameter is p. It is given that the gradient of the curve at P is $\frac{1}{2}$.

   a Show that $p = \dfrac{1}{3p^2 + 2} - 2$. [1]

   b By first using an iterative formula based on the equation in **part a**, determine the coordinates of the point P. Give the result of each iteration to 5 decimal places and each coordinate of P correct to 2 decimal places. [4]

   Cambridge International A Level Mathematics 9709 Paper 31 Q10 June 2015

24 i Prove that $2\operatorname{cosec} 2\theta \tan\theta \equiv \sec^2\theta$. [3]

   ii Hence

      a solve the equation $2\operatorname{cosec} 2\theta \tan\theta = 5$ for $0 < x < \pi$. [3]

      b find the exact value of $\displaystyle\int_0^{\frac{1}{6}\pi} 2\operatorname{cosec} 4x \tan 2x \,dx$. [4]

   Cambridge International AS & A Level Mathematics 9709 Paper 21 Q6 June 2015

25 i Prove the identity $\cos 4\theta + 4\cos 2\theta \equiv 8\cos^4\theta - 3$. [4]

   ii Hence

      a solve the equation $\cos 4\theta + 4\cos 2\theta = 1$ for $-\tfrac{1}{2}\pi \leq \theta \leq \tfrac{1}{2}\pi$. [3]

      b find the exact value of $\displaystyle\int_0^{\frac{1}{4}\pi} \cos^4\theta \,d\theta$. [3]

   Cambridge International A Level Mathematics 9709 Paper 31 Q9 June 2011

26 i By first expanding $\cos(2x + x)$, show that $\cos 3x = 4\cos^3 x - 3\cos x$. [5]

   ii Hence show that $\displaystyle\int_0^{\frac{1}{6}\pi} (2\cos^3 x - \cos x)\,dx = \dfrac{5}{12}$. [5]

   Cambridge International AS & A Level Mathematics 9709 Paper 21 Q8 November 2011

27

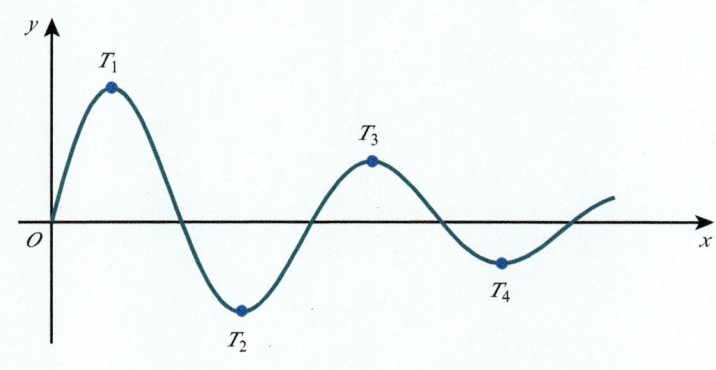

The diagram shows the curve $y = 10e^{-\frac{1}{2}x}\sin 4x$ for $x \geq 0$. The stationary points are labeled $T_1, T_2, T_3, \ldots$ as shown.

   i Find the x-coordinates of T_1 and T_2, giving each x-coordinate correct to 3 decimal places. [6]

   ii It is given that the x-coordinate of T_n is greater than 25. Find the least possible value of n. [4]

   Cambridge International A Level Mathematics 9709 Paper 31 Q10 June 2014

28 The equation of a curve is $y = \dfrac{3x^2}{x^2+4}$. At the point on the curve with positive x-coordinate p, the gradient of the curve is $\dfrac{1}{2}$.

    i Show that $p = \sqrt{\left(\dfrac{48p-16}{p^2+8}\right)}$. [5]

    ii Show by calculation that $2 < p < 3$. [2]

    iii Use an iterative formula based on the equation in **part i** to find the value of p correct to 4 significant figures. Give the result of each iteration to 6 significant figures. [3]

Cambridge International AS & A Level Mathematics 9709 Paper 21 Q6 June 2016

29 **i** By differentiating $\dfrac{1}{\cos\theta}$, show that if $y = \sec\theta$ then $\dfrac{dy}{d\theta} = \tan\theta \sec\theta$. [3]

    ii Hence show that $\dfrac{d^2y}{d\theta^2} = a\sec^3\theta + b\sec\theta$, giving the values of a and b. [4]

    iii Find the exact value of $\displaystyle\int_0^{\frac{1}{4}\pi} (1 + \tan^2\theta - 3\sec\theta\tan\theta)\, d\theta$. [5]

Cambridge International AS & A Level Mathematics 9709 Paper 21 Q8 November 2012

Chapter 7
Further algebra

P3 This chapter is for Pure Mathematics 3 students only.

In this chapter you will learn how to:

- recall an appropriate form for expressing rational functions in partial fractions, and carry out the decomposition, in cases where the denominator is no more complicated than:
  - $(ax + b)(cx + d)(ex + f)$
  - $(ax + b)(cx + d)^2$
  - $(ax + b)(cx^2 + d)$
- use the expansion of $(1 + x)^n$, where n is a rational number and $|x| < 1$.

PREREQUISITE KNOWLEDGE

Where it comes from	What you should be able to do	Check your skills
Pure Mathematics 1 Coursebook, Chapter 6	Equate coefficients of polynomials.	1 Find the value of A, B and C for: a $Ax^2 - 3x + C \equiv 6x^2 + Bx - 9$ b $(2 - A)x^2 + 5x + 2C \equiv 3x^2 - 3Bx + 8$
Pure Mathematics 1 Coursebook, Chapter 6	Expand $(a + b)^n$ where n is a positive integer.	2 Find the first 3 terms, in ascending powers of x, in the expansion of: a $(1 + 2x)^7$ b $(3 - 2x)^5$
Chapter 1	Divide polynomials.	3 Find the quotient and remainder when $x^2 - 8x + 4$ is divided by $x - 3$.

Why do we study algebra?

At IGCSE / O Level we learnt how to add and subtract algebraic fractions. In this chapter we will learn how to do the 'reverse process'. This reverse process is often referred to as splitting a fraction into its partial fractions. In Mathematics it is often easier to deal with two or more simple fractions than it is to deal with one complicated fraction.

In the Pure Mathematics 1 Coursebook, Chapter 6, you learnt how to find the binomial expansion of $(a + b)^n$ for positive integer values of n. After working through this chapter you will be able to expand expressions of the form $(1 + x)^n$ for values of n that are not positive integers (providing $|x| < 1$).

Combining your partial fraction and binomial expansion skills will enable you to obtain series expansions of complicated expressions such as $\dfrac{2x - 1}{2x^2 + 3x - 20}$.

7.1 Improper algebraic fractions

A numerical improper fraction is defined as a fraction where the numerator $\geqslant$ the denominator. For example, $\dfrac{11}{5}$ is an improper fraction. This fraction can be expressed as $2 + \dfrac{1}{5}$, which is the sum of a positive integer and a proper fraction.

So how do we define an **algebraic improper fraction**?

> **KEY POINT 7.1**
>
> The algebraic fraction $\dfrac{P(x)}{Q(x)}$, where $P(x)$ and $Q(x)$ are polynomials in x, is said to be an algebraic improper fraction if the degree of $P(x) \geqslant$ the degree of $Q(x)$.

For example, the fraction $\dfrac{x^3 - 3x^2 + 7}{x - 2}$ is an improper algebraic fraction because the degree of the numerator (3) is greater than the degree of the denominator (1).

FAST FORWARD

In Chapter 8 you will be shown another use for partial fractions: how to integrate rational expressions such as $\dfrac{2x - 1}{2x^2 + 3x - 20}$ by first splitting the expression into partial fractions.

WEB LINK

Explore the *Polynomials and rational functions* station on the Underground Mathematics website.

We can use long division to write the fraction $\dfrac{x^3 - 3x^2 + 7}{x - 2}$ as the sum of a polynomial and a proper algebraic fraction.

$$\begin{array}{r}
x^2 - x - 2 \\
x-2 \overline{\smash{\big)}\, x^3 - 3x^2 + 0x + 7} \\
\underline{x^3 - 2x^2 } \\
-x^2 + 0x \\
\underline{-x^2 + 2x } \\
-2x + 7 \\
\underline{-2x + 4} \\
3
\end{array}$$

$\therefore \dfrac{x^3 - 3x^2 + 7}{x - 2} = x^2 - x - 2 + \dfrac{3}{x - 2}$

EXPLORE 7.1

1 Discuss with your classmates which of the following are improper algebraic fractions.

$\dfrac{1}{2x + 1}$ $\dfrac{x^2 - 4x}{5 - x^2}$ $\dfrac{3x}{x - 5}$ $\dfrac{6x^3 - 2x + 1}{2x^2 - 1}$

$\dfrac{2x - 3}{(x + 2)(x - 1)}$ $\dfrac{x^3 + 2x^2 - 7}{(x + 2)(x + 1)}$ $\dfrac{4x^5 - 1}{3x^7 + 2}$ $\dfrac{2x^4 - 8}{x^2 - 2x - 1}$

2 Write each improper fraction in **question 1** as the sum of a polynomial and a proper fraction.

EXERCISE 7A

1 Express each of the following improper fractions as the sum of a polynomial and a proper fraction.

  a $\dfrac{8x}{2x - 5}$ b $\dfrac{6x + 1}{3x + 2}$ c $\dfrac{4x^3 - 3}{2x + 1}$

  d $\dfrac{x^3 + 4x^2 + 3x - 1}{x^2 + 2x + 3}$ e $\dfrac{7x^3 + 2x^2 - 5x + 1}{x^2 - 5}$ f $\dfrac{x^4 + 2x^2 - 5}{x^2 + 1}$

2 Given that $\dfrac{x^3 + x^2 - 7}{x - 3} \equiv Ax^2 + Bx + C + \dfrac{D}{x - 3}$, find the values of A, B, C and D.

3 Given that $\dfrac{x^4 + 5x^2 - 1}{x + 1} \equiv Ax^3 + Bx^2 + Cx + D + \dfrac{E}{x + 1}$, find the values of A, B, C, D and E.

4 Given that $\dfrac{2x^4 + 3x^3 + 4x^2 + 5x + 6}{x^3 + 2x} \equiv Ax + B + \dfrac{Cx + D}{x^3 + 2x}$, find the values of A, B, C and D.

7.2 Partial fractions

At IGCSE / O Level, we learnt how to add and subtract algebraic fractions. The following table shows three different cases for addition of algebraic fractions.

Case 1	Case 2	Case 3
$\dfrac{5}{x+3} + \dfrac{4}{x-2}$ $= \dfrac{5(x-2) + 4(x+3)}{(x+3)(x-2)}$ $= \dfrac{9x+12}{(x+3)(x-2)}$	$\dfrac{3}{x+1} + \dfrac{2}{(x+1)^2}$ $= \dfrac{3(x+1)+2}{(x+1)^2}$ $= \dfrac{3x+5}{(x+1)^2}$	$\dfrac{2}{x-1} + \dfrac{3}{x^2+5}$ $= \dfrac{2(x^2+5) + 3(x-1)}{(x-1)(x^2+5)}$ $= \dfrac{2x^2 + 3x + 7}{(x-1)(x^2+5)}$
Denominator in answer has: two distinct linear factors.	Denominator in answer has: one linear repeated factor.	Denominator in answer has: one linear factor and one quadratic factor.

In this section we will learn how to do the reverse of this process. We will learn how to split (or **decompose**) a single algebraic fraction into two or more **partial fractions**.

Case 1: Proper fraction where the denominator has distinct linear factors

If the denominator of a proper algebraic fraction has two distinct linear factors, the fraction can be split into partial fractions.

> **KEY POINT 7.2**
>
> $$\frac{px + q}{(ax+b)(cx+d)} \equiv \frac{A}{ax+b} + \frac{B}{cx+d}$$

This rule can be extended for 3 or more linear factors.

> **KEY POINT 7.3**
>
> $$\frac{px + q}{(ax+b)(cx+d)(ex+f)} \equiv \frac{A}{ax+b} + \frac{B}{cx+d} + \frac{C}{ex+f}$$

WORKED EXAMPLE 7.1

Express $\dfrac{2x - 13}{(2x+1)(x-3)}$ in partial fractions.

Answer

$\dfrac{2x-13}{(2x+1)(x-3)} \equiv \dfrac{A}{2x+1} + \dfrac{B}{x-3}$ .... Multiply throughout by $(2x+1)(x-3)$.

$\quad\quad 2x - 13 \equiv A(x-3) + B(2x+1)$ --------(1)

Method 1: Choose appropriate values for x.

Let $x = 3$ in equation (1):

$$2(3) - 13 = A(3-3) + B(6+1)$$
$$-7 = 7B$$
$$B = -1$$

Let $x = -\dfrac{1}{2}$ in equation (1):

$$2\left(-\dfrac{1}{2}\right) - 13 = A\left(-\dfrac{1}{2} - 3\right) + B(-1 + 1)$$
$$-14 = -\dfrac{7}{2}A$$
$$A = 4$$

$$\therefore \dfrac{2x - 13}{(2x+1)(x-3)} \equiv \dfrac{4}{2x+1} - \dfrac{1}{x-3}$$

Method 2: Equating coefficients.

$2x - 13 \equiv A(x-3) + B(2x+1)$ ········· Expand brackets and collect like terms.

$2x - 13 \equiv (A + 2B)x - 3A + B$ ········· Equate the coefficients of x and the constants.

$A + 2B = 2$ ---------(2)
$-3A + B = -13$ --------(3) ········· Solve equations (2) and (3) simultaneously.
$A = 4, B = -1$

$$\therefore \dfrac{2x - 13}{(2x+1)(x-3)} \equiv \dfrac{4}{2x+1} - \dfrac{1}{x-3}$$

Case 2: Proper fraction with repeated linear factor in the denominator

If the denominator has a repeated linear factor, the fraction can be split into partial fractions using the rule:

KEY POINT 7.4

$$\dfrac{px + q}{(ax+b)^2} \equiv \dfrac{A}{ax+b} + \dfrac{B}{(ax+b)^2}$$

WORKED EXAMPLE 7.2

Express $\dfrac{2x - 11}{(x-4)^2}$ in partial fractions.

Answer

$\dfrac{2x - 11}{(x-4)^2} \equiv \dfrac{A}{x-4} + \dfrac{B}{(x-4)^2}$ ········· Multiply throughout by $(x-4)^2$.

$2x - 11 \equiv A(x - 4) + B$ ---------(1)

Let $x = 4$ in equation (1):
$$2(4) - 11 = A(4 - 4) + B$$
$$B = -3$$
Let $x = 0$ in equation (1):
$$2(0) - 11 = A(0 - 4) - 3$$
$$A = 2$$
$$\therefore \frac{2x - 11}{(x - 4)^2} \equiv \frac{2}{x - 4} - \frac{3}{(x - 4)^2}$$

WORKED EXAMPLE 7.3

Express $\dfrac{1 - 9x - 8x^2}{(x - 2)(2x + 3)^2}$ in partial fractions.

Answer

$$\frac{1 - 9x - 8x^2}{(x - 2)(2x + 3)^2} \equiv \frac{A}{x - 2} + \frac{B}{2x + 3} + \frac{C}{(2x + 3)^2}$$ .... Multiply throughout by $(x - 2)(2x + 3)^2$.
$$1 - 9x - 8x^2 \equiv A(2x + 3)^2 + B(x - 2)(2x + 3) + C(x - 2) \quad \text{----(1)}$$

Let $x = 2$ in equation (1):
$$1 - 9(2) - 8(2)^2 = A(4 + 3)^2 + B(2 - 2)(4 + 3) + C(2 - 2)$$
$$-49 = 49A$$
$$A = -1$$

Let $x = -\dfrac{3}{2}$ in equation (1):
$$1 - 9\left(-\frac{3}{2}\right) - 8\left(-\frac{3}{2}\right)^2 = A(-3 + 3)^2 + B\left(-\frac{3}{2} - 2\right)(-3 + 3) + C\left(-\frac{3}{2} - 2\right)$$
$$-\frac{7}{2} = -\frac{7}{2}C$$
$$C = 1$$

Let $x = 0$ in equation (1):

$$1 = 9A - 6B - 2C \quad \text{------------------------(2)}$$ .... Substitute $A = -1$ and $C = 1$ in equation (2).

$$B = -2$$

$$\therefore \frac{1 - 9x - 8x^2}{(x - 2)(2x + 3)^2} \equiv -\frac{1}{x - 2} - \frac{2}{2x + 3} + \frac{1}{(2x + 3)^2}$$

Case 3: Proper fraction with a quadratic factor in the denominator that cannot be factorised

In this syllabus you will only be asked questions where the quadratic factor is of the form $cx^2 + d$.

If the denominator of a proper algebraic fraction has a quadratic factor of the form $cx^2 + d$ that cannot be factorised, the fraction can be split into partial fractions using the following rule.

> **KEY POINT 7.5**
>
> $$\frac{px + q}{(ax + b)(cx^2 + d)} \equiv \frac{A}{ax + b} + \frac{Bx + C}{cx^2 + d}$$

WORKED EXAMPLE 7.4

Express $\dfrac{8 + 2x - x^2}{(x - 1)(x^2 + 2)}$ in partial fractions.

Answer

$\dfrac{8 + 2x - x^2}{(x - 1)(x^2 + 2)} \equiv \dfrac{A}{x - 1} + \dfrac{Bx + C}{x^2 + 2}$ …… Multiply throughout by $(x - 1)(x^2 + 2)$.

$8 + 2x - x^2 \equiv A(x^2 + 2) + Bx(x - 1) + C(x - 1)$ -----(1)

Let $x = 1$ in equation (1):

$8 + 2(1) - (1)^2 = A(1^2 + 2) + B(1)(1 - 1) + C(1 - 1)$

$ 9 = 3A$

$ A = 3$

Let $x = 0$ in equation (1):

$ 8 = 6 - C$

$ C = -2$

Let $x = -1$ in equation (1):

$ 5 = 9 + 2B + 4$

$ B = -4$

$\therefore \dfrac{8 + 2x - x^2}{(x - 1)(x^2 + 2)} \equiv \dfrac{3}{x - 1} + \dfrac{-4x - 2}{x^2 + 2} \equiv \dfrac{3}{x - 1} - \dfrac{4x + 2}{x^2 + 2}$

Case 4: Improper fractions

If an algebraic fraction is improper we must first express it as the sum of a polynomial and a proper fraction and then split the proper fraction into partial fractions.

WORKED EXAMPLE 7.5

Express $\dfrac{2x^2 + 5x - 11}{x^2 + 2x - 3}$ in partial fractions.

Answer

The fraction is improper. Hence it must first be written as the sum of a polynomial and a proper fraction.

$$\begin{array}{r}2\\x^2+2x-3\overline{\smash{\big)}\,2x^2+5x-11}\\\underline{2x^2+4x-6}\\x-5\end{array}$$

$\dfrac{2x^2 + 5x - 11}{x^2 + 2x - 3} \equiv 2 + \dfrac{x - 5}{x^2 + 2x - 3}$

Split the proper fraction $\dfrac{x-5}{x^2 + 2x - 3}$ into partial fractions:

$\dfrac{x-5}{(x+3)(x-1)} \equiv \dfrac{A}{x+3} + \dfrac{B}{x-1}$ .... Multiply throughout by $(x+3)(x-1)$.

$x - 5 \equiv A(x-1) + B(x+3)$ -----(1)

Let $x = 1$ in equation (1):
$\qquad -4 = 4B$
$\qquad B = -1$

Let $x = -3$ in equation (1):
$\qquad -8 = -4A$
$\qquad A = 2$

$\dfrac{x-5}{(x+3)(x-1)} \equiv \dfrac{2}{x+3} - \dfrac{1}{x-1}$

$\therefore \dfrac{2x^2 + 5x - 11}{x^2 + 2x - 3} \equiv 2 + \dfrac{2}{x+3} - \dfrac{1}{x-1}$

Application of partial fractions

Partial fractions can be used to help evaluate the sum of some sequences.

WORKED EXAMPLE 7.6

Given that the algebraic fraction $\dfrac{1}{x(x+1)}$ can be written in partial fractions as $\dfrac{1}{x} - \dfrac{1}{x+1}$, find the sum of the first n terms of the series $\dfrac{1}{1 \times 2} + \dfrac{1}{2 \times 3} + \dfrac{1}{3 \times 4} + \dfrac{1}{4 \times 5} + \ldots$. Hence, state the sum to infinity.

Answer

$S_n = \dfrac{1}{1 \times 2} + \dfrac{1}{2 \times 3} + \dfrac{1}{3 \times 4} + \dfrac{1}{4 \times 5} + \ldots + \dfrac{1}{(n-1)n} + \dfrac{1}{n(n+1)}$

$= \left(\dfrac{1}{1} - \dfrac{1}{\cancel{2}}\right) + \left(\dfrac{\cancel{1}}{\cancel{2}} - \dfrac{1}{\cancel{3}}\right) + \left(\dfrac{\cancel{1}}{\cancel{3}} - \dfrac{1}{\cancel{4}}\right) + \left(\dfrac{\cancel{1}}{\cancel{4}} - \dfrac{1}{\cancel{5}}\right) + \ldots + \left(\dfrac{\cancel{1}}{\cancel{n-1}} - \dfrac{1}{\cancel{n}}\right) + \left(\dfrac{\cancel{1}}{\cancel{n}} - \dfrac{1}{n+1}\right)$

$= 1 - \dfrac{1}{n+1}$

As $n \to \infty$, $\dfrac{1}{n+1} \to 0$

$\therefore S_\infty = 1$

DID YOU KNOW?

The series in Worked example 7.6 is called a telescoping series, because its partial sums only have a fixed number of terms after cancellation.

EXERCISE 7B

1. Express the following proper fractions as partial fractions.

   a $\dfrac{6x-2}{(x-2)(x+3)}$

   b $\dfrac{7x+12}{2x(x-4)}$

   c $\dfrac{15x+13}{(x-1)(3x+1)}$

   d $\dfrac{x-1}{(3x-5)(x-3)}$

   e $\dfrac{6x^2+5x-2}{x(x-1)(2x+1)}$

   f $\dfrac{11x+12}{(2x+3)(x+2)(x-3)}$

2. Express the following proper fractions as partial fractions.

   a $\dfrac{2x}{(x+2)^2}$

   b $\dfrac{11x^2+14x+5}{(2x+1)(x+1)^2}$

   c $\dfrac{x^2-2}{x(x-1)^2}$

   d $\dfrac{36x^2+2x-4}{(2x-3)(2x+1)^2}$

   e $\dfrac{3}{(x+2)(x-2)^2}$

   f $\dfrac{3x+4}{(x+2)(x-1)^2}$

3. Express the following proper fractions as partial fractions.

   a $\dfrac{2x^2-3x+2}{x(x^2+1)}$

   b $\dfrac{3x^2+4x+17}{(2x+1)(x^2+5)}$

   c $\dfrac{2x^2-6x-9}{(3x+5)(2x^2+1)}$

   d $-\dfrac{6x^2-21x+50}{(3x-5)(2x^2+5)}$

4. Express the following improper fractions as partial fractions.

   a $\dfrac{2x^2+3x+4}{(x-1)(x+2)}$

   b $\dfrac{x^2+3}{x^2-4}$

   c $\dfrac{22-17x+21x^2-4x^3}{(x-4)(x^2+1)}$

   d $\dfrac{4x^3+x^2-16x+7}{2x^3-4x^2+2x}$

5. Find the values of A, B, C and D such that $\dfrac{4x^3-9x^2+11x-4}{x^2(2x-1)} \equiv A + \dfrac{B}{x} + \dfrac{C}{x^2} + \dfrac{D}{2x-1}$.

6. a Factorise $2x^3 - 3x^2 - 3x + 2$ completely.

   b Hence express $\dfrac{x^2-13x-5}{2x^3-3x^2-3x+2}$ in partial fractions.

7. a Factorise $2x^3 - 11x^2 + 12x + 9$ completely.

   b Hence express $\dfrac{24-x}{2x^3-11x^2+12x+9}$ in partial fractions.

8. The polynomial $p(x) = 2x^3 + 5x^2 + ax + b$ is exactly divisible by $2x+1$ and leaves a remainder of 9 when divided by $x+2$.

   a Find the value of a and the value of b.

   b Factorise $p(x)$ completely.

   c Express $\dfrac{120}{p(x)}$ in partial fractions.

9 **a** Express $\dfrac{2}{x(x+2)}$ in partial fractions.

b Find an expression for the sum of the first n terms of the series:

$$\dfrac{2}{1\times 3}+\dfrac{2}{2\times 4}+\dfrac{2}{3\times 5}+\dfrac{2}{4\times 6}+\dfrac{2}{5\times 7}+\ldots$$

c Find the sum to infinity of this series.

10 Investigate whether the series $\dfrac{1}{1\times 2\times 3}+\dfrac{1}{2\times 3\times 4}+\dfrac{1}{3\times 4\times 5}+\dfrac{1}{4\times 5\times 6}+\ldots$ is a telescoping series. If it is a telescoping series, find an expression for the sum of the first n terms and the sum to infinity.

EXPLORE 7.2

There is another method that can be used to decompose a proper algebraic fraction where the denominator contains only distinct linear factors. This alternative method is called **Heaviside's cover-up** method. Find out more about this method and use it to express $\dfrac{4x^2+15x+29}{(x-2)(x+1)(x+3)}$ in partial fractions.

7.3 Binomial expansion of $(1+x)^n$ for values of n that are not positive integers

REWIND

In the Pure Mathematics 1 Coursebook, Chapter 6, we learnt that the binomial expansion of $(1+x)^n$, where n is a positive integer is:

$$(1+x)^n = \binom{n}{0} + \binom{n}{1}x + \binom{n}{2}x^2 + \binom{n}{3}x^3 + \ldots + \binom{n}{n}x^n$$

and that this terminating series can also be written in the form:

$$(1+x)^n = 1 + nx + \dfrac{n(n-1)}{2!}x^2 + \dfrac{n(n-1)(n-2)}{3!}x^3 + \ldots + x^n.$$

The **general binomial theorem** can be used for any rational value of n. It states that:

KEY POINT 7.6

$$(1+x)^n = 1 + nx + \dfrac{n(n-1)}{2!}x^2 + \dfrac{n(n-1)(n-2)}{3!}x^3 + \ldots \text{ where } n \text{ is rational and } |x|<1$$

There are two very important differences when n is not a positive integer. These are:

- the series is infinite
- the expansion is only valid for $|x|<1$.

The proof of the general binomial theorem is beyond the scope of this course.

Consider, however, the following two examples where n is not a positive integer $\left(n=-3 \text{ and } n=\dfrac{1}{2}\right)$.

Chapter 7: Further algebra

$$(1+x)^{-3} = 1 + (-3)x + \frac{(-3)(-4)}{2!}x^2 + \frac{(-3)(-4)(-5)}{3!}x^3 + \frac{(-3)(-4)(-5)(-6)}{4!}x^4 + \ldots$$

$$= 1 - 3x + 6x^2 - 10x^3 + 15x^4 + \ldots$$

$$(1+x)^{\frac{1}{2}} = 1 + \left(\frac{1}{2}\right)x + \frac{\left(\frac{1}{2}\right)\left(-\frac{1}{2}\right)}{2!}x^2 + \frac{\left(\frac{1}{2}\right)\left(-\frac{1}{2}\right)\left(-\frac{3}{2}\right)}{3!}x^3 + \frac{\left(\frac{1}{2}\right)\left(-\frac{1}{2}\right)\left(-\frac{3}{2}\right)\left(-\frac{5}{2}\right)}{4!}x^4 + \ldots$$

$$= 1 + \frac{1}{2}x - \frac{1}{8}x^2 + \frac{1}{16}x^3 - \frac{5}{128}x^4 + \ldots$$

We can see that in each case the sequence of coefficients forms an infinite sequence.

When $|x| < 1$, the binomial series converges and has a sum to infinity (in a similar way to a geometric series converging for $|r| < 1$, where r is the common ratio).

WORKED EXAMPLE 7.7

Expand $(1 - 3x)^{-2}$ in ascending powers of x up to and including the term in x^3, and state the range of values of x for which the expansion is valid.

Answer

Use $n = -2$ and replace x by $-3x$ in the binomial formula:

$$(1 - 3x)^{-2} = 1 + (-2)(-3x) + \frac{(-2)(-3)}{2!}(-3x)^2 + \frac{(-2)(-3)(-4)}{3!}(-3x)^3 + \ldots$$

$$= 1 + 6x + 27x^2 + 108x^3 + \ldots$$

Expansion is valid for $|-3x| < 1$

$$\therefore |x| < \frac{1}{3}.$$

DID YOU KNOW?

Isaac Newton is credited with the generalised binomial theorem for any rational power. A poet once wrote that 'Newton's Binomial is as beautiful as the Venus de Milo. The truth is that few people notice it.'

WORKED EXAMPLE 7.8

Find the first four terms in the expansion of $(1 + 2x^2)^{-4}$ and state the range of values of x for which the expansion is valid.

Answer

Use $n = -4$ and replace x by $2x^2$ in the binomial formula:

$$(1 + 2x^2)^{-4} = 1 + (-4)(2x^2) + \frac{(-4)(-5)}{2!}(2x^2)^2 + \frac{(-4)(-5)(-6)}{3!}(2x^2)^3 + \ldots$$

$$= 1 - 8x^2 + 40x^4 - 160x^6 + \ldots$$

Expansion is valid for $|2x^2| < 1$

$$\therefore |x| < \frac{1}{\sqrt{2}}$$

Cambridge International AS & A Level Mathematics: Pure Mathematics 2 & 3

> **WORKED EXAMPLE 7.9**
>
> Find the coefficient of x^3 in the expansion of $\dfrac{x-5}{1+2x}$.
>
> **Answer**
>
> $\dfrac{x-5}{1+2x} = (x-5)(1+2x)^{-1}$
>
> Use $n = -1$ and replace x by $2x$ in the binomial formula:
>
> $(1+2x)^{-1} = 1 + (-1)(2x) + \dfrac{(-1)(-2)}{2!}(2x)^2 + \dfrac{(-1)(-2)(-3)}{3!}(2x)^3 + \ldots$
>
> $\qquad\qquad\quad = 1 - 2x + 4x^2 - 8x^3 + \ldots$
>
> $(x-5)(1+2x)^{-1} = (x-5)(1 - 2x + 4x^2 - 8x^3 + \ldots)$
>
> $\qquad\qquad\qquad\quad = x(1 - 2x + 4x^2 - 8x^3) - 5(1 - 2x + 4x^2 - 8x^3 + \ldots)$
>
> Term in x^3 is $x(4x^2) + (-5)(-8x^3) = 44x^3$.
>
> $\therefore$ The coefficient of x^3 is 44.

EXERCISE 7C

1. Expand the following, in ascending powers of x, up to and including the term in x^3. State the range of values of x for which each expansion is valid.

   a $(1+x)^{-2}$
   b $(1+3x)^{-1}$
   c $(1-2x)^{-4}$
   d $\left(1+\dfrac{x}{2}\right)^{-3}$
   e $\sqrt{1+2x}$
   f $\sqrt[3]{1-3x}$
   g $\dfrac{2}{(1-4x)^2}$
   h $\dfrac{1-3x}{(1+2x)^3}$
   i $\dfrac{1+x}{\sqrt{1-2x}}$

2. Find the first 3 terms, in ascending powers of x, in the expansion of each of the following expressions. State the range of values of x for which each expansion is valid.

   a $(1+x^2)^{-3}$
   b $\sqrt[3]{1-2x^2}$
   c $\left(\sqrt{1-4x^2}\right)^3$

3. Find the first 5 terms, in ascending powers of x, in the expansion of $\dfrac{2+3x}{\sqrt{1-5x^2}}$.

4. Can the binomial expansion be used to expand each of these expressions?

   a $(3x-1)^{-2}$
   b $\sqrt{2x-1}$

   Give reasons for your answers.

5. Expand $(2x-1)^{-3}$ in ascending powers of x, up to and including the term in x^3.

6. Show that the expansion of $(1+x)^{\frac{1}{2}}$ is not valid when $x = 3$.

PS 7. Given that $(1-3x)^{-4} - (1+2x)^{\frac{3}{2}} \approx 9x + kx^2$ for small values of x, find the value of the constant k.

PS 8. Given that $\dfrac{a}{1-x} + \dfrac{b}{1+2x} \approx -3 + 12x$ for small values of x, find the value of a and the value of b.

9 When $(1+ax)^{-3}$, where a is a positive constant, is expanded the coefficients of x and x^2 are equal.

   a Find the value of a.

   b When a has this value, obtain the first 5 terms in the expansion.

10 When $(3-2x)(1+ax)^{\frac{2}{3}}$ is expanded the coefficient of x^2 is -15. Find the two possible values of a.

11 The first 3 terms in the expansion of $(1+ax)^n$ are $1 - 24x + 384x^2$.

   a Find the value of a and the value of n.

   b Hence find the term in x^3.

7.4 Binomial expansion of $(a + x)^n$ for values of n that are not positive integers

To expand $(a + x)^n$, where n is not a positive integer, we take a outside of the brackets:

$$(a+x)^n = a^n \left(1 + \frac{x}{a}\right)^n$$

Hence the expression is:

> **KEY POINT 7.7**
>
> $$(a+x)^n = a^n \left[1 + n\left(\frac{x}{a}\right) + \frac{n(n-1)}{2!}\left(\frac{x}{a}\right)^2 + \frac{n(n-1)(n-2)}{3!}\left(\frac{x}{a}\right)^3 + \cdots\right] \text{ for } \left|\frac{x}{a}\right| < 1$$

WORKED EXAMPLE 7.10

Expand $(2 + x)^{-3}$ in ascending powers of x, up to and including the term in x^3 and state the range of values of x for which the expansion is valid.

Answer

$$(2+x)^{-3} = (2)^{-3}\left(1 + \frac{x}{2}\right)^{-3}$$

$$= (2)^{-3}\left[1 + (-3)\left(\frac{x}{2}\right) + \frac{(-3)(-4)}{2!}\left(\frac{x}{2}\right)^2 + \frac{(-3)(-4)(-5)}{3!}\left(\frac{x}{2}\right)^3 + \ldots\right] \text{ for } \left|\frac{x}{2}\right| < 1$$

$$= \frac{1}{8}\left[1 - \frac{3}{2}x + \frac{3}{2}x^2 - \frac{5}{4}x^3 + \ldots\right]$$

$$= \frac{1}{8} - \frac{3}{16}x + \frac{3}{16}x^2 - \frac{5}{32}x^3 + \ldots \text{ valid for } |x| < 2$$

Cambridge International AS & A Level Mathematics: Pure Mathematics 2 & 3

WORKED EXAMPLE 7.11

Find the coefficient of x^2 in the expansion of $\dfrac{1-3x}{\sqrt{4-x}}$.

Answer

$\dfrac{1-3x}{\sqrt{4-x}} = (1-3x)(4-x)^{-\frac{1}{2}}$ .... Factor the 4 out of $(4-x)^{-\frac{1}{2}}$.

$= (1-3x)(4)^{-\frac{1}{2}}\left(1-\dfrac{x}{4}\right)^{-\frac{1}{2}}$ .... Simplify $(4)^{-\frac{1}{2}}$ to $\dfrac{1}{2}$.

$= \dfrac{1}{2}(1-3x)\left[1+\left(-\dfrac{1}{2}\right)\left(-\dfrac{x}{4}\right)+\dfrac{\left(-\frac{1}{2}\right)\left(-\frac{3}{2}\right)}{2!}\left(-\dfrac{x}{4}\right)^2+\ldots\right]$

$= \dfrac{1}{2}(1-3x)\left[1+\dfrac{1}{8}x+\dfrac{3}{128}x^2+\ldots\right]$

$= \dfrac{1}{2}\left[1+\dfrac{1}{8}x+\dfrac{3}{128}x^2+\ldots\right]-\dfrac{3}{2}x\left[1+\dfrac{1}{8}x+\dfrac{3}{128}x^2+\ldots\right]$

Coefficient of x^2 is $\dfrac{1}{2}\times\dfrac{3}{128}-\dfrac{3}{2}\times\dfrac{1}{8}=-\dfrac{45}{256}$

EXERCISE 7D

1. Expand the following, in ascending powers of x, up to and including the term in x^3. State the range of values of x for which each expansion is valid.

   a $(2+x)^{-2}$
   b $(5-2x)^{-1}$
   c $\sqrt{9-x}$
   d $\sqrt[3]{3x+8}$
   e $\dfrac{4}{(3-x)^3}$
   f $\dfrac{1+2x}{(2x-5)^3}$

2. Find the first 3 terms, in ascending powers of x, in the expansion of each of the following expressions. State the range of values of x for which each expansion is valid.

   a $(2+x^2)^{-2}$
   b $\sqrt[3]{8-3x^2}$
   c $\left(\sqrt{3-x^2}\right)^5$

3. Expand $(1-2x)\sqrt[3]{8-2x}$ in ascending powers of x, up to and including the term in x^3.

4. a Expand $(2-x)^{-1}$ and $(1+3x)^{-2}$ giving the first 3 terms in each expansion.
   b Use your answers to **part a** to find the first 3 terms in the expansion of $(2-x)^{-1}(1+3x)^{-2}$, stating the range of values of x for which this expansion is valid.

PS 5. The first 3 terms in the expansion of $(a-5x)^{-2}$ are $\dfrac{1}{4}+\dfrac{5}{4}x+bx^2$.

Find the value of a, the value of b and the term in x^3.

PS 6. In the expansion of $\sqrt{3+ax}$ where $a \neq 0$, the coefficient of the term in x^2 is 3 times the coefficient of the term in x^3. Find the value of a.

7 **a** Find the first 4 terms in the expansion of $\left(1+\dfrac{2}{x}\right)^{-1}$, where $\left|\dfrac{2}{x}\right| < 1$.

  b Show that $\left(1+\dfrac{2}{x}\right)^{-1} \equiv \dfrac{x}{x+2} \equiv \dfrac{x}{2}\left(1+\dfrac{x}{2}\right)^{-1}$.

  c Find the first 4 terms in the expansion of $\dfrac{x}{2}\left(1+\dfrac{x}{2}\right)^{-1}$, where $\left|\dfrac{x}{2}\right| < 1$.

  d Explain why your expansions are different.

EXPLORE 7.3

1 Given that $f(x) = \dfrac{1}{(1+x)(1+2x)}$, discuss why $f(x)$ can be written in each of the following ways:

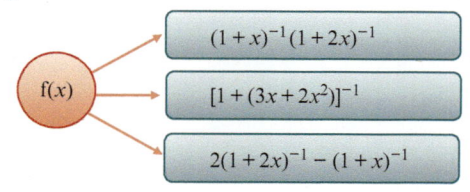

f(x) →
- $(1+x)^{-1}(1+2x)^{-1}$
- $[1+(3x+2x^2)]^{-1}$
- $2(1+2x)^{-1} - (1+x)^{-1}$

2 Use the binomial theorem to expand each of these three expressions in ascending powers of x, up to the term in x^3.

3 Discuss with your classmates which method you would prefer if you were asked for:

  a a linear approximation for $f(x)$
  b a quadratic approximation for $f(x)$
  c a cubic approximation for $f(x)$.

7.5 Partial fractions and binomial expansions

One of the most common reasons for splitting a fraction into partial fractions is so that binomial expansions can be applied.

WORKED EXAMPLE 7.12

Given that $f(x) = \dfrac{10x+1}{(1-2x)(1+x)}$, express $f(x)$ in partial fractions and hence obtain the expansion of $f(x)$ in ascending powers of x, up to and including the term in x^3. State the range of values of x for which the expansion is valid.

Answer

$\dfrac{10x+1}{(1-2x)(1+x)} \equiv \dfrac{A}{1-2x} + \dfrac{B}{1+x}$ .... Multiply throughout by $(1-2x)(1+x)$.

$10x+1 \equiv A(1+x) + B(1-2x)$ ----(1)

Let $x = -1$ in equation (1):

$$-9 = 3B$$
$$B = -3$$

Let $x = \dfrac{1}{2}$ in equation (1):

$$6 = \dfrac{3}{2}A$$
$$A = 4$$

$$\therefore \dfrac{10x+1}{(1-2x)(1+x)} \equiv \dfrac{4}{1-2x} - \dfrac{3}{1+x} \equiv 4(1-2x)^{-1} - 3(1+x)^{-1}$$

Expanding $(1-2x)^{-1}$ gives:

$$(1-2x)^{-1} = 1 + (-1)(-2x) + \dfrac{(-1)(-2)}{2!}(-2x)^2 + \dfrac{(-1)(-2)(-3)}{3!}(-2x)^3 + \ldots$$

$$= 1 + 2x + 4x^2 + 8x^3 + \ldots \quad \text{valid for } |-2x| < 1 \quad \text{i.e.} \quad -\dfrac{1}{2} < x < \dfrac{1}{2}$$

Expanding $(1+x)^{-1}$ gives:

$$(1+x)^{-1} = 1 + (-1)x + \dfrac{(-1)(-2)}{2!}x^2 + \dfrac{(-1)(-2)(-3)}{3!}x^3$$

$$= 1 - x + x^2 - x^3 + \ldots \quad \text{valid for } |x| < 1 \quad \text{i.e.} \quad -1 < x < 1$$

$$\therefore \dfrac{10x+1}{(1-2x)(1+x)} = 4(1 + 2x + 4x^2 + 8x^3 + \ldots) - 3(1 - x + x^2 - x^3 + \ldots)$$

$$= 1 + 11x + 13x^2 + 35x^3 + \ldots$$

The expansion is valid for the range of values of x satisfying both $-\dfrac{1}{2} < x < \dfrac{1}{2}$ and $-1 < x < 1$

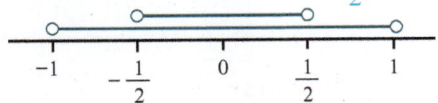

Hence, the expansion is valid for $-\dfrac{1}{2} < x < \dfrac{1}{2}$.

EXERCISE 7E

1 a Express $\dfrac{7x-1}{(1-x)(1+2x)}$ in partial fractions.

  b Hence obtain the expansion of $\dfrac{7x-1}{(1-x)(1+2x)}$ in ascending powers of x, up to and including the term in x^3.

2 a Express $\dfrac{5x^2 + x}{(1-x)^2(1-3x)}$ in partial fractions.

  b Hence obtain the expansion of $\dfrac{5x^2 + x}{(1-x)^2(1-3x)}$ in ascending powers of x, up to and including the term in x^3.

3 **a** Express $\dfrac{7x^2 + 4x + 4}{(1-x)(2x^2+1)}$ in partial fractions.

 b Hence obtain the expansion of $\dfrac{7x^2 + 4x + 4}{(1-x)(2x^2+1)}$ in ascending powers of x, up to and including the term in x^3.

4 **a** Express $\dfrac{19 - 7x - 6x^2}{(2x+1)(2-3x)}$ in partial fractions.

 b Hence obtain the coefficient of x^2 in the expansion of $\dfrac{19 - 7x - 6x^2}{(2x+1)(2-3x)}$.

5 **a** Express $\dfrac{21}{(x-4)(x+3)}$ in partial fractions.

 b Hence obtain the expansion of $\dfrac{24}{(x-4)(x+3)}$ in ascending powers of x, up to and including the term in x^2.

6 **a** Express $\dfrac{6x^2 - 24x + 15}{(x+1)(x-2)^2}$ in partial fractions.

 b Hence obtain the expansion of $\dfrac{6x^2 - 24x + 15}{(x+1)(x-2)^2}$ in ascending powers of x, up to and including the term in x^2.

Checklist of learning and understanding

Partial fractions

- **Case 1:** If the denominator of a proper algebraic fraction has two distinct linear factors, the fraction can be split into partial fractions using the rule $\dfrac{px+q}{(ax+b)(cx+d)} \equiv \dfrac{A}{ax+b} + \dfrac{B}{cx+d}$

- **Case 2:** If the denominator of a proper algebraic fraction has a repeated linear factor, the fraction can be split into partial fractions using the rule $\dfrac{px+q}{(ax+b)^2} \equiv \dfrac{A}{ax+b} + \dfrac{B}{(ax+b)^2}$

- **Case 3:** If the denominator of a proper algebraic fraction has a quadratic factor of the form $cx^2 + d$ that cannot be factorised, the fraction can be split into partial fractions using the rule $\dfrac{px+q}{(ax+b)(cx^2+d)} \equiv \dfrac{A}{ax+b} + \dfrac{Bx+C}{cx^2+d}$

- **Case 4:** If an algebraic fraction is improper we must first express it as the sum of a polynomial and a proper fraction and then split the proper fraction into partial fractions.

Binomial expansion of $(1 + x)^n$ for values of n that are not positive integers

- $(1+x)^n = 1 + nx + \dfrac{n(n-1)}{2!} x^2 + \dfrac{n(n-1)(n-2)}{3!} x^3 + \ldots$ for $|x| < 1$.

Binomial expansion of $(a + x)^n$ for values of n that are not positive integers

- $(a+x)^n = a^n \left[1 + n\left(\dfrac{x}{a}\right) + \dfrac{n(n-1)}{2!}\left(\dfrac{x}{a}\right)^2 + \dfrac{n(n-1)(n-2)}{3!}\left(\dfrac{x}{a}\right)^3 + \ldots \right]$ for $\left|\dfrac{x}{a}\right| < 1$.

END-OF-CHAPTER REVIEW EXERCISE 7

1. Expand $(1-2x)^{-4}$ in ascending powers of x, up to and including the term in x^3, simplifying the coefficients. [4]

2. Expand $\sqrt[3]{1-(6x)}$ in ascending powers of x up to and including the term in x^3, simplifying the coefficients. [4]

   Cambridge International A Level Mathematics 9709 Paper 31 Q1 June 2011

3. Expand $(2-x)(1+2x)^{-\frac{3}{2}}$ in ascending powers of x, up to and including the term in x^2, simplifying the coefficients. [4]

   Cambridge International A Level Mathematics 9709 Paper 31 Q2 November 2016

4. Expand $\dfrac{1+3x}{\sqrt{(1+2x)}}$ in ascending powers of x up to and including the term in x^2, simplifying the coefficients. [4]

   Cambridge International A Level Mathematics 9709 Paper 31 Q2 June 2013

5. Expand $\dfrac{1}{\sqrt{1+3x}}$ in ascending powers of x, up to and including the term in x^3, simplifying the coefficients. [4]

6. Expand $\dfrac{10}{(2-x)^2}$ in ascending powers of x, up to and including the term in x^2, simplifying the coefficients. [4]

7. Expand $\dfrac{1}{\sqrt{4-5x}}$ in ascending powers of x, up to and including the term in x^2, simplifying the coefficients. [4]

8. Express $\dfrac{12}{x^2(2x-3)}$ in the form $\dfrac{A}{x} + \dfrac{B}{x^2} + \dfrac{C}{2x-3}$. [4]

9. Express $\dfrac{8x^2+4x+21}{(x+2)(x^2+5)}$ in the form $\dfrac{A}{x+2} + \dfrac{Bx+C}{x^2+5}$. [4]

10. Express $\dfrac{7x^2-3x+2}{x(x^2+1)}$ in partial fractions. [5]

    Cambridge International A Level Mathematics 9709 Paper 31 Q3 June 2013

11. Given that $\dfrac{9x^3-11x^2+8x-4}{x^2(3x-2)} \equiv A + \dfrac{B}{x} + \dfrac{C}{x^2} + \dfrac{D}{3x-2}$, find the values of A, B, C and D. [5]

12. Express $\dfrac{4x^2-5x+3}{(x+2)(2x-1)}$ in partial fractions. [5]

13. Show that for small values of x^2, $(1-2x^2)^{-2} - (1+6x^2)^{\frac{2}{3}} \approx kx^4$, where the value of the constant k is to be determined. [6]

    Cambridge International A Level Mathematics 9709 Paper 31 Q3 June 2015

14. Let $f(x) = \dfrac{4x^2+12}{(x+1)(x-3)^2}$.

    i Express $f(x)$ in partial fractions. [5]

    ii Hence obtain the expansion of $f(x)$ in ascending powers of x, up to and including the term in x^2. [5]

    Cambridge International A Level Mathematics 9709 Paper 31 Q8 June 2016

15 Let $f(x) = \dfrac{2x^2 - 7x - 1}{(x-2)(x^2+3)}$.

  i Express $f(x)$ in partial fractions. [5]

  ii Hence obtain the expansion of $f(x)$ in ascending powers of x, up to and including the term in x^2. [5]

  Cambridge International A Level Mathematics 9709 Paper 31 Q7 November 2013

16 Let $f(x) = \dfrac{3x}{(1+x)(1+2x^2)}$.

  i Express $f(x)$ in partial fractions. [5]

  ii Hence obtain the expansion of $f(x)$ in ascending powers of x, up to and including the term in x^3. [5]

  Cambridge International A Level Mathematics 9709 Paper 31 Q8 November 2010

Chapter 8
Further calculus

P3 This chapter is for Pure Mathematics 3 students only.

In this chapter you will learn how to:

- use the derivative of $\tan^{-1} x$
- extend the ideas of 'reverse differentiation' to include the integration of $\dfrac{1}{x^2 + a^2}$
- recognise an integrand of the form $\dfrac{k f'(x)}{f(x)}$, and integrate such functions
- use a given substitution to simplify and evaluate either a definite or an indefinite integral
- integrate rational functions by means of decomposition into partial fractions
- recognise when an integrand can usefully be regarded as a product, and use integration by parts.

Chapter 8: Further calculus

PREREQUISITE KNOWLEDGE

Where it comes from	What you should be able to do	Check your skills
Chapter 4	Differentiate e^x, $\ln x$, $\sin x$, $\cos x$ and $\tan x$.	1 Differentiate with respect to x. a $\sin 3x$ b e^{x^2+1} c $\ln(5x-3)$ d $\tan 2x - 5\cos x$
Chapter 5	Integrate e^{ax+b}, $\dfrac{1}{ax+b}$, $\sin(ax+b)$, $\cos(ax+b)$ and $\sec^2(ax+b)$ and use trigonometrical relationships in carrying out integration.	2 Find each of these integrals. a $\displaystyle\int e^{5x+1}\,dx$ b $\displaystyle\int \sin^2 2x\,dx$ c $\displaystyle\int \dfrac{5}{3x-2}\,dx$ d $\displaystyle\int \sec^2(3x)\,dx$
Chapter 7	Express rational functions as partial fractions.	3 Express in partial fractions. a $\dfrac{2x}{(x-1)(x+3)}$ b $\dfrac{4x^2-6}{x(x^2+2)}$ c $\dfrac{5}{x(x-1)^2}$ d $\dfrac{3x^2-3x-3}{x^2-x-2}$

Why do we study calculus?

This chapter will further extend your knowledge and skills in differentiation and integration.

In particular, we will learn about integration by substitution and integration by parts, which can be considered as the 'reverse process' of the chain rule and the product rule for differentiation that we learnt about earlier in this course. These new rules will enable you to integrate many more functions than you could previously.

The skills that we learn in this chapter will be very important when studying Chapter 10. Chapter 10 is about solving differential equations, many of which model real-life situations.

Integration by parts plays a major role in engineering and its application is often found in problems including electric circuits, heat transfer, vibrations, structures, fluid mechanics, air pollution, electromagnetics and digital signal processing.

WEB LINK

Explore the *Chain rule and integration by substitution* and the *Product rule and integration by parts* stations on the Underground Mathematics website.

8.1 Derivative of $\tan^{-1} x$

The function $y = \tan^{-1} x$ can be written as:

$\tan y = x$ Differentiate both sides with respect to x.

$\sec^2 y \dfrac{dy}{dx} = 1$

$\dfrac{dy}{dx} = \dfrac{1}{\sec^2 y}$ Use $\sec^2 y = 1 + \tan^2 y$.

$\dfrac{dy}{dx} = \dfrac{1}{1 + \tan^2 y}$ Use $\tan y = x$.

$\dfrac{dy}{dx} = \dfrac{1}{1 + x^2}$

> **KEY POINT 8.1**
>
> $\dfrac{d}{dx}(\tan^{-1} x) = \dfrac{1}{x^2 + 1}$

WORKED EXAMPLE 8.1

Differentiate with respect to x.

a $\tan^{-1} 3x$

b $\tan^{-1} \sqrt{x}$

c $\tan^{-1}\left(\dfrac{x}{x-2}\right)$

Answer

a $\dfrac{d}{dx}(\tan^{-1} 3x) = \dfrac{1}{(3x)^2 + 1} \times 3$ Chain rule.

   $= \dfrac{3}{9x^2 + 1}$

b $\dfrac{d}{dx}\left(\tan^{-1} \sqrt{x}\right) = \dfrac{1}{\left(\sqrt{x}\right)^2 + 1} \times \dfrac{1}{2} x^{-\frac{1}{2}}$ Chain rule.

   $= \dfrac{1}{x + 1} \times \dfrac{1}{2\sqrt{x}}$

   $= \dfrac{1}{2\sqrt{x}(x + 1)}$

c $\dfrac{d}{dx}\left[\tan^{-1}\left(\dfrac{x}{x-2}\right)\right] = \dfrac{1}{\left(\dfrac{x}{x-2}\right)^2 + 1} \times \dfrac{d}{dx}\left(\dfrac{x}{x-2}\right)$ .... Chain rule and quotient rule.

$\phantom{\dfrac{d}{dx}\left[\tan^{-1}\left(\dfrac{x}{x-2}\right)\right]} = \dfrac{(x-2)^2}{x^2+(x-2)^2} \times \dfrac{(x-2)(1)-(x)(1)}{(x-2)^2}$ .... Simplify.

$\phantom{\dfrac{d}{dx}\left[\tan^{-1}\left(\dfrac{x}{x-2}\right)\right]} = \dfrac{-2}{2x^2-4x+4}$

$\phantom{\dfrac{d}{dx}\left[\tan^{-1}\left(\dfrac{x}{x-2}\right)\right]} = \dfrac{-1}{x^2-2x+2}$

EXERCISE 8A

1 Differentiate with respect to x.

   a $\tan^{-1} 2x$ **b** $\tan^{-1} 5x$ **c** $\tan^{-1}\dfrac{x}{3}$

   d $\tan^{-1}(x-1)$ **e** $\tan^{-1} x^2$ **f** $\tan^{-1}\left(\dfrac{2x}{x+1}\right)$

2 Differentiate with respect to x.

   a $x\tan^{-1} x$ **b** $\dfrac{\tan^{-1} 2x}{x}$ **c** $e^x \tan^{-1} x$

3 Find the equation of the tangent to the curve $y = \tan^{-1}\left(\dfrac{x}{2}\right)$ at the point where $x = 2$.

PS 4 Show that the tangent to the curve $y = \tan^{-1} x$ at the point where $x = -1$ is perpendicular to the normal to the curve at the point where $x = 1$. Find the x-coordinate of the point where this tangent and normal intersect.

8.2 Integration of $\dfrac{1}{x^2+a^2}$

Since integration is the reverse process of differentiation we can say that:

> **KEY POINT 8.2**
>
> $\displaystyle\int \dfrac{1}{x^2+1}\,dx = \tan^{-1} x + c$
>
> This can be extended to:
>
> $\displaystyle\int \dfrac{1}{x^2+a^2}\,dx = \dfrac{1}{a}\tan^{-1}\dfrac{x}{a} + c$

> **REWIND**
>
> In Section 8.1 we learnt that
> $\dfrac{d}{dx}(\tan^{-1} x) = \dfrac{1}{x^2+1}$.

WORKED EXAMPLE 8.2

Find $\displaystyle\int \dfrac{1}{x^2+25}\,dx$.

Answer

$\displaystyle\int \dfrac{1}{x^2+25}\,dx = \dfrac{1}{5}\tan^{-1}\dfrac{x}{5} + c$

WORKED EXAMPLE 8.3

Find $\int \dfrac{1}{2x^2 + 3}\,dx$.

Answer

$$\int \dfrac{1}{2x^2+3}\,dx = \dfrac{1}{2}\int \dfrac{1}{x^2+\frac{3}{2}}\,dx$$

$$= \dfrac{1}{2} \times \sqrt{\dfrac{2}{3}} \tan^{-1}\left(\dfrac{x}{\sqrt{\frac{3}{2}}}\right) + c$$

$$= \dfrac{1}{2}\sqrt{\dfrac{2}{3}} \tan^{-1}\left(\sqrt{\dfrac{2}{3}}\,x\right) + c$$

WORKED EXAMPLE 8.4

Find the exact value of $\int_0^2 \dfrac{1}{x^2+4}\,dx$.

Answer

$$\int_0^2 \dfrac{1}{x^2+4}\,dx = \left[\dfrac{1}{2}\tan^{-1}\dfrac{x}{2}\right]_0^2$$

$$= \left(\dfrac{1}{2}\tan^{-1}1\right) - \left(\dfrac{1}{2}\tan^{-1}0\right)$$

$$= \left(\dfrac{1}{2}\times\dfrac{\pi}{4}\right) - \left(\dfrac{1}{2}\times 0\right)$$

$$= \dfrac{\pi}{8}$$

EXERCISE 8B

1 Find the following integrals.

   a $\displaystyle\int \dfrac{1}{x^2+9}\,dx$ **b** $\displaystyle\int \dfrac{1}{16+x^2}\,dx$ **c** $\displaystyle\int \dfrac{1}{4x^2+1}\,dx$

   d $\displaystyle\int \dfrac{1}{9x^2+16}\,dx$ **e** $\displaystyle\int \dfrac{1}{4x^2+3}\,dx$ **f** $\displaystyle\int \dfrac{1}{2+3x^2}\,dx$

2 Find the exact value of each of these integrals.

   a $\displaystyle\int_0^3 \dfrac{1}{x^2+9}\,dx$ **b** $\displaystyle\int_0^{\frac{1}{2}} \dfrac{2}{4x^2+1}\,dx$ **c** $\displaystyle\int_{-\sqrt{2}}^{\sqrt{2}} \dfrac{1}{3x^2+2}\,dx$

3 The diagram shows part of the curve $y = \dfrac{2}{\sqrt{x^2+1}}$. Find the exact value of the volume of the solid formed when the shaded region is rotated completely about the x-axis.

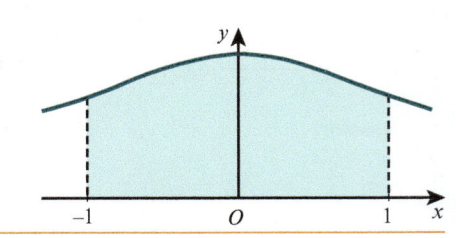

8.3 Integration of $\dfrac{k f'(x)}{f(x)}$

Since integration is the reverse process of differentiation and $\ln x$ only exists for $x > 0$, we can say that:

> **KEY POINT 8.3**
>
> $$\int \frac{f'(x)}{f(x)} \, dx = \ln |f(x)| + c$$
>
> This can be extended to:
>
> $$\int \frac{k f'(x)}{f(x)} \, dx = k \ln |f(x)| + c$$

> **REWIND**
>
> In Chapter 4, we learnt that
> $$\frac{d}{dx}[\ln f(x)] = \frac{f'(x)}{f(x)}.$$

WORKED EXAMPLE 8.5

Find $\displaystyle\int \frac{2x}{x^2 + 3} \, dx$.

Answer

If $f(x) = x^2 + 3$ then $f'(x) = 2x$.

$\therefore \displaystyle\int \frac{2x}{x^2 + 3} \, dx = \ln |x^2 + 3| + c$

WORKED EXAMPLE 8.6

Find $\displaystyle\int \tan 3x \, dx$.

Answer

$\displaystyle\int \tan 3x \, dx = \int \frac{\sin 3x}{\cos 3x} \, dx$

If $f(x) = \cos 3x$ then $f'(x) = -3 \sin 3x$.

$\displaystyle\int \tan 3x \, dx = -\frac{1}{3} \int \frac{-3 \sin 3x}{\cos 3x} \, dx$

$ = -\dfrac{1}{3} \ln |\cos 3x| + c$

Cambridge International AS & A Level Mathematics: Pure Mathematics 2 & 3

WORKED EXAMPLE 8.7

Find $\int \dfrac{3x^2 + 5x}{2x^3 + 5x^2}\, dx$.

Answer

If $f(x) = 2x^3 + 5x^2$ then $f'(x) = 6x^2 + 10x$.

$$\int \dfrac{3x^2 + 5x}{2x^3 + 5x^2}\, dx = \dfrac{1}{2}\int \dfrac{6x^2 + 10x}{2x^3 + 5x^2}\, dx$$

$$= \dfrac{1}{2}\ln\left|2x^3 + 5x^2\right| + c$$

WORKED EXAMPLE 8.8

Find $\displaystyle\int_0^1 \dfrac{2e^x}{1 + e^x}\, dx$.

Answer

If $f(x) = 1 + e^x$ then $f'(x) = e^x$.

$$\int_0^1 \dfrac{2e^x}{1 + e^x}\, dx = 2\int_0^1 \dfrac{e^x}{1 + e^x}\, dx$$

$$= \left[2\ln\left|1 + e^x\right|\right]_0^1 \quad \cdots \text{Substitute limits.}$$

$$= 2\ln(1 + e) - 2\ln 2 \quad \cdots \text{Simplify the logarithms.}$$

$$= 2\ln\left(\dfrac{1 + e}{2}\right)$$

EXERCISE 8C

1 Find the following integrals.

    a $\displaystyle\int \dfrac{6x^2}{x^3 - 1}\, dx$ **b** $\displaystyle\int \dfrac{\cos x}{1 + \sin x}\, dx$ **c** $\displaystyle\int \dfrac{4x - 10}{x^2 - 5x + 1}\, dx$

    d $\displaystyle\int \cot x\, dx$ **e** $\displaystyle\int \dfrac{x}{2 - x^2}\, dx$ **f** $\displaystyle\int \dfrac{\sec^2 x}{1 + \tan x}\, dx$

2 Find the exact value of each of these integrals.

    a $\displaystyle\int_{\frac{1}{6}\pi}^{\frac{1}{4}\pi} \tan x\, dx$ **b** $\displaystyle\int_0^2 \dfrac{3x^2}{x^3 + 2}\, dx$ **c** $\displaystyle\int_0^{\ln 2} \dfrac{e^{2x}}{1 + e^{2x}}\, dx$

    d $\displaystyle\int_1^2 \dfrac{x + 1}{x^2 + 2x - 1}\, dx$ **e** $\displaystyle\int_{\frac{1}{6}\pi}^{\frac{1}{4}\pi} \cot x\, dx$ **f** $\displaystyle\int_0^1 \dfrac{2x + 3}{(x + 1)(x + 2)}\, dx$

3 Show that $\displaystyle\int \dfrac{e^x + e^{-x}}{e^x - e^{-x}}\, dx = \ln(1 - e^{2x}) - x + c$.

4 Show that $\displaystyle\int_{\frac{1}{3}\pi}^{\frac{1}{4}\pi} \dfrac{\sin x \cos x}{1 - 2\cos 2x}\, dx = -\dfrac{1}{8}\ln 2$.

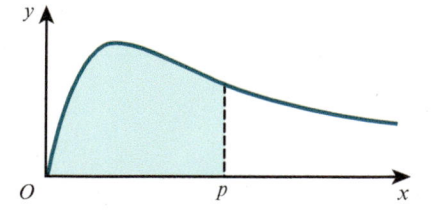

5) The diagram shows part of the curve $y = \dfrac{4x}{x^2 + 1}$. Find the exact value of p for which the shaded region has an area of 4.

8.4 Integration by substitution

Integration by substitution can be considered as the reverse process of differentiation by the chain rule.

This method is used when a simple substitution can be applied that will transform a difficult integral into an easier integral. The integral must be completely rewritten in terms of the new variable.

> **REWIND**
>
> In the Pure Mathematics 1 Coursebook Chapter 7, we learnt the chain rule for differentiation:
> $$\frac{dy}{dx} = \frac{dy}{du} \times \frac{du}{dx}$$

WORKED EXAMPLE 8.9

Use the substitution $u = 2x + 1$ to find $\displaystyle\int 4x\sqrt{2x+1}\, dx$.

Answer

$u = 2x + 1 \Rightarrow x = \dfrac{u - 1}{2}$

$u = 2x + 1 \Rightarrow \dfrac{du}{dx} = 2 \Rightarrow dx = \dfrac{du}{2}$

$\displaystyle\int 4x\sqrt{2x+1}\, dx = \int 4\left(\dfrac{u-1}{2}\right)\sqrt{u}\, \dfrac{du}{2}$ .... Rewrite the integral in terms of u and simplify.

$\phantom{\displaystyle\int 4x\sqrt{2x+1}\, dx} = \displaystyle\int \left(u^{\frac{3}{2}} - u^{\frac{1}{2}}\right) du$ .... Integrate with respect to u.

$\phantom{\displaystyle\int 4x\sqrt{2x+1}\, dx} = \dfrac{2}{5} u^{\frac{5}{2}} - \dfrac{2}{3} u^{\frac{3}{2}} + c$ .... Rewrite in terms of x.

$\phantom{\displaystyle\int 4x\sqrt{2x+1}\, dx} = \dfrac{2}{5}(2x+1)^{\frac{5}{2}} - \dfrac{2}{3}(2x+1)^{\frac{3}{2}} + c$

WORKED EXAMPLE 8.10

Use the substitution $x = 2\tan u$ to find $\displaystyle\int \frac{8}{x^2 + 4}\,dx$.

Answer

$x = 2\tan u \Rightarrow x^2 = 4\tan^2 u$

$x = 2\tan u \Rightarrow \dfrac{dx}{du} = 2\sec^2 u \Rightarrow dx = 2\sec^2 u\,du$

$\displaystyle\int \frac{8}{x^2 + 4}\,dx = \int \left(\frac{8}{4\tan^2 u + 4}\right) 2\sec^2 u\,du$ .... Rewrite the integral in terms of u and simplify.

$\displaystyle\phantom{\int \frac{8}{x^2+4}dx} = \int \frac{16\sec^2 u}{4\tan^2 u + 4}\,du$ .... Use $\tan^2 u + 1 \equiv \sec^2 u$.

$\displaystyle\phantom{\int \frac{8}{x^2+4}dx} = \int \frac{16\sec^2 u}{4\sec^2 u}\,du$ .... Simplify.

$\displaystyle\phantom{\int \frac{8}{x^2+4}dx} = \int 4\,du$ .... Integrate with respect to u.

$\displaystyle\phantom{\int \frac{8}{x^2+4}dx} = 4u + c$ .... Rewrite in terms of x.

$\displaystyle\phantom{\int \frac{8}{x^2+4}dx} = 4\tan^{-1}\left(\frac{x}{2}\right) + c$

WORKED EXAMPLE 8.11

Use the substitution $u = \sin x$ to find $\displaystyle\int \sin^2 2x \cos x\,dx$.

Answer

$u = \sin x \Rightarrow \sin^2 2x = (2\sin x \cos x)^2$

$ = 4\sin^2 x \cos^2 x$

$ = 4u^2(1 - u)^2$

$u = \sin x \Rightarrow \dfrac{du}{dx} = \cos x \Rightarrow dx = \dfrac{du}{\cos x}$

$\displaystyle\int \sin^2 2x \cos x\,dx = \int \sin^2 2x \cos x \frac{du}{\cos x}$ ...... Replace dx by $\dfrac{du}{\cos x}$ and simplify.

$\displaystyle = \int \sin^2 2x\,du$

$\displaystyle = \int (4u^2 - 4u^4)\,du$ ...... Replace $\sin^2 2x$ by $4u^2 - 4u^4$.

$\displaystyle = \frac{4}{3}u^3 - \frac{4}{5}u^5 + c$ ...... Integrate with respect to u.

$\displaystyle = \frac{4}{3}\sin^3 x - \frac{4}{5}\sin^5 x + c$ ...... Rewrite in terms of x.

> **TIP**
>
> When using substitution in definite integrals you must also remember to convert the limits to limits for u.

WORKED EXAMPLE 8.12

Use the substitution $u = x + 1$ to find $\int_0^3 \dfrac{x}{\sqrt{x+1}}\,dx$.

Answer

$u = x + 1 \Rightarrow x = u - 1$

$u = x + 1 \Rightarrow \dfrac{du}{dx} = 1 \Rightarrow dx = du$

Find the new limits for u:

$x = 3 \Rightarrow u = 3 + 1 = 4$

$x = 0 \Rightarrow u = 0 + 1 = 1$

$\displaystyle\int_{x=0}^{x=3} \dfrac{x}{\sqrt{x+1}}\,dx = \int_{u=1}^{u=4} \dfrac{u-1}{\sqrt{u}}\,du$ .... Write the integral in terms of u and simplify.

$\qquad = \displaystyle\int_1^4 \left(u^{\frac{1}{2}} - u^{-\frac{1}{2}}\right) du$ .... Integrate with respect to u.

$\qquad = \left[\dfrac{2}{3} u^{\frac{3}{2}} - 2u^{\frac{1}{2}}\right]_1^4$ .... Evaluate the function at the limits.

$\qquad = \left(\dfrac{16}{3} - 4\right) - \left(\dfrac{2}{3} - 2\right)$

$\qquad = \dfrac{8}{3}$

EXERCISE 8D

1 Given that $I = \displaystyle\int \dfrac{x}{\sqrt{x^2 - 3}}\,dx$, use the substitution $u = x^2 - 3$ to show that $I = \dfrac{1}{2}\displaystyle\int \dfrac{1}{\sqrt{u}}\,du$.

Hence find I.

2 Find these integrals, using the given substitution.

a $\displaystyle\int \dfrac{x}{(x+2)^4}\,dx,\ u = x + 2$

b $\displaystyle\int x\sqrt{1 - 2x^2}\,dx,\ u = 1 - 2x^2$

c $\displaystyle\int \cos x \sin^5 x\,dx,\ u = \sin x$

d $\displaystyle\int e^x \sqrt{e^x + 2}\,dx,\ u = e^x + 2$

e $\displaystyle\int \dfrac{5x}{5x + 1}\,dx,\ u = 5x + 1$

f $\displaystyle\int x\sqrt{3x - 1}\,dx,\ u = 3x - 1$

3 Given that $I = \displaystyle\int_0^1 \dfrac{x^2}{\sqrt{1 - x^2}}\,dx$, use the substitution $x = \sin\theta$ to show that $I = \displaystyle\int_0^{\frac{1}{2}\pi} \sin^2\theta\,d\theta$.

Hence find the exact value of I.

4 Find the exact value of each of these integrals.

a $\displaystyle\int_0^{\frac{1}{2}\pi} \cos^3 x\,dx,\ u = \sin x$

b $\displaystyle\int_0^4 \dfrac{\sqrt{x}}{3 - \sqrt{x}}\,dx,\ u = 3 - \sqrt{x}$

c $\displaystyle\int_1^2 \dfrac{3}{x\left(2 - \sqrt{x}\right)}\,dx,\ u = \sqrt{x}$

d $\displaystyle\int_0^{\sqrt{\pi}} 2x \cos\left(\dfrac{x^2}{4}\right) dx,\ u = \cos\left(\dfrac{x^2}{4}\right)$

e $\quad \int_{\frac{1}{6}\pi}^{\frac{1}{3}\pi} \sec^2 x \tan^3 x \, dx, \; u = \tan x$

f $\quad \int_{-2}^{2} \dfrac{5}{(x^2 + 4)^2} \, dx, \; x = 2\tan\theta$

g $\quad \int_{0}^{1} \dfrac{x^2}{\sqrt{4 - x^2}}, \; x = 2\cos\theta$

h $\quad \int_{0}^{2} \sqrt{x(2 - x)} \, dx, \; x = 2\sin^2\theta$

5 Use the substitution $u = x - 2$ to find the exact value of $\displaystyle\int_{1}^{2} \dfrac{4}{1 + (x - 2)^2} \, dx$.

6 The diagram shows part of the curve $y = 2\sin^2 x \cos^3 x$. Using the substitution $u = \sin x$, find the exact area of the shaded region.

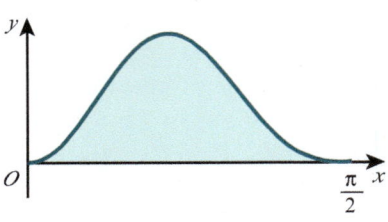

7 The diagram shows part of the curve $y = \dfrac{\ln x}{\sqrt{x}}$. Using the substitution $u = \ln x$, find the exact value of the volume of the solid formed when the shaded region is rotated completely about the x-axis.

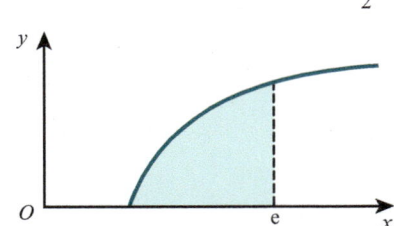

8 The diagram shows part of the curve $y = 3e^{\cos x} \sin x$. The curve has rotational symmetry of order 2 about the origin. Using the substitution $u = \cos x$, find the exact value of the total area bounded by the curve and the x-axis between $x = -\pi$ and $x = \pi$.

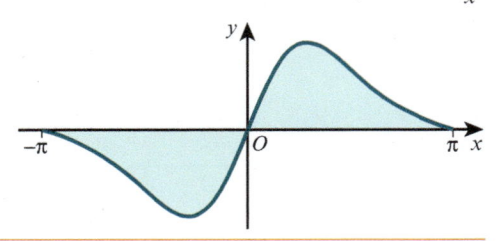

8.5 The use of partial fractions in integration

EXPLORE 8.1

Consider $\displaystyle\int \dfrac{2}{x^2 - 1} \, dx$.

Try using some of the integration techniques that you have learnt so far to find this integral and discuss what happens with your classmates.

In the integral $\displaystyle\int \dfrac{2}{x^2 - 1} \, dx$, the denominator $x^2 - 1$ can be factorised to $(x - 1)(x + 1)$.

Hence the fraction $\dfrac{2}{x^2 - 1}$ can be split into partial fractions.

$\dfrac{2}{x^2 - 1} \equiv \dfrac{A}{x - 1} + \dfrac{B}{x + 1}$ Multiply throughout by $(x - 1)(x + 1)$.

$2 \equiv A(x + 1) + B(x - 1)$ ------ (1)

Let $x = 1$ in equation (1):

$2 = 2A$

$A = 1$

Let $x = -1$ in equation (1):
$$2 = -2B$$
$$B = -1$$
$$\therefore \frac{2}{x^2 - 1} \equiv \frac{1}{x-1} - \frac{1}{x+1}$$

Hence $\int \frac{2}{x^2 - 1} \, dx = \int \left(\frac{1}{x-1} - \frac{1}{x+1} \right) dx$

$$= \ln|x-1| - \ln|x+1| + c$$
$$= \ln \left| \frac{x-1}{x+1} \right| + c$$

WORKED EXAMPLE 8.13

Find $\int \frac{3-x}{(x+1)(x^2+3)} \, dx$.

Answer

First split into partial fractions.

$$\frac{3-x}{(x+1)(x^2+3)} \equiv \frac{A}{x+1} + \frac{Bx+C}{x^2+3}$$ …… Multiply throughout by $(x+1)(x^2+3)$.

$$3 - x \equiv A(x^2 + 3) + Bx(x+1) + C(x+1)$$

Let $x = -1$ $4 = 4A$ $\Rightarrow A = 1$

Let $x = 0$ $3 = 3A + C$ $\Rightarrow C = 0$

Let $x = 1$ $2 = 4A + 2B + 2C$ $\Rightarrow B = -1$

$$\therefore \int \frac{3-x}{(x+1)(x^2+3)} \, dx = \int \left(\frac{1}{x+1} - \frac{x}{x^2+3} \right) dx$$

$$= \int \frac{1}{x+1} \, dx - \frac{1}{2} \int \frac{2x}{x^2+3} \, dx$$

$$= \ln|x+1| - \frac{1}{2} \ln|x^2+3| + c$$

WORKED EXAMPLE 8.14

Find $\int \frac{9}{(x+2)(x-1)^2} \, dx$.

Answer

First split into partial fractions.

$$\frac{9}{(x+2)(x-1)^2} \equiv \frac{A}{x+2} + \frac{B}{x-1} + \frac{C}{(x-1)^2}$$ …… Multiply throughout by $(x+2)(x-1)^2$.

$$9 \equiv A(x-1)^2 + B(x+2)(x-1) + C(x+2)$$

Let $x = 1$ $9 = 3C$ $\Rightarrow C = 3$
Let $x = -2$ $9 = 9A$ $\Rightarrow A = 1$
Let $x = 0$ $9 = A - 2B + 2C$ $\Rightarrow B = -1$

$$\therefore \int \frac{9}{(x+2)(x-1)^2} \, dx = \int \left(\frac{1}{x+2} - \frac{1}{x-1} + \frac{3}{(x-1)^2} \right) dx$$

$$= \int \left(\frac{1}{x+2} - \frac{1}{x-1} + 3(x-1)^{-2} \right) dx$$

$$= \ln|x+2| - \ln|x-1| - 3(x-1)^{-1} + c$$

$$= \ln \left| \frac{x+2}{x-1} \right| - \frac{3}{x-1} + c$$

EXERCISE 8E

1 Find each of these integrals.

  a $\displaystyle\int \frac{1}{x(2-x)} \, dx$ **b** $\displaystyle\int \frac{2x-5}{(x+2)(x-1)} \, dx$ **c** $\displaystyle\int \frac{2(4x-9)}{(2x+1)(x-5)} \, dx$

  d $\displaystyle\int \frac{x^2+2x-5}{(x-3)(x^2+1)} \, dx$ **e** $\displaystyle\int \frac{6x^2-19x+20}{(x+1)(x-2)^2} \, dx$ **f** $\displaystyle\int \frac{2x^2+4x-21}{(x-1)(x+4)} \, dx$

2 Evaluate each of these integrals.

  a $\displaystyle\int_1^2 \frac{5x+13}{(x+2)(x+3)} \, dx$ **b** $\displaystyle\int_0^2 \frac{4x+5}{(2x+1)(x+2)} \, dx$ **c** $\displaystyle\int_1^2 \frac{4-3x}{(2x-1)(x+2)} \, dx$

  d $\displaystyle\int_{-1}^1 \frac{3x^2+x}{(x-2)(x^2+1)} \, dx$ **e** $\displaystyle\int_0^1 \frac{x^2+x+2}{(x+1)(1+x^2)} \, dx$ **f** $\displaystyle\int_{-2}^0 \frac{8}{(2-x)(4+x^2)} \, dx$

  g $\displaystyle\int_2^3 \frac{1-x-2x^2}{x^2(1-x)} \, dx$ **h** $\displaystyle\int_0^1 \frac{1-2x}{(2x+1)(x+1)^2} \, dx$ **i** $\displaystyle\int_1^2 \frac{8x^2-3x-2}{4x^3-3x+1} \, dx$

3 By first dividing the numerator by the denominator, evaluate each of these integrals.

  a $\displaystyle\int_2^3 \frac{x^2+2x+3}{(x-1)(x+2)} \, dx$ **b** $\displaystyle\int_1^3 \frac{2x^2+5x+1}{(x+1)(x+2)} \, dx$ **c** $\displaystyle\int_1^2 \frac{4x^2+2x-5}{(x+1)(2x-1)} \, dx$

4 Show that $\displaystyle\int_{2\sqrt{2}}^{3\sqrt{2}} \frac{x+3\sqrt{2}}{x^2-2} \, dx = \ln 3$.

5 Show that $\displaystyle\int_0^{\frac{1}{2}\pi} \frac{\cos x}{9-\sin^2 x} \, dx = \frac{1}{6} \ln 2$.

6 Show that $\displaystyle\int_0^{\ln 2} \frac{e^{2x}}{(1+e^x)(2e^x+1)} \, dx = \frac{1}{2} \ln \frac{27}{20}$.

8.6 Integration by parts

Consider $\int x e^{2x} \, dx$. The function to be integrated is clearly the product of two simpler functions x and e^{2x}. If we try using a substitution to simplify the integral we will find that it does not help. To integrate this function we require a new technique.

Integrating both sides of the equation for the product rule for differentiation with respect to x and using the fact that integration is the reverse process of differentiation we can say that:

$$uv = \int u \frac{dv}{dx} \, dx + \int v \frac{du}{dx} \, dx$$

Rearranging gives the formula for **integration by parts**:

> **KEY POINT 8.4**
>
> $$\int u \frac{dv}{dx} \, dx = uv - \int v \frac{du}{dx} \, dx$$

> **REWIND**
>
> In Chapter 4, we learnt the product rule for differentiation:
> $$\frac{d}{dx}(uv) = u \frac{dv}{dx} + v \frac{du}{dx}.$$

The following Worked examples show how to use this formula.

WORKED EXAMPLE 8.15

Find $\int x e^{2x} \, dx$.

Answer

$u = x \quad \Rightarrow \quad \frac{du}{dx} = 1$

$\frac{dv}{dx} = e^{2x} \quad \Rightarrow \quad v = \frac{1}{2} e^{2x}$

Substitute into $\int u \frac{dv}{dx} \, dx = uv - \int v \frac{du}{dx} \, dx$:

$$\int x e^{2x} \, dx = \frac{1}{2} x e^{2x} - \int \frac{1}{2} e^{2x} \, dx$$

$$= \frac{1}{2} x e^{2x} - \frac{1}{4} e^{2x} + c$$

It is important to choose our values for u and $\frac{dv}{dx}$ carefully.

Consider trying to find $\int x \ln x \, dx$. If we let $u = x$ and $\frac{dv}{dx} = \ln x$, then finding $\frac{du}{dx}$ is easy but we do not know how to integrate $\ln x$ to find v. Hence we must let $u = \ln x$ and $\frac{dv}{dx} = x$. This is demonstrated in Worked example 8.16.

WORKED EXAMPLE 8.16

Find $\int x \ln x \, dx$.

Answer

$u = \ln x \implies \dfrac{du}{dx} = \dfrac{1}{x}$

$\dfrac{dv}{dx} = x \implies v = \dfrac{1}{2}x^2$

Substitute into $\int u \dfrac{dv}{dx} dx = uv - \int v \dfrac{du}{dx} dx$:

$$\int x \ln x \, dx = \dfrac{1}{2} x^2 \ln x - \int \dfrac{1}{2} x^2 \cdot \dfrac{1}{x} \, dx$$

$$= \dfrac{1}{2} x^2 \ln x - \int \dfrac{1}{2} x \, dx$$

$$= \dfrac{1}{2} x^2 \ln x - \dfrac{1}{4} x^2 + c$$

WORKED EXAMPLE 8.17

Find $\int \ln x \, dx$.

Answer

First write $\int \ln x \, dx$ as $\int 1 \cdot \ln x \, dx$.

$u = \ln x \implies \dfrac{du}{dx} = \dfrac{1}{x}$

$\dfrac{dv}{dx} = 1 \implies v = x$

Substitute into $\int u \dfrac{dv}{dx} dx = uv - \int v \dfrac{du}{dx} dx$:

$$\int \ln x \, dx = x \ln x - \int \dfrac{1}{x} \cdot x \, dx$$

$$= x \ln x - \int 1 \, dx$$

$$= x \ln x - x + c$$

Chapter 8: Further calculus

Sometimes we might need to use integration by parts more than once to integrate a function.

WORKED EXAMPLE 8.18

Find $\int x^2 \sin x \, dx$.

Answer

$u = x^2 \implies \dfrac{du}{dx} = 2x$

$\dfrac{dv}{dx} = \sin x \implies v = -\cos x$

Substitute into $\int u \dfrac{dv}{dx} dx = uv - \int v \dfrac{du}{dx} dx$

$\int x^2 \sin x \, dx = -x^2 \cos x - \int -2x \cos x \, dx$

$\qquad = -x^2 \cos x + \int 2x \cos x \, dx \ ---- (1)$

Use integration by parts a second time to find $\int 2x \cos x \, dx$

$u = 2x \implies \dfrac{du}{dx} = 2$

$\dfrac{dv}{dx} = \cos x \implies v = \sin x$

Substitute into $\int u \dfrac{dv}{dx} dx = uv - \int v \dfrac{du}{dx} dx$

$\int 2x \cos x \, dx = 2x \sin x - \int 2 \sin x \, dx$

$\qquad = 2x \sin x + 2 \cos x + c$

Substituting in (1) gives:

$\int x^2 \sin x \, dx = -x^2 \cos x + 2x \sin x + 2 \cos x + c$

WORKED EXAMPLE 8.19

Given that $I = \int \dfrac{\ln x}{x} dx$, use integration by parts to show that $I = \dfrac{1}{2}(\ln x)^2 + c$.

Answer

$u = \ln x \implies \dfrac{du}{dx} = \dfrac{1}{x}$

$\dfrac{dv}{dx} = \dfrac{1}{x} \implies v = \ln x$

Substitute into $\int u \dfrac{dv}{dx} dx = uv - \int v \dfrac{du}{dx} dx$

$$I = (\ln x)^2 - \int \dfrac{\ln x}{x} dx \quad \text{Rearrange.}$$

$$2I = (\ln x)^2 + k \quad \text{Divide both sides by 2.}$$

$$I = \dfrac{1}{2}(\ln x)^2 + \dfrac{k}{2}$$

$$\therefore I = \dfrac{1}{2}(\ln x)^2 + c$$

WORKED EXAMPLE 8.20

Find the exact value of $\int_0^{\frac{1}{2}\pi} 4x \sin 2x \, dx$.

Answer

$u = 4x \quad \Rightarrow \dfrac{du}{dx} = 4$

$\dfrac{dv}{dx} = \sin 2x \quad \Rightarrow v = -\dfrac{1}{2}\cos 2x$

Substitute into $\int u \dfrac{dv}{dx} dx = uv - \int v \dfrac{du}{dx} dx$

$$\int 4x \sin 2x \, dx = -2x \cos 2x + \int 2 \cos 2x \, dx$$

$$= -2x \cos 2x + \sin 2x + c$$

$$\therefore \int_0^{\frac{1}{2}\pi} 4x \sin 2x \, dx = \Big[-2x \cos 2x + \sin 2x \Big]_0^{\frac{1}{2}\pi}$$

$$= (-\pi \cos \pi + \sin \pi) - (0 + \sin 0)$$

$$= \pi$$

EXERCISE 8F

1 Find each of these integrals.

a $\int 3xe^x \, dx$
b $\int x \cos x \, dx$
c $\int x \ln 2x \, dx$

d $\int x \sin 2x \, dx$
e $\int x^3 \ln x \, dx$
f $\int \dfrac{\ln x}{\sqrt{x}} dx$

2 Evaluate each of these integrals.

a $\int_0^{\frac{1}{6}\pi} x \cos 3x \, dx$
b $\int_1^2 x \ln 2x \, dx$
c $\int_0^1 xe^{2x} \, dx$

d $\int_1^3 \ln x \, dx$
e $\int_0^1 (2-x)e^{-3x} \, dx$
f $\int_{-\frac{1}{5}}^0 (1 - 3xe^{-5x}) \, dx$

3 Use integration by parts twice to find the exact value of each of these integrals.

a $\displaystyle\int_1^2 (\ln x)^2 \, dx$

b $\displaystyle\int_0^{\frac{1}{4}\pi} x^2 \cos 2x \, dx$

c $\displaystyle\int_0^\pi x^2 \sin x \, dx$

d $\displaystyle\int_0^e x(\ln x)^2 \, dx$

e $\displaystyle\int_{-\infty}^0 x^2 e^x \, dx$

f $\displaystyle\int_0^\pi e^x \sin x \, dx$

4 Find the exact value of the area for each of these shaded regions.

a

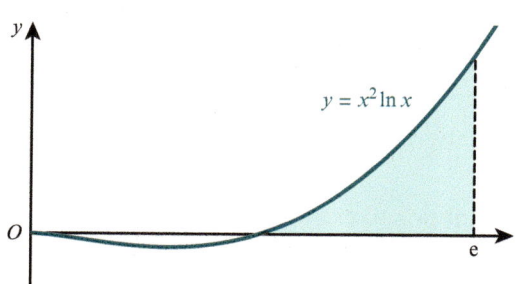

$y = x^2 \ln x$

b

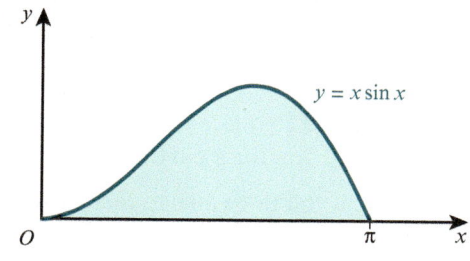

$y = x \sin x$

c

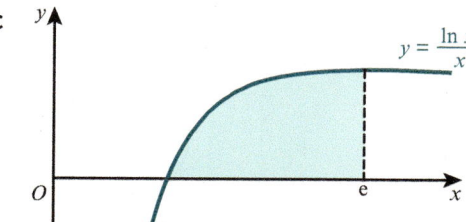

$y = \dfrac{\ln x}{x}$

d

$y = x^2 e^{-x}$

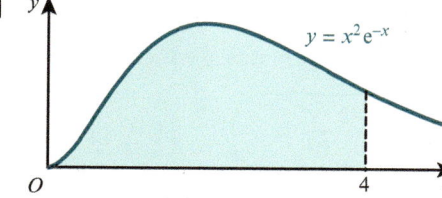

5

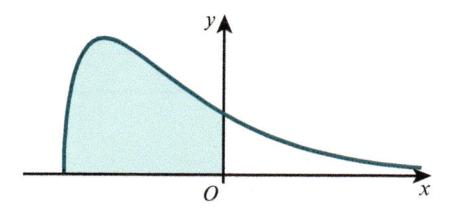

The diagram shows the curve $y = e^{-x}\sqrt{x+2}$. Find the exact value of the volume of the solid formed when the shaded region is rotated completely about the x-axis.

8.7 Further integration

Some integrals can be found using different integration techniques.

The activities in Explore 8.2 look at why our answers to an indefinite integral might 'appear' to be different when using different integration techniques when they are actually equivalent.

EXPLORE 8.2

1 Nadia and Richard are asked to find $\displaystyle\int x(x-2)^7 \, dx$.

Nadia uses the substitution $u = x - 2$ and her answer is: $\dfrac{1}{9}(x-2)^9 + \dfrac{1}{4}(x-2)^8 + c$

Richard uses integration by parts and his answer is: $\dfrac{1}{8}x(x-2)^8 - \dfrac{1}{72}(x-2)^9 + c$

Use each of their methods to show that they are both correct and then use algebra to explain why their answers are equivalent.

2 a Find $\int \dfrac{4x+1}{2x^2+x-1}\,dx$ using:

    i the substitution $u = 2x^2 + x - 1$

    ii partial fractions.

b Suggest another method for finding $\int \dfrac{4x+1}{2x^2+x-1}\,dx$.

c Use algebra to show that your two answers to **part a** are equivalent.

Explore 8.3 is designed to make you think about the different integration techniques that might be available to you for a particular integral.

This activity expects you to be able to recognise the appropriate substitution to use when integrating by substitution.

EXPLORE 8.3

Discuss with your classmates which integration methods can be used to find each of these integrals.

- $\int \dfrac{2x+3}{x^2+3x-10}\,dx$
- $\int \dfrac{3x^2}{\sqrt{x^3+1}}\,dx$
- $\int x\,e^{x^2}\,dx$
- $\int \dfrac{2x}{x+5}\,dx$
- $\int \ln 5x\,dx$
- $\int x\sqrt{2x+1}\,dx$
- $\int \dfrac{3}{(2x+1)^5}\,dx$
- $\int \sqrt{3x-2}\,dx$
- $\int \dfrac{2x+3}{x^2+3x-10}\,dx$
- $\int \cos^5 x \sin x\,dx$
- $\int x \sin 4x\,dx$
- $\int \dfrac{\sin x}{3+\cos x}\,dx$

Find each integral.

Checklist of learning and understanding

Integrating $\dfrac{k f'(x)}{f(x)}$

- $\int \dfrac{k f'(x)}{f(x)}\,dx = k \ln |f(x)| + c$

Integration by substitution

- Substitutions can sometimes be used to simplify the form of a function so that its integral can be easily recognised. When using a substitution, we must ensure that the integral is completely rewritten in terms of the new variable before integrating.

Integrating rational functions

- Some rational functions can be split into partial fractions that can then be integrated.

Integration by parts

- $\int u \dfrac{dv}{dx}\,dx = uv - \int v \dfrac{du}{dx}\,dx$

Chapter 8: Further calculus

END-OF-CHAPTER REVIEW EXERCISE 8

1. Find the exact value of $\int_0^{\frac{1}{2}} x e^{-2x} \, dx$. [5]

   Cambridge International A Level Mathematics 9709 Paper 31 Q2 June 2016

2. Find the exact value of $\int_1^4 \dfrac{\ln x}{\sqrt{x}} \, dx$. [5]

   Cambridge International A Level Mathematics 9709 Paper 31 Q3 November 2013

3. Use the substitution $u = 1 + 3\tan x$ to find the exact value of $\int_0^{\frac{1}{4}\pi} \dfrac{\sqrt{(1 + 3\tan x)}}{\cos^2 x} \, dx$. [5]

   Cambridge International A Level Mathematics 9709 Paper 31 Q2 June 2014

4.
   i. Prove that $\cot \theta + \tan \theta \equiv 2 \operatorname{cosec} 2\theta$. [3]

   ii. Hence show that $\int_{\frac{1}{6}\pi}^{\frac{1}{3}\pi} \operatorname{cosec} 2\theta \, d\theta = \dfrac{1}{2} \ln 3$. [4]

   Cambridge International A Level Mathematics 9709 Paper 31 Q5 November 2013

5. Let $I = \int_0^1 \dfrac{x^2}{\sqrt{(4 - x^2)}} \, dx$.

   i. Using the substitution $x = 2 \sin \theta$, show that $I = \int_0^{\frac{1}{6}\pi} 4 \sin^2 \theta \, d\theta$. [3]

   ii. Hence find the exact value of I. [4]

   Cambridge International A Level Mathematics 9709 Paper 31 Q5 November 2010

6. The integral I is defined by $I = \int_0^2 4t^3 \ln(t^2 + 1) \, dt$.

   i. Use the substitution $x = t^2 + 1$ to show that $I = \int_1^5 (2x - 2) \ln x \, dx$. [3]

   ii. Hence find the exact value of I. [5]

   Cambridge International A Level Mathematics 9709 Paper 31 Q7 June 2011

7.
   i. Express $\dfrac{2}{(x + 1)(x + 3)}$ in partial fractions. [2]

   ii. Using your answer to **part i**, show that $\left(\dfrac{2}{(x + 1)(x + 3)}\right)^2 \equiv \dfrac{1}{(x + 1)^2} - \dfrac{1}{x + 1} + \dfrac{1}{x + 3} + \dfrac{1}{(x + 3)^2}$. [2]

   iii. Hence show that $\int_0^1 \dfrac{4}{(x + 1)^2 (x + 3)^2} \, dx = \dfrac{7}{12} - \ln \dfrac{3}{2}$. [5]

   Cambridge International A Level Mathematics 9709 Paper 31 Q8 June 2010

8.
   a. Find $\int (4 + \tan^2 2x) \, dx$. [3]

   b. Find the exact value of $\int_{\frac{1}{4}\pi}^{\frac{1}{2}\pi} \dfrac{\sin\left(x + \frac{1}{6}\pi\right)}{\sin x} \, dx$. [5]

   Cambridge International A Level Mathematics 9709 Paper 31 Q5 June 2015

9   **i**   Prove the identity $\tan 2\theta - \tan\theta \equiv \tan\theta \sec 2\theta$. [4]

   ii   Hence show that $\int_0^{\frac{1}{6}\pi} \tan\theta \sec 2\theta \, d\theta = \frac{1}{2}\ln\frac{3}{2}$. [4]

Cambridge International A Level Mathematics 9709 Paper 31 Q5 November 2016

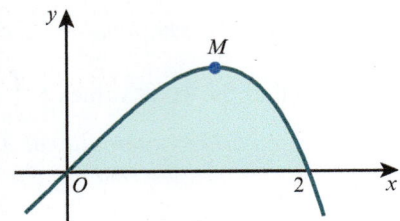

10

The diagram shows part of the curve $y = (2x - x^2)e^{\frac{1}{2}x}$ and its maximum point M.

   i   Find the exact x-coordinate of M. [4]

   ii   Find the exact value of the area of the shaded region bounded by the curve and the positive x-axis. [5]

Cambridge International A Level Mathematics 9709 Paper 31 Q7 November 2016

11   By first expressing $\dfrac{4x^2 + 5x + 3}{2x^2 + 5x + 2}$ in partial fractions, show that $\int_0^4 \dfrac{4x^2 + 5x + 3}{2x^2 + 5x + 2}\,dx = 8 - \ln 9$. [10]

Cambridge International A Level Mathematics 9709 Paper 31 Q9 June 2012

12

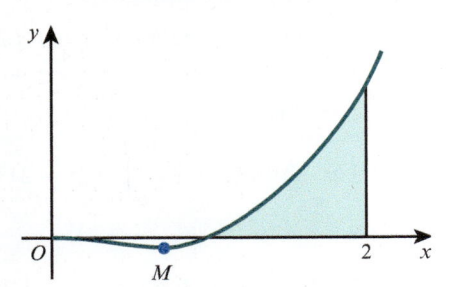

The diagram shows the curve $y = x^3 \ln x$ and its minimum point M.

   i   Find the exact coordinates of M. [5]

   ii   Find the exact area of the shaded region bounded by the curve, the x-axis and the line $x = 2$. [5]

Cambridge International A Level Mathematics 9709 Paper 31 Q9 November 2010

13

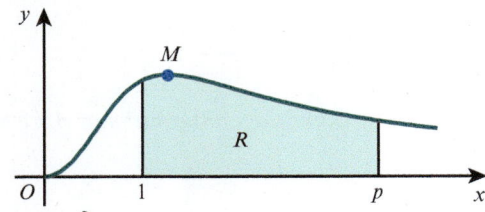

The diagram shows the curve $y = \dfrac{x^2}{1 + x^3}$ for $x \geq 0$, and its maximum point M. The shaded region R is enclosed by the curve, the x-axis and the lines $x = 1$ and $x = p$.

   i   Find the exact value of the x-coordinate of M. [4]

   ii   Calculate the value of p for which the area of R is equal to 1. Give your answer correct to 3 significant figures. [6]

Cambridge International A Level Mathematics 9709 Paper 31 Q10 November 2015

CROSS-TOPIC REVIEW EXERCISE 3

This exercise is for Pure Mathematics 3 students only.

1. Expand $(1 - 3x)^{-5}$ in ascending powers of x, up to and including the term in x^3. [3]

2. Evaluate $\int_0^1 \frac{1}{3x^2 + 1} \, dx$. [3]

3. Expand $\frac{32}{(x + 2)^3}$ in ascending powers of x, up to and including the term in x^3. [4]

4. Expand $\frac{1}{\sqrt{4 - 2x}}$ in ascending powers of x, up to and including the term in x^2. [4]

5. Evaluate $\int_0^{\frac{1}{2}\pi} x^2 \sin x \, dx$. [5]

6. Expand $\sqrt{\frac{1 + 2x}{1 - x}}$ in ascending powers of x, up to and including the term in x^3. [5]

7. Show that $\int_0^1 (x + 2)e^{-2x} \, dx = \frac{5}{4} - \frac{7}{4e^2}$. [5]

8.
   i. Expand $\frac{1}{\sqrt{(1 - 4x)}}$ in ascending powers of x, up to and including the term in x^2, simplifying the coefficients. [3]

   ii. Hence find the coefficient of x^2 in the expansion of $\frac{1 + 2x}{\sqrt{(4 - 16x)}}$. [2]

   Cambridge International A Level Mathematics 9709 Paper 31 Q2 June 2012

9. When $(1 + ax)^{-2}$, where a is a positive constant, is expanded in ascending powers of x, the coefficients of x and x^3 are equal.

   i. Find the exact value of a. [4]

   ii. When a has this value, obtain the expansion up to and including the term in x^2, simplifying the coefficients. [3]

   Cambridge International A Level Mathematics 9709 Paper 31 Q4 November 2012

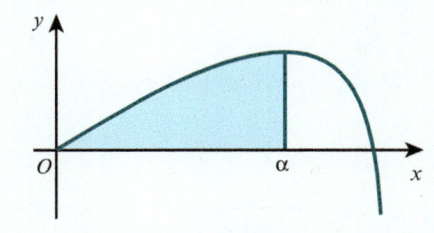

10.

    The diagram shows the curve $y = 8 \sin \frac{1}{2}x - \tan \frac{1}{2}x$ for $0 \leq x < \pi$. The x-coordinate of the maximum point is α and the shaded region is enclosed by the curve and the lines $x = \alpha$ and $y = 0$.

    i. Show that $\alpha = \frac{2}{3}\pi$. [3]

    ii. Find the exact value of the area of the shaded region. [4]

    Cambridge International A Level Mathematics 9709 Paper 31 Q5 June 2012

11 Show that $\displaystyle\int_0^5 \frac{5-3x}{(x+1)(3x+1)}\,dx = 4\ln\frac{4}{3}$. [7]

12 Let $I = \displaystyle\int_0^{\frac{1}{2}} \frac{4x^2}{\sqrt{1-x^2}}\,dx$.

    a Using the substitution $x = \sin\theta$, show that $I = \displaystyle\int_0^{\frac{\pi}{6}} 4\sin^2\theta\,d\theta$. [3]

    b Hence show that $I = \dfrac{\pi}{3} - \dfrac{\sqrt{3}}{2}$. [4]

13 It is given that $\displaystyle\int_1^a x\ln x\,dx = 30$, where $a > 1$.

    a Show that $a = \sqrt{\dfrac{119}{2\ln a - 1}}$. [5]

    b Use the iterative formula $a_{n+1} = \sqrt{\dfrac{119}{2\ln a_n - 1}}$ to determine the value of a correct to 2 decimal places. Give the result of each iteration correct to 4 decimal places. [3]

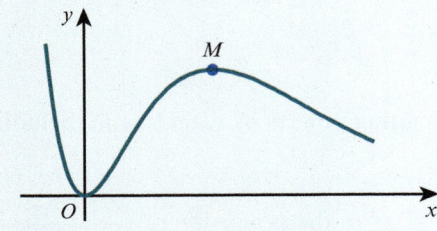

14

The diagram shows the curve $y = x^2 e^{2-x}$ and its maximum point M.

    i Show that the x-coordinate of M is 2. [3]

    ii Find the exact value of $\displaystyle\int_0^2 x^2 e^{2-x}\,dx$. [6]

Cambridge International A Level Mathematics 9709 Paper 31 Q9 June 2015

15 Let $f(x) = \dfrac{12 + 8x - x^2}{(2-x)(4+x^2)}$.

    i Express $f(x)$ in the form $\dfrac{A}{2-x} + \dfrac{Bx+C}{4+x^2}$. [4]

    ii Show that $\displaystyle\int_0^1 f(x)\,dx = \ln\left(\dfrac{25}{2}\right)$. [5]

Cambridge International A Level Mathematics 9709 Paper 31 Q8 November 2011

16 It is given that $\displaystyle\int_1^a \ln(2x)\,dx = 1$, where $a > 1$.

    i Show that $a = \dfrac{1}{2}\exp\left(1 + \dfrac{\ln 2}{a}\right)$, where $\exp(x)$ denotes e^x. [6]

    ii Use the iterative formula $a_{n+1} = \dfrac{1}{2}\exp\left(1 + \dfrac{\ln 2}{a_n}\right)$ to determine the value of a correct to 2 decimal places. Give the result of each iteration to 4 decimal places. [3]

Cambridge International A Level Mathematics 9709 Paper 31 Q6 November 2014

17 i Express $\dfrac{4 + 12x + x^2}{(3 - x)(1 + 2x)^2}$ in partial fractions. [5]

ii Hence obtain the expansion of $\dfrac{4 + 12x + x^2}{(3 - x)(1 + 2x)^2}$ in ascending powers of x, up to and including the term in x^2. [5]

Cambridge International A Level Mathematics 9709 Paper 31 Q9 June 2014

18 Let $f(x) = \dfrac{x^2 - 8x + 9}{(1 - x)(2 - x)^2}$.

i Express $f(x)$ in partial fractions. [5]

ii Hence obtain the expansion of $f(x)$ in ascending powers of x, up to and including the term in x^2. [5]

Cambridge International A Level Mathematics 9709 Paper 31 Q9 November 2014

19 Let $f(x) = \dfrac{5x - 2}{(x - 1)(2x^2 - 1)}$.

a Express $f(x)$ in partial fractions. [5]

b Hence obtain the expansion of $f(x)$ in ascending powers of x, up to and including the term in x^3. [5]

20 a Given that $\dfrac{x^3 - 2}{x^2(2x - 1)} \equiv A + \dfrac{B}{x} + \dfrac{C}{x^2} + \dfrac{D}{2x - 1}$, find the values of the constants A, B, C and D. [5]

b Hence show that $\displaystyle\int_1^2 \dfrac{x^3 - 2}{x^2(2x - 1)} \, dx = \dfrac{3}{2} - 2\ln\dfrac{9}{4} + \dfrac{1}{4}\ln 3$. [5]

21

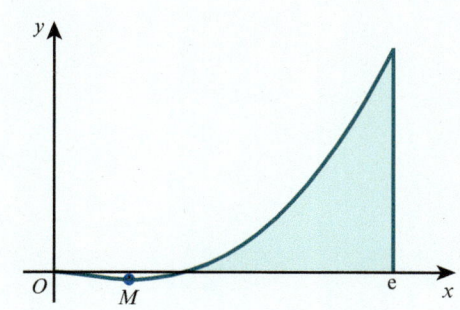

The diagram shows the curve $y = x^2 \ln x$ and its minimum point M.

i Find the exact values of the coordinates of M. [5]

ii Find the exact value of the area of the shaded region bounded by the curve, the x-axis and the line $x = e$. [5]

Cambridge International A Level Mathematics 9709 Paper 31 Q9 November 2011

22 i Show that $\displaystyle\int_2^4 4x \ln x \, dx = 56 \ln 2 - 12$. [5]

ii Use the substitution $u = \sin 4x$ to find the exact value of $\displaystyle\int_0^{\frac{1}{24}\pi} \cos^3 4x \, dx$. [5]

Cambridge International A Level Mathematics 9709 Paper 31 Q8 June 2013

23 **i** Express $4\cos\theta + 3\sin\theta$ in the form $R\cos(\theta - \alpha)$, where $R > 0$ and $0 < \alpha < \frac{1}{2}\pi$. Give the value of α correct to 4 decimal places. [3]

ii Hence

a solve the equation $4\cos\theta + 3\sin\theta = 2$ for $0 < \theta < 2\pi$, [4]

b find $\int \dfrac{50}{(4\cos\theta + 3\sin\theta)^2}\, d\theta$. [3]

Cambridge International A Level Mathematics 9709 Paper 31 Q9 June 2013

Chapter 9
Vectors

P3 This chapter is for Pure Mathematics 3 students only.

In this chapter you will learn how to:

- use standard notations for vectors, in two dimensions and three dimensions
- add and subtract vectors, multiply a vector by a scalar and interpret these operations geometrically
- calculate the magnitude of a vector and find and use unit vectors
- use displacement vectors and position vectors
- find the vector equation of a line
- find whether two lines are parallel, intersect or are skew
- find the common point of two intersecting lines
- find and use the scalar product of two vectors.

Cambridge International AS & A Level Mathematics: Pure Mathematics 2 & 3

PREREQUISITE KNOWLEDGE

Where it comes from	What you should be able to do	Check your skills
IGCSE / O Level Mathematics	Recall and apply: • Pythagoras' theorem • simple trigonometry to solve problems in two and three dimensions.	1 The diagram shows a cuboid with $CD = 5$ cm, $DH = 6$ cm and $EH = 3$ cm. Find the angle between AG and AH.
IGCSE / O Level Mathematics Pure Mathematics 1 Coursebook, Chapter 3	Show familiarity with the following areas of coordinate geometry in two dimensions: • Use and apply the general equation of a straight line in gradient/intercept form. • Calculate the coordinates of the midpoint of a line. • Find and interpret the gradients of parallel and perpendicular lines.	2 Three points A, B and C are such that $A(1, 7)$, $B(7, 3)$ and $C(0, -6)$. a Find the equation of the perpendicular bisector of the line AB. Give your answer in the form $y = mx + c$. b Write down the equation of the line parallel to AB that passes through the point C.
IGCSE / O Level Mathematics	Divide a quantity in a given ratio. Recall and apply properties of prisms, other solids and plane shapes, such as trapezia and parallelograms.	3 The diagram shows a cuboid with $CD = 5$ cm, $DH = 6$ cm and $EH = 3$ cm. The point M lies on DH such that $DM:MH = 1:2$. The point N is the midpoint of EF. a Write down the length of MH. b Find the **exact** length of NH.
IGCSE / O Level Mathematics	Translate points in 2-dimensional space using column vectors.	4 The point P has coordinates $(3, 9)$. The vector $\overrightarrow{PQ} = \begin{pmatrix} -1 \\ 3 \end{pmatrix}$. Write down the coordinates of the point Q.

Chapter 9: Vectors

Why are vectors important?

Vectors are used to model many situations in the world around us. They are used by engineers to ensure that buildings are designed safely. They are used in navigation to ensure that planes and boats, for example, arrive at their destinations in differing winds or water currents. There are countless examples of the use of vectors to model the movement of objects in mathematics and physics. Displacement, velocity and acceleration are all examples of vector quantities. Vectors are essential to describing the position and location of objects. A boat lost at sea can be found if its bearing and distance from each of two fixed points is known. These are just a few examples. Our world is 3-dimensional, as far as we know, and in this chapter we will be studying vectors in two and three dimensions. As the mathematics of vectors can be applied to all vectors, the study of vector geometry and algebra in general is important.

It is helpful if you have a good understanding of simple coordinate geometry before you start working with vectors because you will be able to make comparisons between the coordinate geometry that you have previously studied and the vector geometry introduced in this chapter.

Recall that a vector has both **magnitude** (size) and **direction**.

You should have some knowledge of how to work with vectors in 2-dimensional space from your previous studies. You will be reminded of this here, before being introduced to vectors in 3-dimensional space.

> **WEB LINK**
>
> Explore the *Vector geometry* station on the Underground Mathematics website.

9.1 Displacement or translation vectors

A movement through a certain distance and in a given direction is called a **displacement** or a **translation**. The magnitude or length of a displacement represents the distance moved.

Examples of displacement or translation vectors

- Sami walks 100 m in a straight line on a bearing of 045° from place A to place B. Jan is also standing at place A. Jan walks directly to Sami. Jan therefore also walks for 100 m on a bearing of 045° from place A. Jan follows the same displacement vector as Sami.
- The diagram shows two arrows.

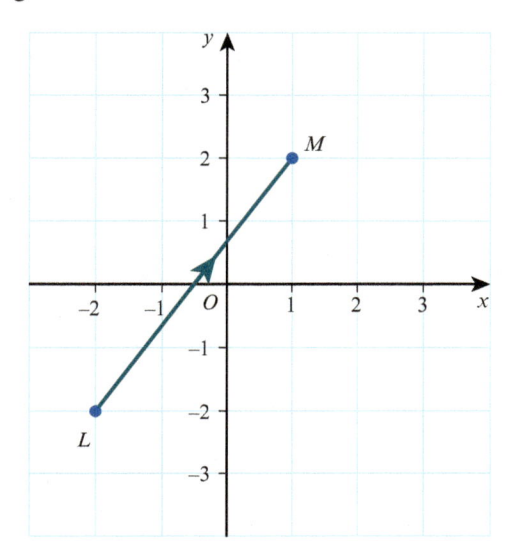

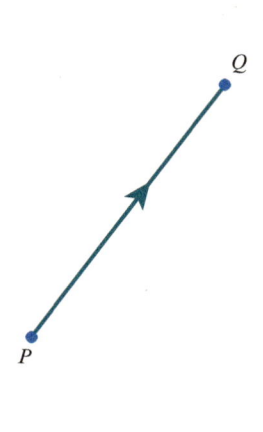

The arrow from L to M is the same length and has been drawn in the same direction as the arrow from P to Q.

This means that the points L and P have been displaced or translated by the same vector to M and Q, respectively.

The arrow LM and the arrow PQ represent the same displacement.

Notation using upper case

A displacement from L to M is written as $\overrightarrow{LM}$.

The order of the letters is important as $\overrightarrow{ML}$ would represent a displacement from M to L. The arrow above the upper case letters always goes from left to right. Reverse the letters to show the opposite direction.

In the previous diagram, $\overrightarrow{LM} = \overrightarrow{PQ}$. Even though the arrows are drawn on different parts of the page, the displacement shown by each is the same.

> **TIP**
> In some books **LM** is used for $\overrightarrow{LM}$.

> **TIP**
> When you write your vectors using the correct notation, you are less likely to make errors.

Components of a displacement in two dimensions

Any displacement can be represented by the displacement in the x-direction followed by the displacement in the y-direction. These are called the components of the displacement.

In the following example, the displacement vector $\overrightarrow{AB} = \begin{pmatrix} 3 \\ 1 \end{pmatrix}$.

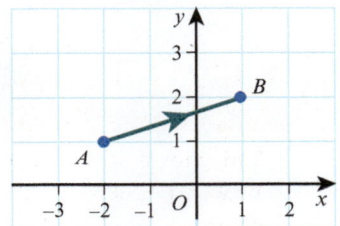

3 is the x-component and 1 is the y-component of the displacement. As 3 and 1 are both positive, the movement is to the right and up, respectively.

Components of a displacement in three dimensions

In a 3-dimensional system of coordinates, a point P is located using three coordinates, (x, y, z). The axis for the third coordinate, the z-coordinate, intersects with the x- and y-axes at the origin and the three axes are perpendicular to each other. The axes may be rotated, as shown.

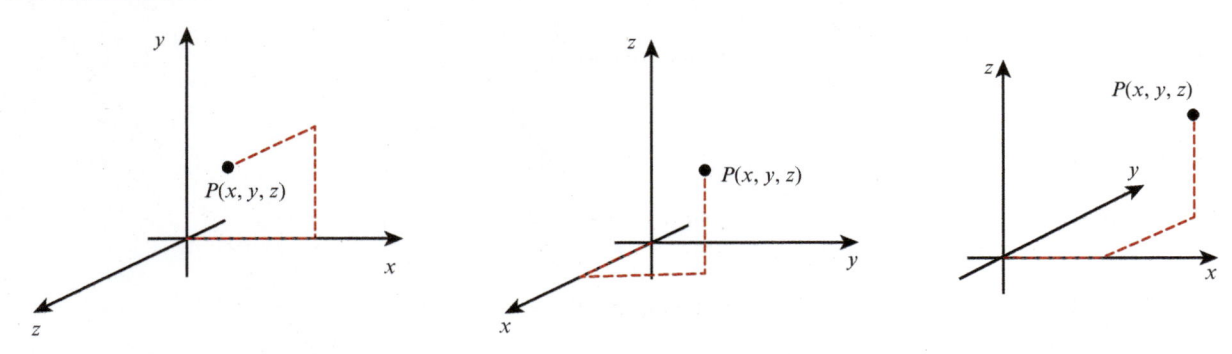

A displacement vector in a 3-dimensional system needs to have three components.

For example, the point $P(1, 3, 7)$ translates to the point $Q(5, 3, 8)$ by the displacement vector $\overrightarrow{PQ} = \begin{pmatrix} 4 \\ 0 \\ 1 \end{pmatrix}$.

Notation using lower case

Sometimes lower case letters are used for vectors. In text, these are in bold type. For example, **a** or **b**.

When written with a pen or pencil, they should be underlined with a bar or a short, wavy line instead. For example, a or b. Sometimes the bar or wavy line is written over the top of the lower case letter.

This diagram shows three vectors, **a**, **b** and **c**.

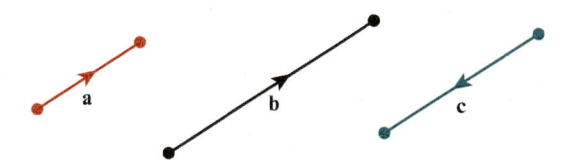

The vectors **a** and **b** have the same direction but not the same magnitude or length. Vectors **a** and **b** are parallel but not equal. In fact, **a** is half the length of **b**. When the length of **a** is doubled, the vectors are equal. In other words **b** = 2**a**. When **a** is scaled so that it is the same length as **b**, then the vectors are equal.

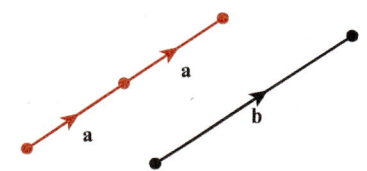

Vector **c** is $\frac{3}{4}$ of the length of vector **b**, but it has opposite direction.
In other words **c** = $-\frac{3}{4}$**b**.

A quantity (usually a number) used to scale a vector is called a **scalar**. So in '2**a**' the vector **a** is being multiplied by the scalar 2.

Parallel vectors

Two vectors are therefore **parallel** provided that one is a scalar multiple of the other. For example, when **a** = λ**b**, where λ is some scalar, then **a** and **b** are parallel.

Combining displacements (adding and subtracting)

Starting at point A, make the displacement $\overrightarrow{AB} = \begin{pmatrix} 4 \\ 2 \end{pmatrix}$ followed by the displacement $\overrightarrow{BC} = \begin{pmatrix} 3 \\ 7 \end{pmatrix}$.
What is the result?

The overall displacement shown by $\overrightarrow{AC}$, is $\begin{pmatrix} 7 \\ 9 \end{pmatrix}$. This is usually called the **resultant vector**.

- The x-component of the resultant vector is easily found by adding the x-components of $\overrightarrow{AB}$ and $\overrightarrow{BC}$.
- The y-component of the resultant vector is easily found by adding the y-components of $\overrightarrow{AB}$ and $\overrightarrow{BC}$.

> **TIP**
>
> This may also be written, for example, as 4**c** = −3**b**.
>
> Can you write an expression for **b** in terms of **c**?

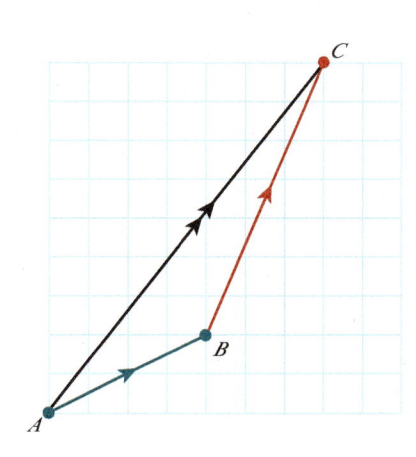

This operation is called **vector addition**.

> **KEY POINT 9.1**
>
> The formula for vector addition:
>
> $\overrightarrow{AB} + \overrightarrow{BC} = \overrightarrow{AC}$

The displacement $\overrightarrow{AC}$ followed by the displacement $\overrightarrow{CA}$ must return you to point A.

In other words $\overrightarrow{AC} + \overrightarrow{CA} = \begin{pmatrix} 0 \\ 0 \end{pmatrix}$, as there is no overall displacement.

$\begin{pmatrix} 7 \\ 9 \end{pmatrix} + \begin{pmatrix} -7 \\ -9 \end{pmatrix} = \begin{pmatrix} 0 \\ 0 \end{pmatrix}$, so $\overrightarrow{CA} = \begin{pmatrix} -7 \\ -9 \end{pmatrix}$.

As -7 and -9 are both negative, the movement is to the left and down, respectively.

Alternatively, we may write $\begin{pmatrix} 7 \\ 9 \end{pmatrix} - \begin{pmatrix} 7 \\ 9 \end{pmatrix} = \begin{pmatrix} 0 \\ 0 \end{pmatrix}$. In other words $\overrightarrow{AC} - \overrightarrow{AC} = \begin{pmatrix} 0 \\ 0 \end{pmatrix}$.

It is reasonable, therefore, to state that $\overrightarrow{CA} = -\overrightarrow{AC}$.

> **KEY POINT 9.2**
>
> $\overrightarrow{CA} = -\overrightarrow{AC}$ is always true.

Taking the negative of a vector changes the direction of the vector and does not change its length.

Vectors $\overrightarrow{AC}$ and $\overrightarrow{CA}$ have **opposite direction**.

As $\mathbf{a} - \mathbf{b}$ is the same as $\mathbf{a} + -\mathbf{b}$, **vector subtraction** is defined as the addition of the negative of a vector. The displacement $\mathbf{a} - \mathbf{b}$ is therefore the displacement $\mathbf{a}$ followed by the displacement $-\mathbf{b}$.

> **EXPLORE 9.1**
>
> $\mathbf{a}$, $\mathbf{b}$ and $\mathbf{c}$ are displacement vectors such that $\mathbf{c} = \mathbf{a} + \mathbf{b}$.
>
> Using diagrams to help you, show that $\mathbf{a} = \mathbf{c} - \mathbf{b}$ and that $\mathbf{a} + \mathbf{b} = \mathbf{b} + \mathbf{a}$.

> **KEY POINT 9.3**
>
> $\mathbf{a} + \mathbf{b} = \mathbf{b} + \mathbf{a}$ means that the addition of vectors is commutative.

Chapter 9: Vectors

WORKED EXAMPLE 9.1

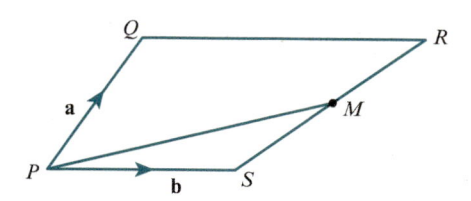

$PQRS$ is a quadrilateral. M is the midpoint of RS.
$\overrightarrow{PQ} = \mathbf{a}$, $\overrightarrow{PS} = \mathbf{b}$ and $\overrightarrow{PM} = \frac{1}{4}(2\mathbf{a} + 5\mathbf{b})$.

a Find an expression in terms of **a** and **b** for $\overrightarrow{SM}$.
   Give your answer in its simplest form.

b Find an expression for $\overrightarrow{QR}$.

c Describe quadrilateral $PQRS$, justifying your answer.

Answer

a $\overrightarrow{SM} = \overrightarrow{SP} + \overrightarrow{PM}$ .... Substitute $\overrightarrow{SP} = -\overrightarrow{PS}$.

$\overrightarrow{SM} = -\overrightarrow{PS} + \overrightarrow{PM}$ or $\overrightarrow{PM} - \overrightarrow{PS}$ .... Substitute given vectors in **a** and **b**.

$\overrightarrow{SM} = \frac{1}{4}(2\mathbf{a} + 5\mathbf{b}) - \mathbf{b}$ .... Expand and simplify.

$\overrightarrow{SM} = \frac{1}{2}\mathbf{a} + \frac{5}{4}\mathbf{b} - \mathbf{b} = \frac{1}{2}\mathbf{a} + \frac{1}{4}\mathbf{b}$

b $\overrightarrow{QR} = \overrightarrow{QP} + \overrightarrow{PS} + \overrightarrow{SR}$ .... Use $\overrightarrow{SR} = 2\overrightarrow{SM}$ and $\overrightarrow{QP} = -\overrightarrow{PQ}$.

$\overrightarrow{QR} = -\overrightarrow{PQ} + \overrightarrow{PS} + 2\overrightarrow{SM}$ .... Substitute known vectors in **a** and **b**.

$\overrightarrow{QR} = -\mathbf{a} + \mathbf{b} + 2\left(\frac{1}{2}\mathbf{a} + \frac{1}{4}\mathbf{b}\right)$ .... Expand and simplify.

$\overrightarrow{QR} = \frac{3}{2}\mathbf{b}$

c $\overrightarrow{QR} = \frac{3}{2}\overrightarrow{PS}$ so the lines QR and PS are parallel and $PQRS$ is a trapezium.

Magnitude of a vector

The magnitude of a vector is the length of the line segment representing the vector. This is found using the components of the displacement and Pythagoras' theorem.

Notation: The magnitude of a vector **p** is written as $|\mathbf{p}|$. This is also known as the modulus of the vector.

> **TIP**
> A diagram can often be useful in vector questions.

Cambridge International AS & A Level Mathematics: Pure Mathematics 2 & 3

WORKED EXAMPLE 9.2

a Find the magnitude of each of the vectors $\mathbf{a} = \begin{pmatrix} 4 \\ 3 \end{pmatrix}$, $\mathbf{b} = \begin{pmatrix} -3 \\ 4 \end{pmatrix}$ and $\mathbf{a} - \mathbf{b}$.

b $\mathbf{p} = \begin{pmatrix} 5 \\ 0 \\ 8 \end{pmatrix}$ and $\mathbf{q} = \begin{pmatrix} -1 \\ 8 \\ -3 \end{pmatrix}$. Find the magnitude of the vector $\mathbf{p} + \mathbf{q}$.

Answer

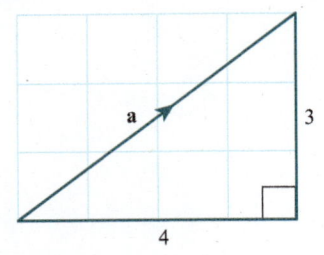

a

Draw a diagram showing the x- and y-components.

$|\mathbf{a}| = \sqrt{4^2 + 3^2} = \sqrt{25} = 5$

Apply Pythagoras' theorem.

Again, draw a diagram to show the x- and y-components.

$|\mathbf{b}| = \sqrt{(-3)^2 + 4^2} = \sqrt{25} = 5$

Apply Pythagoras' theorem.

The diagram has been given to help you to see the components of the vector $\mathbf{a} - \mathbf{b}$.

$|\mathbf{a} - \mathbf{b}| = \left| \begin{pmatrix} 7 \\ -1 \end{pmatrix} \right| = \sqrt{7^2 + (-1)^2} = \sqrt{49 + 1} = \sqrt{50} = 5\sqrt{2}$

Apply Pythagoras' theorem. (Notice that $|\mathbf{a} - \mathbf{b}|$ is **not** the same as $|\mathbf{a}| - |\mathbf{b}|$.)

b $\mathbf{p} + \mathbf{q} = \begin{pmatrix} 5 \\ 0 \\ 8 \end{pmatrix} + \begin{pmatrix} -1 \\ 8 \\ -3 \end{pmatrix} = \begin{pmatrix} 4 \\ 8 \\ 5 \end{pmatrix}$ and

It is not a good idea to try to draw a diagram here.

$\left| \begin{pmatrix} 4 \\ 8 \\ 5 \end{pmatrix} \right| = \sqrt{4^2 + 8^2 + 5^2} = \sqrt{105}$

Apply Pythagoras' theorem in 3 dimensions.

Chapter 9: Vectors

Unit vectors

A **unit vector** is any vector of magnitude or length 1.

Notation: $\hat{\mathbf{a}}$ (read as **a** hat) is the unit vector in the direction of **a**.

To find a unit displacement vector in a particular direction:

- find the magnitude of the displacement
- then divide the components of the vector by the magnitude.

This means that, for a vector $\mathbf{p} = \begin{pmatrix} p_1 \\ p_2 \\ p_3 \end{pmatrix}$, the corresponding unit vector is $\hat{\mathbf{p}} = \begin{pmatrix} \frac{p_1}{|\mathbf{p}|} \\ \frac{p_2}{|\mathbf{p}|} \\ \frac{p_3}{|\mathbf{p}|} \end{pmatrix}$.

WORKED EXAMPLE 9.3

Find the unit vector in the direction $\begin{pmatrix} \sqrt{5} \\ 2 \end{pmatrix}$.

Answer

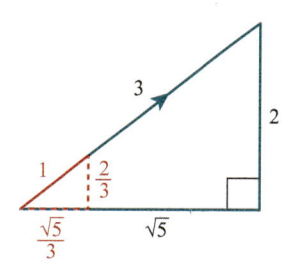

The diagram shows that, geometrically, we are finding lengths of similar triangles when finding components of unit vectors.

$\left| \begin{pmatrix} \sqrt{5} \\ 2 \end{pmatrix} \right| = \sqrt{(\sqrt{5})^2 + 2^2} = \sqrt{9} = 3$

Apply Pythagoras' theorem to find the magnitude of the given vector.

Unit vector in the direction $\begin{pmatrix} \sqrt{5} \\ 2 \end{pmatrix}$ is $\begin{pmatrix} \frac{\sqrt{5}}{3} \\ \frac{2}{3} \end{pmatrix}$.

Divide each component by the magnitude.

WORKED EXAMPLE 9.4

Points A, B and C are such that $\overrightarrow{AB} = \begin{pmatrix} 3 \\ 4 \\ 1 \end{pmatrix}$ and $\overrightarrow{BC} = \begin{pmatrix} 11 \\ -2 \\ -1 \end{pmatrix}$.

Find the unit displacement vector in the direction $\overrightarrow{AC}$.

Answer

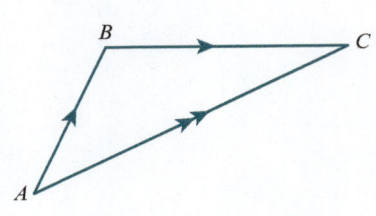

$\overrightarrow{AC} = \overrightarrow{AB} + \overrightarrow{BC}$ .... $\overrightarrow{AC}$ is the resultant vector formed from $\overrightarrow{AB} + \overrightarrow{BC}$.

$\overrightarrow{AC} = \begin{pmatrix} 3 \\ 4 \\ 1 \end{pmatrix} + \begin{pmatrix} 11 \\ -2 \\ -1 \end{pmatrix} = \begin{pmatrix} 14 \\ 2 \\ 0 \end{pmatrix}$ .... Substitute column vectors and sum components.

$|\overrightarrow{AC}| = \sqrt{14^2 + 2^2 + 0^2} = \sqrt{200} = 10\sqrt{2}$ .... Now find the magnitude by applying Pythagoras' theorem.

Unit displacement vector in the direction of $\overrightarrow{AC}$ is .... Divide each component by the magnitude of $\overrightarrow{AC}$.

$\dfrac{1}{10\sqrt{2}}\begin{pmatrix} 14 \\ 2 \\ 0 \end{pmatrix}$.

EXERCISE 9A

1 The diagram shows the vectors $\overrightarrow{AB}$ and $\overrightarrow{BC}$ on 1 centimetre squared paper.

   Using 1 cm as 1 unit of displacement:

   a write $\overrightarrow{AB}$ and $\overrightarrow{BC}$ as column vectors

   b write down the column vector $\overrightarrow{AC}$.

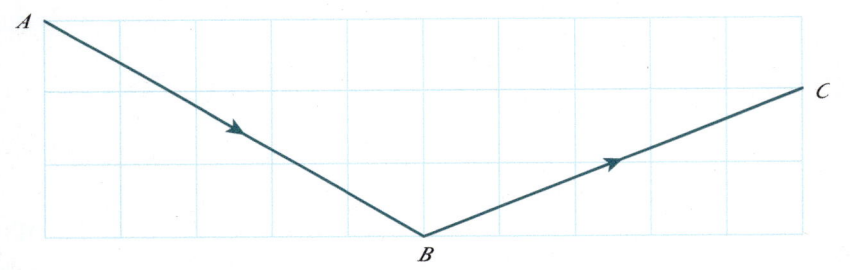

2 The diagram shows the vectors $\overrightarrow{DE}$ and $\overrightarrow{DF}$ on 1 centimetre squared paper.

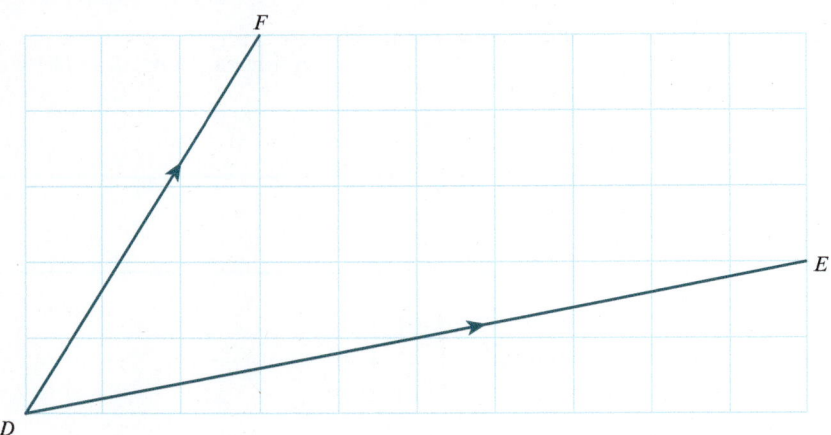

   Using 1 cm as 1 unit of displacement:

   a find the column vector $\overrightarrow{EF}$

   b show that $\overrightarrow{DF} - \overrightarrow{DE} = \overrightarrow{EF}$.

3

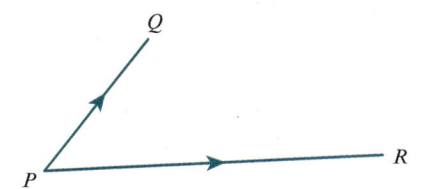

Justify the statement $\vec{QR} = \vec{PR} - \vec{PQ}$.

P 4 In triangle ABC, X and Y are the midpoints of AB and AC, respectively.

$\vec{AX} = \mathbf{a}$ and $\vec{AY} = \mathbf{b}$.

  a Use a vector method to prove that BC is parallel to XY.

  b $|\vec{XY}| = k|\vec{BC}|$

  Write down the value of the constant k.

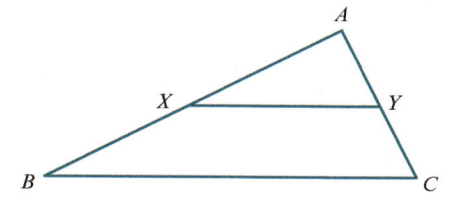

PS 5 The diagram shows a cuboid, $ABCDEFGH$.

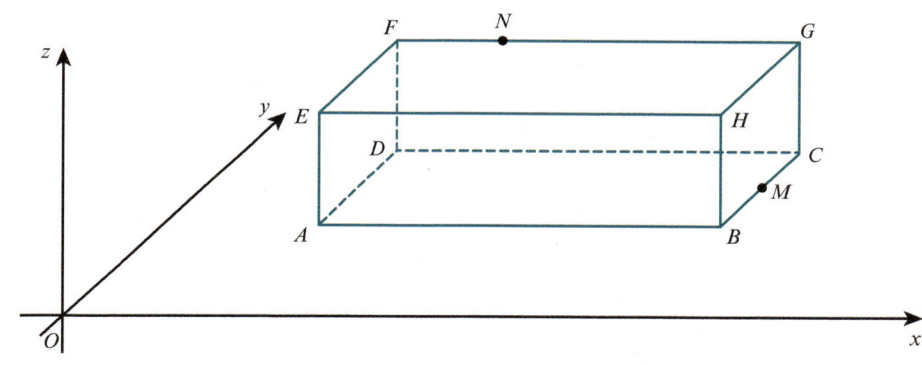

M is the midpoint of BC and the point N is on FG such that $FN:NG$ is $1:3$.

Given that $\vec{AG} = \begin{pmatrix} 12 \\ 4 \\ 2 \end{pmatrix}$, find the displacement vector:

  a $\vec{AM}$ \hspace{2cm} b $\vec{AN}$.

6 $ABCDEFGH$ is a regular octagon. $\vec{AB} = \mathbf{p}$, $\vec{AH} = \mathbf{q}$, $\vec{HG} = \mathbf{r}$ and $\vec{HC} = \mathbf{s}$.

  a Using the vectors $\mathbf{p}$, $\mathbf{q}$, $\mathbf{r}$ and $\mathbf{s}$, write down expressions for:

  i $\vec{BC}$ \hspace{2cm} ii $\vec{EB}$.

  b Explain why $\mathbf{s} = k\mathbf{p}$ and find the exact value of k.

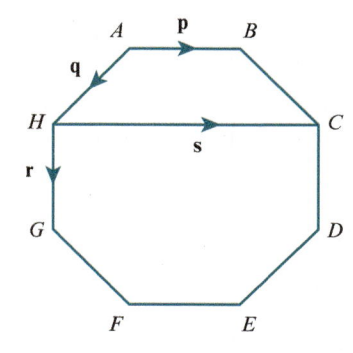

P 7 Use a vector method to show that, when the diagonals of a quadrilateral bisect one another, then the opposite sides are parallel and equal in length.

9.2 Position vectors

A **position vector** is simply a means of locating a point in space relative to an origin, usually O. This means a position vector is fixed in space. Position vectors are sometimes called location vectors.

Up to this point, the displacement vectors we have been working with have been free vectors. This means they are **not** fixed to any particular starting point or origin. Many free vectors can represent the same displacement.

Position vectors are different because they start at a given point or origin. This fixed starting point means there is only one arrow that represents each vector.

Since we can apply the same mathematics to all vectors, the results derived for displacement vectors (free vectors) also apply to position vectors.

> **REWIND**
>
> Look back at vectors $\overrightarrow{LM}$ and $\overrightarrow{PQ}$ at the start of this chapter.

The fixed starting point is often called O, as in the origin of a system of Cartesian coordinates.

So a point with coordinates $A(1, 2, 3)$ can be represented by the position vector $\overrightarrow{OA} = \begin{pmatrix} 1 \\ 2 \\ 3 \end{pmatrix}$.

Position vector of any point in a plane

Consider this case in 2-dimensional space.

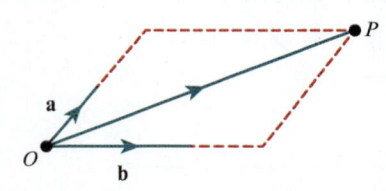

Given that **a** and **b** are position vectors relative to the origin O and that P is any other point in the same plane as **a** and **b**, then the position vector of P can be written using **a** and **b**.

This is because $\overrightarrow{OP}$ is some scalar multiple of **a** plus some scalar multiple of **b**:

$\overrightarrow{OP} = m\mathbf{a} + n\mathbf{b}$, where m and n are scalars.

All the other vectors in this plane can be written using combinations of the vectors **a** and **b**.

Cartesian components

When the vectors **a** and **b** are

- perpendicular to each other

and

- unit vectors

the plane looks like the usual set of x- and y-axes (called the Oxy plane) we use for Cartesian coordinates.

In this case, the unit vector in the x-direction is called **i** and the unit vector in the y-direction is called **j**.

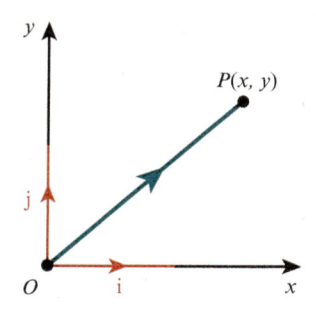

$\overrightarrow{OP} = x\mathbf{i} + y\mathbf{j}$, where x and y are scalars.

In this case, $\overrightarrow{OP}$ is some scalar multiple of $\mathbf{i}$ plus some scalar multiple of $\mathbf{j}$:

The point P with coordinates $(4, 5)$ has position vector $\begin{pmatrix} 4 \\ 5 \end{pmatrix}$, which can also be written as $4\mathbf{i} + 5\mathbf{j}$.

This can be extended to three dimensions.

The unit vector in the z-direction is called $\mathbf{k}$.

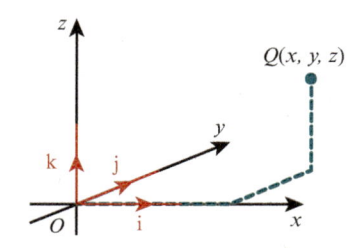

In this case, $\overrightarrow{OQ}$ is some scalar multiple of $\mathbf{i}$ plus some scalar multiple of $\mathbf{j}$ plus some scalar multiple of $\mathbf{k}$:

$$\overrightarrow{OQ} = x\mathbf{i} + y\mathbf{j} + z\mathbf{k}, \text{ where } x, y \text{ and } z \text{ are scalars.}$$

The point Q with coordinates $(4, 2, 3)$ has position vector $\begin{pmatrix} 4 \\ 2 \\ 3 \end{pmatrix}$, which can also be written as $4\mathbf{i} + 2\mathbf{j} + 3\mathbf{k}$.

WORKED EXAMPLE 9.5

The position vectors of the points A, B and C are given by $\mathbf{a}$, $\mathbf{b}$ and $\mathbf{c}$, respectively, where

$$\mathbf{a} = \begin{pmatrix} 2 \\ 5 \\ -3 \end{pmatrix}, \mathbf{b} = \begin{pmatrix} 0 \\ 4 \\ 6 \end{pmatrix} \text{ and } \mathbf{c} = \begin{pmatrix} -4 \\ 2 \\ 24 \end{pmatrix}.$$

a Write $\mathbf{a}$, $\mathbf{b}$ and $\mathbf{c}$ in terms of $\mathbf{i}$, $\mathbf{j}$ and $\mathbf{k}$.

b Giving your answers in terms of $\mathbf{i}$, $\mathbf{j}$ and $\mathbf{k}$, find the vectors representing:

　i $\overrightarrow{AB}$　　　　　　　　　　ii $\overrightarrow{AC}$

c Describe the relationship between the points A, B and C, justifying your answer.

Answer

a $\mathbf{a} = 2\mathbf{i} + 5\mathbf{j} - 3\mathbf{k}$, $\mathbf{b} = 4\mathbf{j} + 6\mathbf{k}$ and $\mathbf{c} = -4\mathbf{i} + 2\mathbf{j} + 24\mathbf{k}$.

b **i** A simple sketch should help you to calculate the answer correctly.

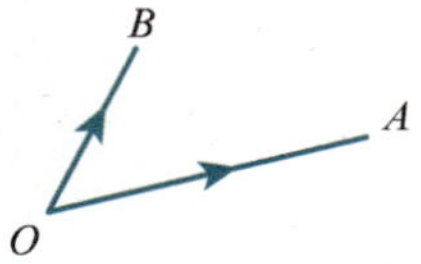

$\overrightarrow{AB} = \overrightarrow{AO} + \overrightarrow{OB} = \overrightarrow{OB} - \overrightarrow{OA}$ .... Substitute the Cartesian form.

$\overrightarrow{AB} = \mathbf{b} - \mathbf{a} = 4\mathbf{j} + 6\mathbf{k} - (2\mathbf{i} + 5\mathbf{j} - 3\mathbf{k})$ .... Expand and simplify.

$\overrightarrow{AB} = -2\mathbf{i} - \mathbf{j} + 9\mathbf{k}$

ii $\overrightarrow{AC} = \overrightarrow{AO} + \overrightarrow{OC} = \overrightarrow{OC} - \overrightarrow{OA}$ .... Substitute the Cartesian form.

$\overrightarrow{AC} = \mathbf{c} - \mathbf{a} = -4\mathbf{i} + 2\mathbf{j} + 24\mathbf{k} - (2\mathbf{i} + 5\mathbf{j} - 3\mathbf{k})$ ...... Expand and simplify.

$\overrightarrow{AC} = -6\mathbf{i} - 3\mathbf{j} + 27\mathbf{k}$

c $\overrightarrow{AC} = 3\overrightarrow{AB}$ and so these vectors are parallel. Since the vectors also have point A in common, then A, B and C must be **collinear** (so they lie on the same straight line).

WORKED EXAMPLE 9.6

The diagram shows a cube $OABCDEFG$ with sides of length $4\,\text{cm}$. Unit vectors $\mathbf{i}$, $\mathbf{j}$ and $\mathbf{k}$ are parallel to OC, OA and OD, respectively. The point M lies on FB such that $FM = 1\,\text{cm}$, N is the midpoint of OC and P is the centre of the square face $DEFG$.

Express each of the vectors $\overrightarrow{NP}$ and $\overrightarrow{MP}$ in terms of $\mathbf{i}$, $\mathbf{j}$ and $\mathbf{k}$.

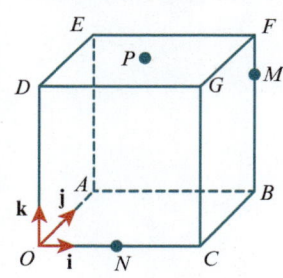

The cube is of side $4\,\text{cm}$ and P' is the centre of the square $AOCB$. $\overrightarrow{NP}$ is parallel to the y-direction and so the first part of the vector $\overrightarrow{NP}$ is $2\mathbf{j}$.

P is directly above this point P' and so the second component $P'P$ is parallel to the z-direction and is $+4\mathbf{k}$.

Answer

Using the displacement of P from N and M:

$\overrightarrow{NP} = 2\mathbf{j} + 4\mathbf{k}$

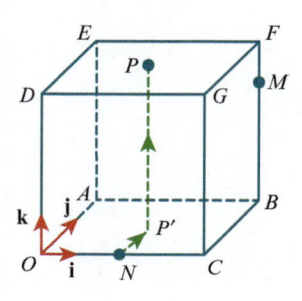

$\overrightarrow{MP} = -2\mathbf{i} - 2\mathbf{j} + \mathbf{k}$

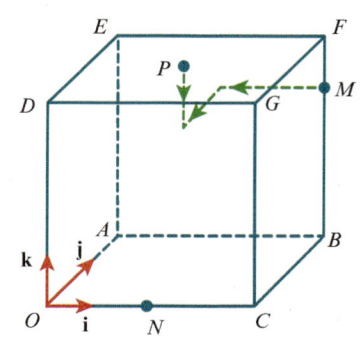

Starting with the x-direction, move $2\mathbf{i}$ left. This is the opposite to the positive direction indicated by the red arrow, so the first component is $-2\mathbf{i}$.

In the y-direction, again we are moving in the negative direction, so the second component is $-2\mathbf{j}$.

In the z-direction, we must move a distance equal to MF. This is 1 unit and so the third component is $+\mathbf{k}$.

Or we can use position vectors:

$\overrightarrow{NP} = \overrightarrow{OP} - \overrightarrow{ON}$

$\overrightarrow{OP}$ is 2 units in x-direction, 2 units in y-direction and 4 units in the z-direction and $\overrightarrow{ON}$ is 2 units in x-direction.

$\overrightarrow{OP} = 2\mathbf{i} + 2\mathbf{j} + 4\mathbf{k}$ and $\overrightarrow{ON} = 2\mathbf{i}$

Subtract vectors.

$\overrightarrow{NP} = \overrightarrow{OP} - \overrightarrow{ON} = 2\mathbf{i} + 2\mathbf{j} + 4\mathbf{k} - 2\mathbf{i}$

Simplify.

$\overrightarrow{NP} = 2\mathbf{j} + 4\mathbf{k}$

$\overrightarrow{MP} = \overrightarrow{OP} - \overrightarrow{OM}$

$\overrightarrow{OM}$ is 4 units in x-direction, 4 units in y-direction and 3 units in z-direction.

$\overrightarrow{OP} = 2\mathbf{i} + 2\mathbf{j} + 4\mathbf{k}$ and $\overrightarrow{OM} = 4\mathbf{i} + 4\mathbf{j} + 3\mathbf{k}$

Subtract vectors.

$\overrightarrow{MP} = \overrightarrow{OP} - \overrightarrow{OM} = 2\mathbf{i} + 2\mathbf{j} + 4\mathbf{k} - (4\mathbf{i} + 4\mathbf{j} + 3\mathbf{k})$

Expand and simplify.

$\overrightarrow{MP} = -2\mathbf{i} - 2\mathbf{j} + \mathbf{k}$

EXERCISE 9B

1 Relative to an origin O, the position vectors of the points A and B are given by $\overrightarrow{OA} = 5\mathbf{i} + 3\mathbf{j}$ and $\overrightarrow{OB} = 5\mathbf{i} + \lambda\mathbf{j} + \mu\mathbf{k}$.

   a In the case where $\lambda = 0$ and $\mu = 7$, find the unit vector in the direction of $\overrightarrow{AB}$.

   b Given that triangle OAB is right-angled with area $5\sqrt{34}$ and $\mu = 10$, find the value of λ.

2 The diagram shows a cuboid $OABCDEFG$ with a horizontal base $OABC$. The cuboid has a length OA of 8 cm, a width AB of 4 cm and a height BF of d cm. The point M is the midpoint of CE and the point N is the point on GF such that $GN = 6$ cm. The unit vectors $\mathbf{i}$, $\mathbf{j}$ and $\mathbf{k}$ are parallel to OA, OC and OD, respectively. It is given that $\overrightarrow{OM} = 4\mathbf{i} + 2\mathbf{j} + \mathbf{k}$.

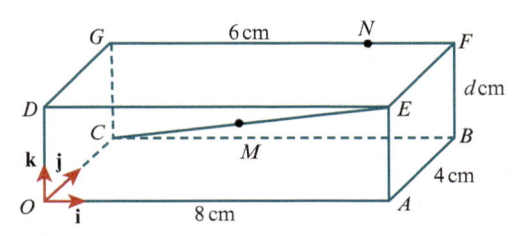

   a Write down the height of the cuboid.

   b Find $\overrightarrow{ON}$.

   PS c Find the unit vector in the direction $\overrightarrow{MN}$.

3 The position vectors of the points A, B, C and D are given by **a**, **b**, **c** and **d**, respectively, where

$$\mathbf{a} = \begin{pmatrix} 2 \\ 2 \\ 0 \end{pmatrix}, \mathbf{b} = \begin{pmatrix} 13 \\ 5 \\ 4 \end{pmatrix}, \mathbf{c} = \begin{pmatrix} 5 \\ -3 \\ 4 \end{pmatrix} \text{ and } \mathbf{d} = \begin{pmatrix} -6 \\ -6 \\ 0 \end{pmatrix}.$$

- **a**
  i Find $|\overrightarrow{AD}|, |\overrightarrow{BC}|, |\overrightarrow{AB}|$ and $|\overrightarrow{DC}|$.
  ii Deduce that $ABCD$ is a parallelogram.
- **b** Find the coordinates of
  - **i** the point M, the midpoint of the line AB
  - **ii** the point P such that $|\overrightarrow{BP}| = \frac{1}{3}|\overrightarrow{BD}|$.

4 Relative to an origin O, the position vector of A is $\mathbf{i} - 2\mathbf{j} + 5\mathbf{k}$ and the position vector of B is $3\mathbf{i} + 4\mathbf{j} + \mathbf{k}$.
- **a** Find the magnitude of $\overrightarrow{AB}$.
- **b** Use Pythagoras' theorem to show that OAB is a right-angled triangle.
- **c** Find the exact area of triangle OAB.

PS 5 The position vectors of the points A, B and C are given by

$$\mathbf{a} = 5\mathbf{i} + \mathbf{j} + 2\mathbf{k}, \ \mathbf{b} = 7\mathbf{i} + 4\mathbf{j} - \mathbf{k} \text{ and } \mathbf{c} = \mathbf{i} + \mathbf{j} + q\mathbf{k}.$$

Given that the length of AB is equal to the length of AC, find the exact possible values of the constant q.

6 The diagram shows a triangular prism, $OABCDE$. The uniform cross-section of the prism, OAE, is a right-angled triangle with base 24 cm and height 7 cm. The length, AB, of the prism is 20 cm. The unit vectors **i**, **j** and **k** are parallel to $\overrightarrow{OA}, \overrightarrow{OC}$ and $\overrightarrow{OE}$, respectively. The point N divides the length of the line DB in the ratio $2:3$.
- **a** Find the magnitude of $\overrightarrow{DB}$.
- **b** Find $\overrightarrow{ON}$.

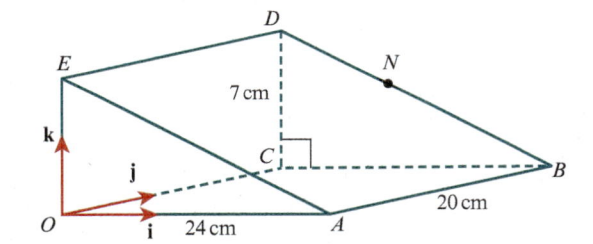

PS 7 Three points O, P and Q are such that $\overrightarrow{OP} = 2\mathbf{i} + 5\mathbf{j} + a\mathbf{k}$ and $\overrightarrow{OQ} = \mathbf{i} + (1+a)\mathbf{j} - 3\mathbf{k}$.

Given that the magnitude of $\overrightarrow{OP}$ is equal to the magnitude of $\overrightarrow{OQ}$, find the value of the constant a.

PS 8 Relative to an origin O, the position vectors of the points P and Q are given by

$$\overrightarrow{OP} = \begin{pmatrix} -6k \\ -2 \\ 8(1+k) \end{pmatrix} \text{ and } \overrightarrow{OQ} = \begin{pmatrix} 2k+13 \\ -8 \\ -32k \end{pmatrix}.$$

Given that OPQ is a straight line:
- **a** find the value of the constant k
- **b** write each of $\overrightarrow{OP}$ and $\overrightarrow{OQ}$ in the form $x\mathbf{i} + y\mathbf{j} + z\mathbf{k}$
- **c** find the magnitude of the vector $\overrightarrow{PQ}$.

9 An ant has an in-built vector navigation system! It is able to go on complex journeys to find food and then return directly home (to its starting point) by the shortest route.

An ant leaves home and its journey is represented by the following list of displacements:

$$\mathbf{i} + 3\mathbf{j} + 6\mathbf{k} \qquad \begin{pmatrix} 1 \\ 6 \\ 0 \end{pmatrix} \qquad 10\mathbf{i} \qquad \begin{pmatrix} -3 \\ -1 \\ -5 \end{pmatrix} \qquad 4\mathbf{i} - \mathbf{k}$$

The ant finds food at this point.

What single displacement takes the ant home?

One unit of displacement is 10 cm.

What is the shortest distance home from the point where the ant finds food?

EXPLORE 9.2

A system of objects with the properties:

- can be represented by a column of numbers
- can be combined using addition
- can be multiplied by a number to give another object in the same system

is called a **vector space**.

Do quadratic expressions of the form $ax^2 + bx + c$, where a, b and c are real constants, form a vector space? Explain your answer.

9.3 The scalar product

We have seen how to add and subtract vectors and also how to multiply a vector by a scalar. It is also possible to multiply two vectors. It is possible to find two different products by multiplying a pair of vectors. The first product gives an answer that is a vector. This is called the vector product or cross product (for example, the cross product of the vectors **a** and **b** is **a** × **b**). The cross product is not required for Pure Mathematics 3. The second product gives an answer that is a scalar. This is called the **scalar product** or **dot product** (the dot product of the vectors **a** and **b**, for example, is **a** · **b**).

Introductory ideas

The angle between two vectors is defined to be the angle made

- either when the direction of each vector is away from a point
- or when the direction of each vector is towards a point.

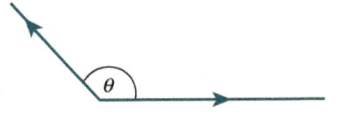

Directions both away from a point.   Directions both towards a point.

As vectors have magnitude **and** direction we can think of the scalar product as being multiplication in relation to a particular direction.

A useful way to understand this is to think about the work done by a person when they pull an object along a flat surface using a rope:

work done by the person = force needed to pull the object × distance the object moves

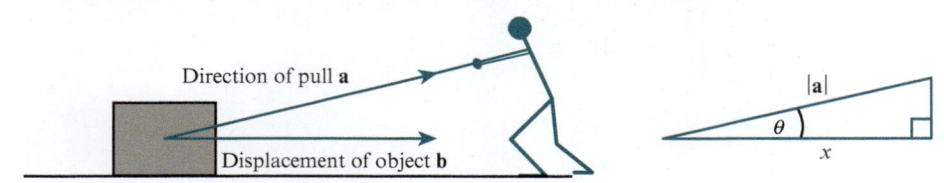

The diagram shows vector **a** representing the pull and vector **b** representing the displacement of the object.

The direction of the pull is not in the same as the direction of the displacement. This means that only some of the force the person uses to pull the object makes it move. In fact, only the x-component of the pull is in the direction of the displacement.

$|\mathbf{a}|$ is the magnitude of the pull. Using the angle made between the two directions, we can see that the magnitude of the x-component of the pull is $x = |\mathbf{a}| \cos \theta$.

$|\mathbf{b}|$ is the magnitude of the displacement, which is a measure of the distance moved.

Applying this information to the formula, we can see that the

work done by the person $= \bigl(|\mathbf{a}| \cos \theta \bigr) |\mathbf{b}|$.

Finding the scalar (number) that is work done by the person is one application of the scalar product.

KEY POINT 9.4

In general, the geometric definition of the scalar product of two vectors is:

$\mathbf{a} \cdot \mathbf{b} = |\mathbf{a}||\mathbf{b}| \cos \theta$

This formula allows us to calculate the value that represents the impact of one vector on the other, taking account of direction.

WORKED EXAMPLE 9.7

$\mathbf{p} = \begin{pmatrix} 8 \\ -6 \\ -10 \end{pmatrix}$, $\mathbf{q} = \begin{pmatrix} 4 \\ -3 \\ 0 \end{pmatrix}$ and the angle, θ, between the two vectors is 45°. Find the value of the scalar product, $\mathbf{p} \cdot \mathbf{q}$.

Answer

$\mathbf{p} \cdot \mathbf{q} = |\mathbf{p}||\mathbf{q}| \cos \theta$

$|\mathbf{p}| = \sqrt{8^2 + (-6)^2 + (-10)^2} = 10\sqrt{2}$ and $|\mathbf{q}| = \sqrt{4^2 + (-3)^2 + 0^2} = 5$

$\mathbf{p} \cdot \mathbf{q} = \bigl(10\sqrt{2}\bigr)(5) \cos 45° = \bigl(10\sqrt{2}\bigr)(5)\left(\dfrac{\sqrt{2}}{2}\right) = 50$

Special case 1: Parallel vectors

When two parallel vectors, **a** and **b**, have the same direction (not opposite directions), $\theta = 0°$ or 0 radians.

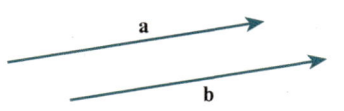

Therefore $\quad\quad\quad\quad\quad\quad \mathbf{a}\cdot\mathbf{b} = |\mathbf{a}||\mathbf{b}|\cos 0$

$\quad\quad\quad\quad\quad\quad\quad\quad\quad\quad \mathbf{a}\cdot\mathbf{b} = |\mathbf{a}||\mathbf{b}|$

This also means that, for example: $\mathbf{a}\cdot\mathbf{a} = |\mathbf{a}||\mathbf{a}|$ or $|\mathbf{a}|^2$

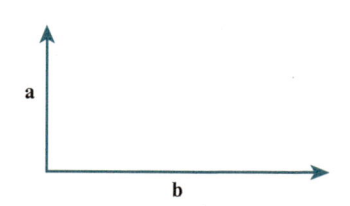

Special case 2: Perpendicular vectors

When two vectors, **a** and **b**, are perpendicular, $\theta = 90°$ or $\dfrac{\pi}{2}$ radians.

$\mathbf{a}\cdot\mathbf{b} = |\mathbf{a}||\mathbf{b}|\cos 90°$

$\mathbf{a}\cdot\mathbf{b} = 0$

> **KEY POINT 9.5**
>
> This is very important. When two vectors are perpendicular, their scalar product is zero. It is also true that when the scalar product is zero, the two vectors are perpendicular.

> **EXPLORE 9.3**
>
> Show that the scalar product has the following properties:
>
> $\mathbf{a}\cdot\mathbf{b} = \mathbf{b}\cdot\mathbf{a}$
>
> $\mathbf{a}\cdot m\mathbf{b} = m\mathbf{a}\cdot\mathbf{b} = m(\mathbf{a}\cdot\mathbf{b})$, where m is a scalar
>
> $\mathbf{a}\cdot(\mathbf{b}+\mathbf{c}) = \mathbf{a}\cdot\mathbf{b} + \mathbf{a}\cdot\mathbf{c}$
>
> $(\mathbf{a}+\mathbf{b})\cdot(\mathbf{c}+\mathbf{d}) = \mathbf{a}\cdot\mathbf{c} + \mathbf{a}\cdot\mathbf{d} + \mathbf{b}\cdot\mathbf{c} + \mathbf{b}\cdot\mathbf{d}$

Scalar product for component form

When we write $\mathbf{a} = a_1\mathbf{i} + a_2\mathbf{j} + a_3\mathbf{k}$ and $\mathbf{b} = b_1\mathbf{i} + b_2\mathbf{j} + b_3\mathbf{k}$ the properties of the scalar product give an algebraic method of finding the same value that we found using $|\mathbf{a}||\mathbf{b}|\cos\theta$.

$\mathbf{a}\cdot\mathbf{b} = (a_1\mathbf{i} + a_2\mathbf{j} + a_3\mathbf{k})(b_1\mathbf{i} + b_2\mathbf{j} + b_3\mathbf{k})$

$\quad\quad = a_1\mathbf{i}(b_1\mathbf{i} + b_2\mathbf{j} + b_3\mathbf{k}) + a_2\mathbf{j}(b_1\mathbf{i} + b_2\mathbf{j} + b_3\mathbf{k}) + a_3\mathbf{k}(b_1\mathbf{i} + b_2\mathbf{j} + b_3\mathbf{k})$

$\quad\quad = a_1 b_1 \mathbf{i}\cdot\mathbf{i} + a_1 b_2 \mathbf{i}\cdot\mathbf{j} + a_1 b_3 \mathbf{i}\cdot\mathbf{k}$

$\quad\quad\quad + a_2 b_1 \mathbf{j}\cdot\mathbf{i} + a_2 b_2 \mathbf{j}\cdot\mathbf{j} + a_2 b_3 \mathbf{j}\cdot\mathbf{k}$

$\quad\quad\quad\quad + a_3 b_1 \mathbf{k}\cdot\mathbf{i} + a_3 b_2 \mathbf{k}\cdot\mathbf{j} + a_3 b_3 \mathbf{k}\cdot\mathbf{k}$

However, as **i** is parallel to **i**, **j** is parallel to **j** and **k** is parallel to **k** then:

$\mathbf{i}\cdot\mathbf{i} = \mathbf{j}\cdot\mathbf{j} = \mathbf{k}\cdot\mathbf{k} = 1$

Also, as **i**, **j** and **k** are perpendicular to each other, scalar products such as $\mathbf{i}\cdot\mathbf{j}$, $\mathbf{j}\cdot\mathbf{k}$, etc. are all 0.

$\mathbf{a}\cdot\mathbf{b} = a_1 b_1 \mathbf{i}\cdot\mathbf{i} + \cancel{a_1 b_2 \mathbf{i}\cdot\mathbf{j}} + \cancel{a_1 b_3 \mathbf{i}\cdot\mathbf{k}}$

$\quad\quad\quad + \cancel{a_2 b_1 \mathbf{j}\cdot\mathbf{i}} + a_2 b_2 \mathbf{j}\cdot\mathbf{j} + \cancel{a_2 b_3 \mathbf{j}\cdot\mathbf{k}}$

$\quad\quad\quad\quad + \cancel{a_3 b_1 \mathbf{k}\cdot\mathbf{i}} + \cancel{a_3 b_2 \mathbf{k}\cdot\mathbf{j}} + a_3 b_3 \mathbf{k}\cdot\mathbf{k}$

KEY POINT 9.6

In general, the algebraic definition of the scalar product of two vectors is:

$\mathbf{a} \cdot \mathbf{b} = a_1b_1 + a_2b_2 + a_3b_3$

When written in column form, $\mathbf{a} = \begin{pmatrix} a_1 \\ a_2 \\ a_3 \end{pmatrix}$ and $\mathbf{b} = \begin{pmatrix} b_1 \\ b_2 \\ b_3 \end{pmatrix}$, so it is perfectly valid to use the components of column vectors in the algebraic form of the scalar product $\mathbf{a} \cdot \mathbf{b} = a_1b_1 + a_2b_2 + a_3b_3$.

We now have a way to find the dot product when we do not know the angle between the two vectors.

Sometimes the two versions of the scalar product are written together like this:

$\mathbf{a} \cdot \mathbf{b} = a_1b_1 + a_2b_2 + a_3b_3 = |\mathbf{a}||\mathbf{b}|\cos\theta$

This can be useful as it allows us to find the angle between the two vectors when we know the components of the vectors.

WORKED EXAMPLE 9.8

$\mathbf{p} = 8\mathbf{i} - 6\mathbf{j} - 10\mathbf{k}$ and $\mathbf{q} = 4\mathbf{i} - 3\mathbf{j}$

Show that the angle, θ, between the two vectors is 45°.

Answer

$\mathbf{p} \cdot \mathbf{q} = p_1q_1 + p_2q_2 + p_3q_3$ .... Find the value of the scalar product $\mathbf{p} \cdot \mathbf{q}$.

$\mathbf{p} \cdot \mathbf{q} = 8(4) + (-6)(-3) + (-10)(0) = 32 + 18 = 50$ .... Find the magnitude of $\mathbf{p}$ and of $\mathbf{q}$ using Pythagoras' theorem.

$|\mathbf{p}| = \sqrt{8^2 + (-6)^2 + (-10)^2} = 10\sqrt{2}$

$|\mathbf{q}| = \sqrt{4^2 + (-3)^2 + 0^2} = 5$ .... Use $\mathbf{p} \cdot \mathbf{q} = |\mathbf{p}||\mathbf{q}|\cos\theta$.

$50 = (10\sqrt{2})(5)\cos\theta$ .... Make $\cos\theta$ the subject.

$\dfrac{50}{50\sqrt{2}} = \cos\theta$ .... Solve for θ.

$\theta = \cos^{-1}\left(\dfrac{1}{\sqrt{2}}\right) = 45°$

WORKED EXAMPLE 9.9

Relative to the origin O, the position vectors of the points A and B are given by

$$\overrightarrow{OA} = \begin{pmatrix} 5 \\ -2 \\ 4 \end{pmatrix} \text{ and } \overrightarrow{OB} = \begin{pmatrix} -1 \\ 3 \\ -1 \end{pmatrix}.$$

Find angle OBA.

Answer

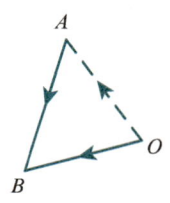

(sketch of triangle with vertices A, O, B)	$\overrightarrow{OB}$ and $\overrightarrow{AB}$ are the two directions towards the point B, so these are the vectors we need to use.
Find $\overrightarrow{OB} \cdot \overrightarrow{AB}$.	First of all, find the column vector $\overrightarrow{AB}$.
$\overrightarrow{AB} = \overrightarrow{OB} - \overrightarrow{OA} = \begin{pmatrix} -1 \\ 3 \\ -1 \end{pmatrix} - \begin{pmatrix} 5 \\ -2 \\ 4 \end{pmatrix} = \begin{pmatrix} -6 \\ 5 \\ -5 \end{pmatrix}$	Using the given vector $\overrightarrow{OB} = \begin{pmatrix} -1 \\ 3 \\ -1 \end{pmatrix}$, find the value of $\overrightarrow{OB} \cdot \overrightarrow{AB}$.
$\overrightarrow{OB} \cdot \overrightarrow{AB} = (-1)(-6) + (3)(5) + (-1)(-5) = 26$	Find the magnitude of $\overrightarrow{OB}$ and of $\overrightarrow{AB}$ using Pythagoras' theorem.
$\lvert \overrightarrow{AB} \rvert = \left\lVert \begin{pmatrix} -6 \\ 5 \\ -5 \end{pmatrix} \right\rVert = \sqrt{(-6)^2 + 5^2 + (-5)^2} = \sqrt{86}$	
$\lvert \overrightarrow{OB} \rvert = \left\lVert \begin{pmatrix} -1 \\ 3 \\ -1 \end{pmatrix} \right\rVert = \sqrt{(-1)^2 + 3^2 + (-1)^2} = \sqrt{11}$	Use $\overrightarrow{OB} \cdot \overrightarrow{AB} = \lvert \overrightarrow{OB} \rvert \lvert \overrightarrow{AB} \rvert \cos OBA$
$26 = \sqrt{86}\sqrt{11} \cos OBA$	Make $\cos OBA$ the subject.
$\dfrac{26}{\sqrt{86}\sqrt{11}} = \cos OBA$	Solve for angle OBA.
Angle $OBA = 32.3°$	

TIP

A sketch will help you to make sure you choose a correct pair of directions for your scalar product.

EXERCISE 9C

1 Find which of the following pairs of position vectors are perpendicular to one another.

For any position vectors that are not perpendicular, find the acute angle between them.

a $\mathbf{a} = \begin{pmatrix} 2 \\ 8 \\ -2 \end{pmatrix}$, $\mathbf{b} = \begin{pmatrix} 7 \\ -1 \\ 3 \end{pmatrix}$

b $\mathbf{c} = \begin{pmatrix} 5 \\ 12 \\ 13 \end{pmatrix}$, $\mathbf{d} = \begin{pmatrix} -3 \\ -1 \\ 3 \end{pmatrix}$

c $\mathbf{e} = \begin{pmatrix} 4 \\ -9 \\ -2 \end{pmatrix}$, $\mathbf{f} = \begin{pmatrix} -4 \\ -2 \\ 1 \end{pmatrix}$

2 Relative to the origin O, the position vectors of the points A and B are given by $\overrightarrow{OA} = \begin{pmatrix} -5 \\ 0 \\ 3 \end{pmatrix}$ and $\overrightarrow{OB} = \begin{pmatrix} 1 \\ 7 \\ 2 \end{pmatrix}$.

Find angle BOA.

3 Given that the vectors $6\mathbf{i} - 2\mathbf{j} + 5\mathbf{k}$ and $a\mathbf{i} + 4\mathbf{j} - 2\mathbf{k}$ are perpendicular, find the value of the constant a.

4 Relative to the origin O, the position vectors of the points P and Q are given by

$\overrightarrow{OP} = \begin{pmatrix} 5k \\ -3 \\ 7k+9 \end{pmatrix}$, $\overrightarrow{OQ} = \begin{pmatrix} k \\ k+2 \\ -1 \end{pmatrix}$ where k is a constant.

a Find the values of k for which $\overrightarrow{OP}$ and $\overrightarrow{OQ}$ are perpendicular to one another.

b Given that $k = 2$, find the angle, θ, between $\overrightarrow{OP}$ and $\overrightarrow{OQ}$.

5 The diagram shows a cuboid $OABCDEFG$ in which $OC = 6\,\text{cm}$, $OA = 4\,\text{cm}$ and $OD = 3\,\text{cm}$. Unit vectors $\mathbf{i}$, $\mathbf{j}$ and $\mathbf{k}$ are parallel to OC, OA and OD respectively. The point M lies on FB such that $FM = 1\,\text{cm}$, N is the midpoint of OC and P is the centre of the rectangular face $DEFG$.

Using a scalar product, find angle NPM.

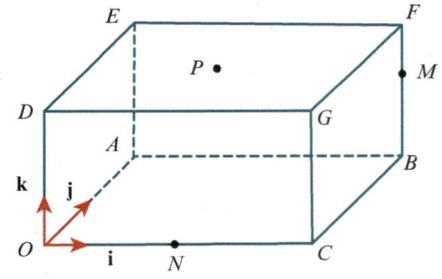

6 The position vector $\mathbf{a}$ is such that $\mathbf{a} = 4\mathbf{i} - 8\mathbf{j} + \mathbf{k}$. Find $\mathbf{a} \cdot \mathbf{j}$ and hence find the angle that $\mathbf{a}$ makes with the positive y-axis.

7 Explain why $\mathbf{a} \cdot \mathbf{b} \cdot \mathbf{c}$ has no meaning.

8 The diagram shows a cube $OABCDEFG$ with side length 4 units. The unit vectors $\mathbf{i}$, $\mathbf{j}$ and $\mathbf{k}$ are parallel to $\overrightarrow{OA}$, $\overrightarrow{OC}$ and $\overrightarrow{OD}$, respectively. The point M is the midpoint of GF and $AN:NB$ is $1:3$.

a Find each of the vectors $\overrightarrow{OM}$ and $\overrightarrow{NG}$.

b Find the angle between the directions $\overrightarrow{OM}$ and $\overrightarrow{NG}$.

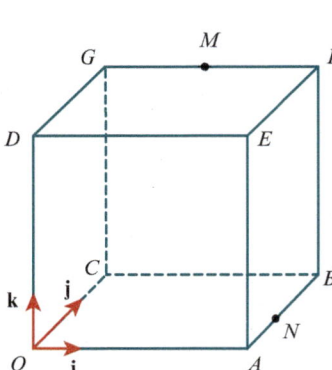

PS 9 The diagram shows a triangular prism, $OABCDE$. The uniform cross-section of the prism, OAE, is a right-angled triangle with base 77 cm and height 36 cm. The length, AB, of the prism is 60 cm. The unit vectors $\mathbf{i}$, $\mathbf{j}$ and $\mathbf{k}$ are parallel to $\overrightarrow{OA}$, $\overrightarrow{OC}$ and $\overrightarrow{OE}$, respectively. The point N divides the length of the line DB in the ratio $4:1$ and M is the midpoint of DE.

Use a scalar product to find angle MAN.

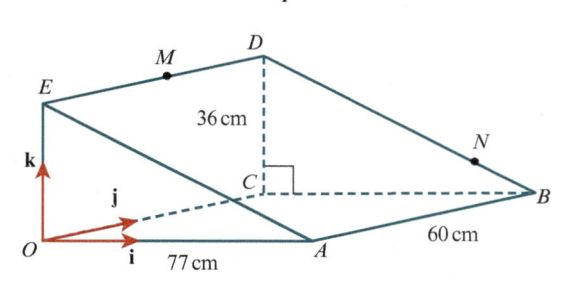

PS 10 The diagram shows a pyramid $VOABC$. The base of the pyramid, $OABC$, is a rectangle. The unit vectors **i**, **j** and **k** are parallel to $\overrightarrow{OA}, \overrightarrow{OC}$ and $\overrightarrow{OV}$, respectively. The position vectors of the points A, B, C and V are given by

$$\overrightarrow{OA} = 6\mathbf{i}, \quad \overrightarrow{OC} = 3\mathbf{j} \text{ and } \overrightarrow{OV} = 9\mathbf{k}.$$

The points M and N are the midpoints of AB and BV, respectively.

a Find the angle between the directions of $\overrightarrow{AN}$ and $\overrightarrow{OC}$.

b Find the vector $\overrightarrow{MN}$.

c The point P lies on OV and is such that PNM is a right angle. Find the position vector of the point P.

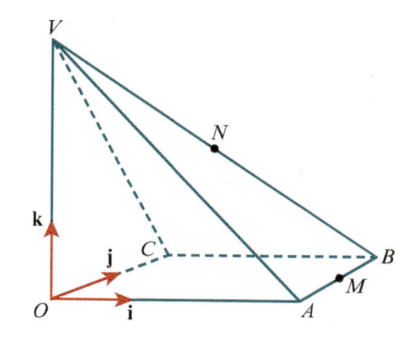

9.4 The vector equation of a line

We already know that, in 2-dimensional space, a line can be represented by an equation of the form $y = mx + c$ or $ax + by = c$. In 3-dimensional space, we need a new approach because introducing the third dimension gives an equation of the form $ax + by + cz = d$, which is actually the Cartesian equation of a plane. One way around this problem is to write the equation of the line using vectors.

The equation of a line in vector form is valid in both 2- and 3-dimensional space.

So what information do we need to be able to do this?

When a line passes through a fixed given point and continues in a known direction then it is uniquely positioned in space.

Therefore, to form the vector equation of a straight line, we need:

- a point on the line, $A(a_1, a_2, a_3)$, and a direction vector, $\mathbf{b} = \begin{pmatrix} b_1 \\ b_2 \\ b_3 \end{pmatrix}$, that is parallel to the line.

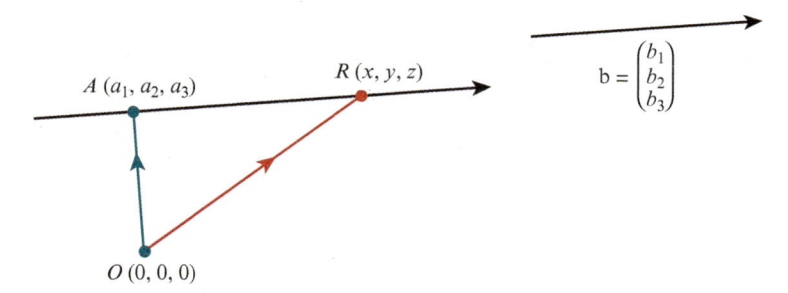

R is any point on the line and so $\overrightarrow{OR}$ represents the position vector of any point on the line.

To reach R, starting at O, take the path $\overrightarrow{OR} = \overrightarrow{OA} + \overrightarrow{AR}$.

$\overrightarrow{OA}$ is a vector we know. To move from A to R, we need to travel in the direction of **b**.

Multiplying **b** by different scalars changes the position of R on the line, so simply gives us different points on the line. Using the parameter t to represent the variable scalar

$$\overrightarrow{OR} = \overrightarrow{OA} + t\mathbf{b}$$

> **TIP**
>
> Sometimes Greek letters such as λ and μ are used as parameters.

> **KEY POINT 9.7**
>
> Using lower case notation for vectors, the vector equation of the line is:
>
> $\mathbf{r} = \mathbf{a} + t\mathbf{b}$

- Or two points on the line, $A(a_1, a_2, a_3)$ and $C(c_1, c_2, c_3)$. One can be used as the fixed given point and the direction can be found from the displacement between the two points.

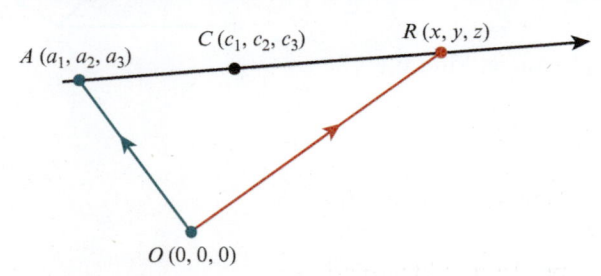

The situation is almost the same. In this case, however, we do not know the direction of the line. We can find the direction using the position vectors of the points A and C.

Using the parameter t as before,

$\overrightarrow{OR} = \overrightarrow{OA} + t\overrightarrow{AC}$.

> **TIP**
>
> Any known point can be used for the position vector, so you could also use
> $\overrightarrow{OR} = \overrightarrow{OC} + t\overrightarrow{AC}$

> **KEY POINT 9.8**
>
> Using lower case notation for vectors, the vector equation of a line is:
>
> $\mathbf{r} = \mathbf{a} + t(\mathbf{c} - \mathbf{a})$

In each case, the structure is:

general point on line = position of **known** point + some way along a **known** direction

Parametric form

$R(x, y, z)$ represents any point on the line.

The position vector of this point can be written as $\mathbf{r} = \begin{pmatrix} x \\ y \\ z \end{pmatrix}$ or $\mathbf{r} = x\mathbf{i} + y\mathbf{j} + z\mathbf{k}$.

In the equation $\mathbf{r} = \mathbf{a} + t\mathbf{b}$, using the position vectors in component form:

$\begin{pmatrix} x \\ y \\ z \end{pmatrix} = \begin{pmatrix} a_1 \\ a_2 \\ a_3 \end{pmatrix} + t \begin{pmatrix} b_1 \\ b_2 \\ b_3 \end{pmatrix}$ or $x\mathbf{i} + y\mathbf{j} + z\mathbf{k} = a_1\mathbf{i} + a_2\mathbf{j} + a_3\mathbf{k} + t(b_1\mathbf{i} + b_2\mathbf{j} + b_3\mathbf{k})$

Simplifying the right-hand side of this:

$\begin{pmatrix} x \\ y \\ z \end{pmatrix} = \begin{pmatrix} a_1 + tb_1 \\ a_2 + tb_2 \\ a_3 + tb_3 \end{pmatrix}$ or $\begin{aligned} x\mathbf{i} + y\mathbf{j} + z\mathbf{k} &= a_1\mathbf{i} + tb_1\mathbf{i} + a_2\mathbf{j} + tb_2\mathbf{j} + a_3\mathbf{k} + tb_3\mathbf{k} \\ x\mathbf{i} + y\mathbf{j} + z\mathbf{k} &= (a_1 + tb_1)\mathbf{i} + (a_2 + tb_2)\mathbf{j} + (a_3 + tb_3)\mathbf{k} \end{aligned}$

Comparing the components:

$x = a_1 + tb_1$

$y = a_2 + tb_2$

$z = a_3 + tb_3$

These three equations are called the **parametric equations** of the line.

Chapter 9: Vectors

WORKED EXAMPLE 9.10

A line passes through the point with position vector $3\mathbf{i} - 2\mathbf{j} + 4\mathbf{k}$ and is parallel to the vector $\mathbf{i} - \mathbf{j} + \mathbf{k}$.

Find:

a the vector equation of the line

b the parametric equations of the line.

Answer

Using t as the parameter:

a General point = $\mathbf{r}$ .... General point on line =

Known point = $3\mathbf{i} - 2\mathbf{j} + 4\mathbf{k}$ .... position of **known** point

Known direction = $\mathbf{i} - \mathbf{j} + \mathbf{k}$ .... + some way along a **known** direction.

$$\mathbf{r} = \begin{pmatrix} 3 \\ -2 \\ 4 \end{pmatrix} + t \begin{pmatrix} 1 \\ -1 \\ 1 \end{pmatrix}$$

b $\mathbf{r} = \begin{pmatrix} 3 \\ -2 \\ 4 \end{pmatrix} + t \begin{pmatrix} 1 \\ -1 \\ 1 \end{pmatrix}$ or $\mathbf{r} = 3\mathbf{i} - 2\mathbf{j} + 4\mathbf{k} + t(\mathbf{i} - \mathbf{j} + \mathbf{k})$ .... Write $\mathbf{r}$ as a column vector and collect together the x-, y- and z-components of the right-hand side.

$$\begin{pmatrix} x \\ y \\ z \end{pmatrix} = \begin{pmatrix} 3 + t \\ -2 - t \\ 4 + t \end{pmatrix} \text{ or }$$

$x\mathbf{i} + y\mathbf{j} + z\mathbf{k} = (3 + t)\mathbf{i} + (-2 - t)\mathbf{j} + (4 + t)\mathbf{k}$ .... Compare components.

$x = 3 + t \quad y = -2 - t \quad z = 4 + t$

WORKED EXAMPLE 9.11

The position vectors of the points A and B are given by

$\mathbf{a} = 3\mathbf{i} + 2\mathbf{j} - 7\mathbf{k}$ and $\mathbf{b} = \mathbf{i} - 4\mathbf{j} + 8\mathbf{k}$,

respectively.

a Write down the vector $\overrightarrow{AB}$.

b Find the vector equation of the line AB.

Answer

a $\overrightarrow{AB} = \mathbf{b} - \mathbf{a}$ .... Substitute for $\mathbf{a}$ and $\mathbf{b}$.

$\overrightarrow{AB} = \mathbf{i} - 4\mathbf{j} + 8\mathbf{k} - (3\mathbf{i} + 2\mathbf{j} - 7\mathbf{k})$ .... Expand and simplify.

$\overrightarrow{AB} = -2\mathbf{i} - 6\mathbf{j} + 15\mathbf{k}$

b Using $\mathbf{a} = 3\mathbf{i} + 4\mathbf{j} - 7\mathbf{k}$ as the known point, t as the parameter and $\overrightarrow{AB}$ from **part a** as the known direction:

General point = $\mathbf{r}$ .... General point on line =

Known point = $3\mathbf{i} + 4\mathbf{j} - 7\mathbf{k}$ .... position of **known** point

Known direction = $-2\mathbf{i} - 6\mathbf{j} + 15\mathbf{k}$ ..... + some way along a **known** direction.

$\mathbf{r} = 3\mathbf{i} + 2\mathbf{j} - 7\mathbf{k} + t\overrightarrow{AB}$

$\mathbf{r} = 3\mathbf{i} + 2\mathbf{j} - 7\mathbf{k} + t(-2\mathbf{i} - 6\mathbf{j} + 15\mathbf{k})$

> ⏮ **REWIND**
>
> Recall, you could also use $\mathbf{b} = \mathbf{i} - 4\mathbf{j} + 8\mathbf{k}$ as the known point.

🔍 KEY POINT 9.9

The angle between two lines is the angle between the two direction vectors of the lines.

WORKED EXAMPLE 9.12

The vector equation of the line L_1 is $\mathbf{r} = 3\mathbf{i} + \mathbf{j} - \mathbf{k} + \lambda(\mathbf{i} - 8\mathbf{j} + 5\mathbf{k})$.

a The line L_2 is parallel to the line L_1 and passes through the point $A(0, 2, 5)$.

Write down the vector equation of the line L_2.

b Find the angle between the line L_1 and the line $\mathbf{r} = 4\mathbf{i} + 3\mathbf{j} - 2\mathbf{k} + \mu(5\mathbf{i} - \mathbf{j} + 2\mathbf{k})$.

Answer

a Using μ as the parameter,

General point is $\mathbf{r}$ .... General point on line =

Known point is $2\mathbf{j} + 5\mathbf{k}$ .... position of **known** point

The direction of the line L_1 is $\mathbf{i} - 8\mathbf{j} + 5\mathbf{k}$ .... + some way along a **known** direction point.

The line L_2 is therefore
$\mathbf{r} = 2\mathbf{j} + 5\mathbf{k} + \mu(\mathbf{i} - 8\mathbf{j} + 5\mathbf{k})$

b Find the angle between
$\mathbf{d}_1 = \mathbf{i} - 8\mathbf{j} + 5\mathbf{k}$ and $\mathbf{d}_2 = 5\mathbf{i} - \mathbf{j} + 2\mathbf{k}$.

$\mathbf{d}_1 \cdot \mathbf{d}_2 = (1)(5) + (-8)(-1) + 5(2) = 23$

.... Find the scalar product $\mathbf{d}_1 \cdot \mathbf{d}_2$

.... Now find the magnitude of $\mathbf{d}_1$ and $\mathbf{d}_2$ using Pythagoras' theorem

$|\mathbf{d}_1| = \sqrt{1^2 + (-8)^2 + (5)^2} = 3\sqrt{10}$

$|\mathbf{d}_2| = \sqrt{5^2 + (-1)^2 + 2^2} = \sqrt{30}$ .... Use $\mathbf{d}_1 \cdot \mathbf{d}_2 = |\mathbf{d}_1||\mathbf{d}_2|\cos\theta$.

$23 = 3\sqrt{10}\sqrt{30} \cos\theta$ .... Make $\cos\theta$ the subject.

$\cos\theta = \dfrac{23}{3\sqrt{10}\sqrt{30}}$ .... Solve for θ.

$\cos^{-1}\left(\dfrac{23}{3\sqrt{10}\sqrt{30}}\right) = 63.727\ldots° = 63.7°$

EXERCISE 9D

1. Write down the vector equation for the line through the point A that is parallel to vector $\mathbf{b}$ when:

   a $A(0, -1, 5)$ and $\mathbf{b} = 2\mathbf{i} + 6\mathbf{j} - \mathbf{k}$

   b $A(0, 0, 0)$ and $\mathbf{b} = 7\mathbf{i} - \mathbf{j} - \mathbf{k}$

   c $A(7, 2, -3)$ and $\mathbf{b} = 3\mathbf{i} - 4\mathbf{k}$.

2. Write each of your answers to **question 1** in parametric form.

3. Show that the line through the points with position vectors $9\mathbf{i} + 2\mathbf{j} - 5\mathbf{k}$ and $\mathbf{i} + 7\mathbf{j} + \mathbf{k}$ is parallel to the line $\mathbf{r} = \lambda(16\mathbf{i} - 10\mathbf{j} - 12\mathbf{k})$.

4. Find the parametric equations of each of the lines.

   a $\mathbf{r} = 2\mathbf{i} + 13\mathbf{j} + \mathbf{k} + t(\mathbf{i} + \mathbf{j} - \mathbf{k})$

   b $\mathbf{r} = 10\mathbf{j} + t(2\mathbf{i} + 5\mathbf{j})$

   c $\mathbf{r} = \mathbf{i} - 3\mathbf{j} + t(2\mathbf{i} + 3\mathbf{j} + 4\mathbf{k})$

5. a Find the parametric equations of the line L through the points $A(0, 4, -2)$ and $B(1, 1, 6)$.

   b Hence find the coordinates of the point where the line L crosses the Oxy plane.

6. The line L_1 has vector equation $\mathbf{r} = 2\mathbf{i} - 3\mathbf{j} + \mathbf{k} + \lambda(6\mathbf{i} + \mathbf{j} + 3\mathbf{k})$ and the line L_2 has parametric equations $x = \mu + 4$, $y = \mu - 7$, $z = 3\mu$.

   a Show that L_1 and L_2 are not parallel.

   b Find the angle between the line L_1 and the line L_2.

7. The vector equation of the line L_1 is given by

   $$\begin{pmatrix} x \\ y \\ z \end{pmatrix} = \begin{pmatrix} -3 \\ 0 \\ 8 \end{pmatrix} + t \begin{pmatrix} 4 \\ -1 \\ -3 \end{pmatrix}.$$

   a Find the vector equation of the line L_2 that is parallel to L_1 and which passes through the point $A(5, -3, 2)$.

   b Show that $A(5, -3, 2)$ is the foot of the perpendicular from the point $B(4, -7, 2)$ to the line L_2.

8. The points A, B and C are position vectors relative to the origin O, given by

   $$\overrightarrow{OA} = \begin{pmatrix} 3 \\ 1 \\ 5 \end{pmatrix}, \overrightarrow{OB} = \begin{pmatrix} 0 \\ -1 \\ 2 \end{pmatrix} \text{ and } \overrightarrow{OC} = \begin{pmatrix} 1 \\ 2 \\ 3 \end{pmatrix}.$$

   a Find a vector equation for the line L passing through A and B.

   b The line through C, perpendicular to L, meets L at the point N. Find the exact coordinates of N.

9.5 Intersection of two lines

In 2-dimensional space, two lines are either parallel or they must intersect at a point.

In 3-dimensional space it might be that:

- they are parallel
- or they are not parallel and intersect
- or they are not parallel and do not intersect.

Lines that are not parallel and do not intersect are called **skew**.

Parallel lines have the same direction vectors and so if a pair of lines is parallel, this can easily be seen by examining the vector equations of the lines.

Non-parallel intersecting lines

In 2-dimensional space, to find the point of intersection of a pair of non-parallel lines using their Cartesian equations ($ax + by = c$), we solve the equations of the lines simultaneously. In other words, we put the two equations equal to each other.

The process is the same in 3-dimensional space and using the vector equations of the lines.

WORKED EXAMPLE 9.13

Given that the lines with vector equations

$\mathbf{r} = \mathbf{i} - 2\mathbf{j} - 4\mathbf{k} + \lambda(-\mathbf{i} - \mathbf{j} + \mathbf{k})$

and $\mathbf{r} = 2\mathbf{i} - 2\mathbf{j} + \mathbf{k} + \mu(\mathbf{i} + 2\mathbf{j} - 7\mathbf{k})$

intersect at the point A, find the coordinates of A.

Answer

The lines are equal at the point of intersection, so write the vector equation in parametric form so that the components can be equated.

$\mathbf{r} = \mathbf{i} - 2\mathbf{j} - 4\mathbf{k} + \lambda(-\mathbf{i} - \mathbf{j} + \mathbf{k})$ $\mathbf{r} = 2\mathbf{i} - 2\mathbf{j} + \mathbf{k} + \mu(\mathbf{i} + 2\mathbf{j} - 7\mathbf{k})$

$x = 1 - \lambda$ $x = 2 + \mu$

$y = -2 - \lambda$ $y = -2 + 2\mu$

$z = -4 + \lambda$ $z = 1 - 7\mu$

Now equate the components.

$1 - \lambda = 2 + \mu$ .... (1)

$-2 - \lambda = -2 + 2\mu$ .... (2)

$-4 + \lambda = 1 - 7\mu$ .... (3)

Solve these equations to find the value of λ and of μ:

From (1): $\lambda = -1 - \mu$

Substitute into (3): $-4 - 1 - \mu = 1 - 7\mu$

$6\mu = 6$ $\mu = 1$

When $\mu = 1$: $\quad\quad\quad\quad\quad\quad\quad \lambda = -1 - 1 = -2$

Check:

$\lambda = -2$	$\mu = 1$
$x = 1 - (-2) = 3$	$x = 2 + 1 = 3$
$y = -2 - (-2) = 0$	$y = -2 + 2 = 0$
$z = -4 - 2 = -6$	$z = 1 - 7 = -6$

The lines intersect at the point $A(3, 0, -6)$.

> **TIP**
>
> It is good to check that the value of λ and of μ that we have found give us the same point.

Skew lines

Skew lines have no point of intersection.

When two lines are skew, they are not parallel and they do not intersect.

WORKED EXAMPLE 9.14

Show that the lines with vector equations

$\quad \mathbf{r} = 4\mathbf{i} + 7\mathbf{j} + 5\mathbf{k} + \lambda(-2\mathbf{i} + \mathbf{j} + 3\mathbf{k})$

$\quad \mathbf{r} = \mathbf{i} + \mathbf{j} + 10\mathbf{k} + \mu(-6\mathbf{i} + \mathbf{j} + 4\mathbf{k})$

do not intersect.

Answer

As $-2\mathbf{i} + \mathbf{j} + 3\mathbf{k}$ is not a scalar multiple of $-6\mathbf{i} + \mathbf{j} + 4\mathbf{k}$ the lines are not parallel.

Writing the equations in parametric form:

$\quad \mathbf{r} = 4\mathbf{i} + 7\mathbf{j} + 5\mathbf{k} + \lambda(-2\mathbf{i} + \mathbf{j} + 3\mathbf{k}) \quad\quad \mathbf{r} = \mathbf{i} + \mathbf{j} + 10\mathbf{k} + \mu(-6\mathbf{i} + \mathbf{j} + 4\mathbf{k})$

$\quad x = 4 - 2\lambda \quad\quad\quad\quad\quad\quad\quad\quad\quad\quad\quad x = 1 - 6\mu$

$\quad y = 7 + \lambda \quad\quad\quad\quad\quad\quad\quad\quad\quad\quad\quad\quad y = 1 + \mu$

$\quad z = 5 + 3\lambda \quad\quad\quad\quad\quad\quad\quad\quad\quad\quad\quad z = 10 + 4\mu$

Now equate the components.

$\quad 4 - 2\lambda = 1 - 6\mu \quad \text{------(1)}$

$\quad 7 + \lambda = 1 + \mu \quad \text{-------(2)}$

$\quad 5 + 3\lambda = 10 + 4\mu \quad \text{------(3)}$

Attempt to solve these equations to find the value of λ and of μ that should give a consistent point:

From (2): $\quad\quad\quad\quad\quad\quad \lambda = \mu - 6$

Substitute into (3): $\quad\quad 5 + 3(\mu - 6) = 10 + 4\mu$

$\quad\quad\quad\quad\quad\quad\quad\quad\quad\quad \mu = -23$

When $\mu = -23$: $\quad\quad\quad\quad \lambda = -23 - 6 = -29$

Check:

$\lambda = -29$	$\mu = -23$
$x = 4 - 2(-29) = 62$	$x = 1 - 6(-23) = 139$
$y = 7 - 29 = -22$	$y = 1 + (-23) = -22$
$z = 5 + 3(-29) = -82$	$z = 10 + 4(-23) = -82$

The values of λ and μ do not give a consistent point, so there is no point of intersection and the lines are skew.

EXERCISE 9E

1 Find whether the following pairs of lines are parallel, intersecting or skew.

State the position vector of the common point of any lines that intersect.

a $x = 3\lambda + 4$ $\qquad$ $x = \mu + 1$
 $y = 5\lambda + 1$ $\qquad$ $y = \mu + 4$
 $z = 4\lambda - 3$ $\qquad$ $z = 2\mu + 5$

b $\mathbf{r} = \mathbf{i} - 5\mathbf{j} + 4\mathbf{k} + \lambda(3\mathbf{i} - 4\mathbf{j} + 9\mathbf{k})$ $\qquad$ $\mathbf{r} = \mathbf{i} - 5\mathbf{j} + 4\mathbf{k} + \lambda(-6\mathbf{i} + 8\mathbf{j} - 18\mathbf{k})$

c $\mathbf{r} = 7\mathbf{i} + 2\mathbf{j} + 6\mathbf{k} + \lambda(\mathbf{i} + 3\mathbf{j} + 9\mathbf{k})$ $\qquad$ $\mathbf{r} = 4\mathbf{j} + \mathbf{k} + \mu(8\mathbf{i} + \mathbf{j} + 14\mathbf{k})$

d $\mathbf{r} = 2\mathbf{i} - 3\mathbf{j} + 4\mathbf{k} + \lambda(6\mathbf{i} + 4\mathbf{j} - 2\mathbf{k})$ $\qquad$ $\mathbf{r} = 4\mathbf{i} - 9\mathbf{j} + 2\mathbf{k} + \mu(-\mathbf{i} - 8\mathbf{j} - \mathbf{k})$

> **TIP**
>
> - Always carry out the check with both λ and μ as it is easy to make a mistake.
> - Re-read the question to check whether you were asked to find the position vector or the coordinates of the point of intersection and give your answer in the correct form.

PS 2 Given that the lines with vector equations

$\mathbf{r} = 2\mathbf{i} + 9\mathbf{j} + \mathbf{k} + \lambda(\mathbf{i} - 4\mathbf{j} + 5\mathbf{k})$

and $\mathbf{r} = 11\mathbf{i} + 9\mathbf{j} + p\mathbf{k} + \mu(-\mathbf{i} - 2\mathbf{j} + 16\mathbf{k})$

intersect at the point P, find the value of p and the position vector of the point P.

3 Three lines, L_1, L_2 and L_3, have vector equations

$\mathbf{r} = 16\mathbf{i} - 4\mathbf{j} - 6\mathbf{k} + \lambda(-12\mathbf{i} + 4\mathbf{j} + 3\mathbf{k})$

$\mathbf{r} = 16\mathbf{i} + 28\mathbf{j} + 15\mathbf{k} + \mu(8\mathbf{i} + 8\mathbf{j} + 5\mathbf{k})$

$\mathbf{r} = \mathbf{i} + 9\mathbf{j} + 3\mathbf{k} + v(4\mathbf{i} - 12\mathbf{j} - 8\mathbf{k})$

The 3 points of intersection of these lines form an acute-angled triangle.

For this triangle, find:

a the position vector of each of the three vertices

b the size of each of the interior angles

c the length of each side.

4 The line L_1 passes through the points $(3, 7, 9)$ and $(-1, 3, 4)$.

a Find a vector equation of the line L_1.

b The line L_2 has vector equation

$\mathbf{r} = \mathbf{i} + 2\mathbf{j} + \mathbf{k} + \mu(3\mathbf{j} + 2\mathbf{k})$.

Show that L_1 and L_2 do not intersect.

5 The point A has coordinates $(1, 0, 5)$ and the point B has coordinates $(-1, 2, 9)$.

   a Find the vector $\overrightarrow{AB}$.

   b Write down a vector equation of the line AB.

   c Find the acute angle between the line AB and the line L with vector equation $\mathbf{r} = \mathbf{i} + 3\mathbf{j} + 4\mathbf{k} + \mu(-\mathbf{i} - 2\mathbf{j} + 3\mathbf{k})$.

   d Find the point of intersection of the line AB and the line L.

Checklist of learning and understanding

- For position vectors $\overrightarrow{OA} = \mathbf{a}$ and $\overrightarrow{OB} = \mathbf{b}$, $\overrightarrow{AB} = \mathbf{b} - \mathbf{a}$.
- The magnitude (length or size) is worked out using Pythagoras' theorem:
  When $\mathbf{a} = a_1\mathbf{i} + a_2\mathbf{j} + a_3\mathbf{k}$, $|\mathbf{a}| = \sqrt{a_1^2 + a_2^2 + a_3^2}$.
- Two vectors are parallel if one is a scalar multiple of the other.
- Problem solving with angles will involve the use of the scalar product.
  When $\mathbf{a} = a_1\mathbf{i} + a_2\mathbf{j} + a_3\mathbf{k}$ and $\mathbf{b} = b_1\mathbf{i} + b_2\mathbf{j} + b_3\mathbf{k}$

  $\mathbf{a} \cdot \mathbf{b} = a_1b_1 + a_2b_2 + a_3b_3 = |\mathbf{a}||\mathbf{b}|\cos\theta$
- Vector equation of a line: $\mathbf{r} = \mathbf{a} + t\mathbf{b}$
- To find the point of intersection of two intersecting lines, solve the parametric equations of the lines simultaneously.

END-OF-CHAPTER REVIEW EXERCISE 9

1 Relative to the origin O, the position vectors of points A, B and C are given by

$$\overrightarrow{OA} = \begin{pmatrix} 2 \\ 3 \\ 7 \end{pmatrix}, \overrightarrow{OB} = \begin{pmatrix} 3m \\ 1 \\ 1 \end{pmatrix} \text{ and } \overrightarrow{OC} = \begin{pmatrix} 4 \\ -m \\ -m(m+1) \end{pmatrix}.$$

   a It is given that $\overrightarrow{AB} = \overrightarrow{OC}$. Find the angle OAB. [4]

   b Find the vector equation of the line AC. [3]

2 With respect to the origin O, the position vectors of the points A and B are given by

$$\overrightarrow{OA} = \begin{pmatrix} -2 \\ 0 \\ 6 \end{pmatrix} \text{ and } \overrightarrow{OB} = \begin{pmatrix} 1 \\ -1 \\ 4 \end{pmatrix}.$$

   a Find whether or not the vectors $\overrightarrow{OA}$ and $\overrightarrow{OB}$ are perpendicular. [2]

   b i Write down the vector $\overrightarrow{AB}$. [1]

       ii Find a vector equation of the line AB. [2]

   c The vector equation of the line L is given by

       $\mathbf{r} = 7\mathbf{i} - 12\mathbf{j} + 7\mathbf{k} + \mu(-3\mathbf{i} + 10\mathbf{j} - 5\mathbf{k})$.

      Show that the lines AB and L intersect and find the position vector of the point of intersection. [5]

3

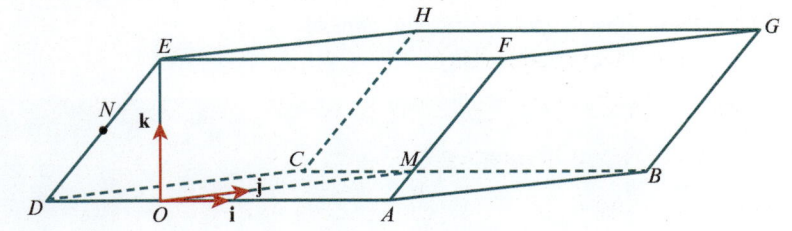

The diagram shows a prism, $ABCDEFGH$, with a parallelogram-shaped uniform cross-section.

The point E is such that OE is the height of the parallelogram. The point M is such that $\overrightarrow{OM}$ is parallel to $\overrightarrow{DC}$ and N is the midpoint of DE. The side OD has a length of 5 units. The unit vectors $\mathbf{i}$, $\mathbf{j}$ and $\mathbf{k}$ are parallel to $\overrightarrow{OA}$, $\overrightarrow{OM}$ and $\overrightarrow{OE}$, respectively.

The position vectors of the points A, E, C and M are given by $\overrightarrow{OA} = 9\mathbf{i}$, $\overrightarrow{OM} = 15\mathbf{j}$ and $\overrightarrow{OE} = 12\mathbf{k}$.

   a Express the vectors $\overrightarrow{AH}$ and $\overrightarrow{NH}$ in terms of $\mathbf{i}$, $\mathbf{j}$ and $\mathbf{k}$. [2]

   b Use a vector method to find angle AHN. [5]

   c Write down a vector equation of the line AH. [2]

4   The position vectors of A, B and C relative to an origin O are given by

$$\overrightarrow{OA} = \begin{pmatrix} 6 \\ 3 \\ 2 \end{pmatrix}, \overrightarrow{OB} = \begin{pmatrix} 2 \\ n \\ -1 \end{pmatrix} \text{ and } \overrightarrow{OC} = \begin{pmatrix} 8 \\ 9 \\ 0 \end{pmatrix}$$

where n is a constant.

   a Find the value of n for which $|\overrightarrow{AB}| = |\overrightarrow{CB}|$. [4]

   b In this case, use a scalar product to find angle ABC. [3]

5   **a** Given the vectors $8\mathbf{i} - 2\mathbf{j} + 5\mathbf{k}$ and $\mathbf{i} + 2\mathbf{j} + p\mathbf{k}$ are perpendicular, find the value of the constant p. [3]

   b The line L_1 passes through the point $(-3, 1, 5)$ and is parallel to the vector $7\mathbf{i} - \mathbf{j} - \mathbf{k}$.

      i Write down a vector equation of the line L_1. [2]

     ii The line L_2 has vector equation

$$\mathbf{r} = \mathbf{i} - 2\mathbf{j} + 2\mathbf{k} + \mu(\mathbf{i} + 8\mathbf{j} - 3\mathbf{k}).$$

        Show that L_1 and L_2 do not intersect. [4]

6   The origin O and the points A, B and C are such that $OABC$ is a rectangle. With respect to O, the position vectors of the points A and B are $-4\mathbf{i} + p\mathbf{j} - 6\mathbf{k}$ and $-10\mathbf{i} - 2\mathbf{j} - 10\mathbf{k}$.

   a Find the value of the positive constant p. [3]

   b Find a vector equation of the line AC. [3]

   c Show that the line AC and the line, L, with vector equation

$$\mathbf{r} = 3\mathbf{i} + 7\mathbf{j} + \mathbf{k} + \mu(-4\mathbf{i} - 4\mathbf{j} - 3\mathbf{k})$$

     intersect and find the position vector of the point of intersection. [5]

   d Find the acute angle between the lines AC and L. [4]

7   Relative to the origin O, the points A, B, C and D have position vectors

$$\overrightarrow{OA} = 4\mathbf{i} + 2\mathbf{j} - \mathbf{k}, \overrightarrow{OB} = 2\mathbf{i} - 2\mathbf{j} + 5\mathbf{k}, \overrightarrow{OC} = 2\mathbf{j} + 7\mathbf{k}, \overrightarrow{OD} = -6\mathbf{i} + 22\mathbf{j} + 9\mathbf{k}.$$

   a Use a scalar product to show that angle ABC is a right angle. [3]

   b Show that $\overrightarrow{AD} = k\overrightarrow{BC}$, where k is a constant, and explain what this means. [2]

   c The point E is the midpoint of the line AD. Find a vector equation of the line EC. [4]

8   With respect to the origin O, the points A, B, C, D have position vectors given by

$$\overrightarrow{OA} = -3\mathbf{i} + \mathbf{j} + 8\mathbf{k}, \overrightarrow{OB} = -10\mathbf{i} + 2\mathbf{j} + 15\mathbf{k}, \overrightarrow{OC} = 5\mathbf{i} - 2\mathbf{j} - 2\mathbf{k}, \overrightarrow{OD} = \mathbf{i} + 6\mathbf{j} + 10\mathbf{k}.$$

   a Calculate the acute angle between the lines AB and CD. [4]

   b Prove that the lines AB and CD intersect. [5]

   c The point E has position vector $5\mathbf{i} + 3\mathbf{j} + 4\mathbf{k}$. Show that the perpendicular distance from E to the line CD is equal to $\sqrt{5}$. [5]

9   $PQRS$ is a rhombus. The coordinates of the vertices P, Q and S are $(9, 2, 4)$, $(-0.5, 6, 6.5)$ and $(4.5, -4, -3.5)$, respectively.

   a   Find the vectors $\overrightarrow{PQ}$ and $\overrightarrow{PS}$. [2]

   b   Find the coordinates of R. [3]

   c   Show that $PQRS$ is in fact a square and find the length of the side of the square. [3]

   d   The point T is the centre of the square. Find the coordinates of T. [3]

   e   The point V has coordinates $(5, 17.5, -13.5)$.

      i   Find a vector equation of the line VT. [3]

      ii Verify that T is the foot of the perpendicular from V to PR. [2]

      iii Describe the solid $VPQRS$. [1]

10 The points P and Q have coordinates $(0, 19, -1)$ and $(-6, 26, -11)$, respectively.

   The line L has vector equation

   $$\mathbf{r} = 3\mathbf{i} + 9\mathbf{j} + 2\mathbf{k} + \lambda(3\mathbf{i} - 10\mathbf{j} + 3\mathbf{k}).$$

   a   Show that the point P lies on the line L. [1]

   b   Find the magnitude of $\overrightarrow{PQ}$. [2]

   c   Find the obtuse angle between PQ and L. [3]

   d   Calculate the perpendicular distance from Q to the line L. [5]

11 The points A and B have coordinates $(7, 1, 6)$ and $(10, 5, 1)$, respectively.

   a   Write down a vector equation of the line AB. [3]

   b   The point P lies on the line AB. The point Q has coordinates $(0, -5, 7)$.

      Given that PQ is perpendicular to AB, find a vector equation of the line PQ. [6]

   c   Find the shortest distance from Q to the line AB. [1]

12 The line L_1 has vector equation

   $$\mathbf{r} = 3\mathbf{i} + 2\mathbf{j} + 5\mathbf{k} + \lambda(4\mathbf{i} + 2\mathbf{j} + 3\mathbf{k}).$$

   The points $A(3, p, 5)$ and $B(q, 0, 2)$, where p and q are constants, lie on the line L_1.

   a   Find the value of p and the value of q. [2]

   The line L_2 has vector equation

   $$\mathbf{r} = 3\mathbf{j} + \mathbf{k} + \mu(7\mathbf{i} + \mathbf{j} + 7\mathbf{k}).$$

   b   Show that L_1 and L_2 intersect and find the position vector of the point of intersection. [5]

   c   Find the acute angle between L_1 and L_2. [4]

Chapter 10
Differential equations

P3 This chapter is for Pure Mathematics 3 students only.

In this chapter you will learn how to:

- formulate a simple statement involving a rate of change as a differential equation
- find, by integration, a general form of solution for a first order differential equation in which the variables are separable
- use an initial condition to find a particular solution
- interpret the solution of a differential equation in the context of a problem being modelled by the equation.

Cambridge International AS & A Level Mathematics: Pure Mathematics 2 & 3

PREREQUISITE KNOWLEDGE

Where it comes from	What you should be able to do	Check your skills
IGCSE / O Level Mathematics	Express direct and inverse variation in algebraic terms and use this form of expression to find unknown quantities.	1 The weight, w newtons, of a sphere is directly proportional to the cube of its radius, r cm. When the radius of the sphere is 60 cm, its weight is 4320 newtons. Find an equation connecting w and r.
IGCSE / O Level Mathematics	Understand and apply • laws of indices • laws of logarithms	2 $\ln y - \ln(20 - y) = \dfrac{x}{4} + \ln 3$ Find and simplify an expression for y in terms of x.
Chapter 5	Integrate • standard functions • using substitution • using parts	3 Find: a $\displaystyle\int \dfrac{1}{3x-1}\,dx$ b $\displaystyle\int \dfrac{\sin x}{\cos x}\,dx$ c $\displaystyle\int x e^{-x}\,dx.$
Chapter 8	Decompose a compound fraction with denominator • $(ax+b)(cx+d)(ex+f)$ • $(ax+b)(cx+d)^2$ • $(ax+b)(cx^2+d)$ into its partial fractions and integrate the result.	4 Find $\displaystyle\int \dfrac{2}{(x-2)(3x+1)}\,dx.$
Chapter 1	Sketch graphs of standard functions e.g. involving $\ln f(x)$ and $e^{f(x)}$ (including understanding end behaviour as $x \to \pm\infty$).	5 a Sketch the graph of $y = e^{x^2}$. b $\dfrac{12e^x}{3+e^x} \to k$ as $x \to \infty$. Find the value of the constant k.

What is a differential equation?

An equation that has a derivative such as $\dfrac{dy}{dx}$ or $\dfrac{d^2y}{dx^2}$ as one of its terms is called a **differential equation**. When the equation has the first derivative $\dfrac{dy}{dx}$ as one of its terms, it is called a **first order** differential equation. For example, the equation $\dfrac{dy}{dx} = 3x^2$ is an alternative way of representing the relationship $y = x^3 + C$, where C is a constant.

To solve a differential equation, we integrate to find the direct relationship between x and y, such as $y = x^3 + C$. The direct equation $y = x^3 + C$ is called the **general solution** of the differential equation $\dfrac{dy}{dx} = 3x^2$. To find a **particular solution**, we must have further information that will enable us to find the value of C.

The solution of an algebraic equation is a number or set of numbers. The solution of a differential equation is a function.

Chapter 10: Differential equations

In this chapter, we will learn the initial skills needed to master the solution of one type of first order differential equations. These simple skills will be the foundation for further study of this subject.

You might wonder about the usefulness of such seemingly abstract equations in the real world. In fact, differential equations are excellent mathematical models for many real-life situations and are essential to solve many problems. First order differential equations can model rates of change, in other words, the rate at which one quantity is increasing or decreasing with respect to another quantity. There are many examples in the physical world that can be modelled using such relationships. A few examples are the rate of increase of the number of bacteria in a culture, simple kinematics problems, the rate of decrease of the value of a car and the rate at which a hot liquid cools.

10.1 The technique of separating the variables

Solving a differential equation such as $\frac{dy}{dx} = 6(x+1)^2$ can be done using an anti-differentiation approach.

We know, from the Pure Mathematics 1 Coursebook, Chapter 9, that if we integrate both sides with respect to x, we have:

$$\int \frac{dy}{dx} \, dx = \int 6(x+1)^2 \, dx$$

$$y = \frac{6(x+1)^3}{3} + C$$

$$y = 2(x+1)^3 + C$$

Notice that the first line on the left-hand side is $\int \frac{dy}{dx} \, dx = \int 6(x+1)^2 \, dx$ and from anti-differentiation we can see that, if we integrate $\frac{dy}{dx}$ with respect to x we obtain an expression for y.

This means that $\int \frac{dy}{dx} \, dx = \ldots$ is equivalent to $\int dy = \ldots$, which is $y = \ldots$

This result is generally true.

> **KEY POINT 10.1**
>
> Integrating an expression of the form $f(y) \frac{dy}{dx}$ with respect to x is equivalent to integrating $f(y)$ with respect to y.
>
> Therefore $\int f(y) \frac{dy}{dx} \, dx$ is equivalent to $\int f(y) \, dy$.

The differential equation $\frac{dy}{dx} = 6(x+1)^2$ is easy to solve because $\frac{dy}{dx}$ is equal to a function that we know how to integrate. This is not always the case.

How do we solve the differential equation $\frac{dy}{dx} = xy$ when $y > 0$?

As it is written, we cannot integrate both sides with respect to x and find a solution.

Separating the variables in the expression xy means that we are then able to integrate.

$$\frac{dy}{dx} = xy \qquad \text{Separate the variables } x \text{ and } y.$$

$$\frac{1}{y}\frac{dy}{dx} = x \qquad \text{Form integrals of both sides with respect to } x.$$

$$\int \frac{1}{y}\frac{dy}{dx}\,dx = \int x\,dx$$

$$\int \frac{1}{y}\,dy = \int x\,dx \qquad \text{Integrate each side.}$$

$$\ln y + A = \frac{x^2}{2} + B \qquad \text{The modulus of } y \text{ is not required as we know that } y > 0.$$

$$\ln y = \frac{x^2}{2} + C \qquad \text{The constants of integration can be combined as one term and we only need to write a constant on one side.}$$

$$y = e^{\frac{x^2}{2} + C}$$

$$y = De^{\frac{x^2}{2}} \quad \text{where } D = e^C \qquad \text{This is the general solution of the differential equation as we do not know the value of the \textbf{arbitrary constant} } C.$$

When we have further information, we can find the particular solution of the differential solution given by that information.

For example, solve $\frac{dy}{dx} = xy$ given that $y = 2$ when $x = 0$.

Using this information:

$$\ln y = \frac{x^2}{2} + C \text{ and so } \ln 2 = \frac{0^2}{2} + C, \; C = \ln 2$$

$$y = e^{\frac{x^2}{2} + \ln 2} \qquad \text{This is the \textbf{particular} solution of the differential equation.}$$

$$y = e^{\frac{x^2}{2}} \times e^{\ln 2} \qquad \text{Note that you often need to write your answer as an expression for } y \text{ in terms of } x. \text{ This commonly means using laws of indices or logarithms.}$$

$$y = 2e^{\frac{x^2}{2}}$$

By separating the variables we are rearranging the differential equation to the form $f(y)\frac{dy}{dx} = g(x)$.

$$f(y)\frac{dy}{dx} = g(x) \qquad \text{(Note: Other variables may be used such as } v \text{ and } t.)$$

$$\int f(y)\frac{dy}{dx}\,dx = \int g(x)\,dx$$

$$\int f(y)\,dy = \int g(x)\,dx$$

So, in the case where the derivative is $\dfrac{dy}{dx}$:

- form *an expression in x only* on one side of the equation
- form *an expression in* $y \times \dfrac{dy}{dx}$ on the other side of the equation.

This gives us a form of the equation that can be integrated on both sides, one side with respect to x and the other side with respect to y.

WORKED EXAMPLE 10.1

The variables x and y satisfy the differential equation $\dfrac{dy}{dx} = \dfrac{1+x}{2y}$.

It is given that $y = 3$ when $x = 0$.

Solve the differential equation and obtain an expression for y in terms of x.

Answer

Even though the question does not ask for it directly, you are expected to use the initial condition that you have been given to find a particular solution.

$\dfrac{dy}{dx} = \dfrac{1+x}{2y}$ — Separate the variables x and y.

Here we have moved the $2y$. It was not essential to move the 2, we could have moved the y only. This would be just as valid as it would also give expressions that can be integrated.

$2y \dfrac{dy}{dx} = 1 + x$ — Form integrals of both sides with respect to x.

$\int 2y \dfrac{dy}{dx} dx = \int (1+x) dx$

$\int 2y \, dy = \int (1+x) dx$ — Integrate each side.

$y^2 = x + \dfrac{x^2}{2} + C$ — $y = 3$ when $x = 0$.

$9 = 0 + \dfrac{0^2}{2} + C, \quad C = 9$

$y^2 = x + \dfrac{x^2}{2} + 9$

$y = \left(x + \dfrac{x^2}{2} + 9\right)^{\frac{1}{2}}$ — You can check this by differentiating.

WORKED EXAMPLE 10.2

Find the general solution of the differential equation $(x^2 + 1)y^3 + x^2 \dfrac{dy}{dx} = 0$.

Answer

$(x^2 + 1)y^3 + x^2 \dfrac{dy}{dx} = 0$ .... Separate the variables x and y.

$x^2 \dfrac{dy}{dx} = -(x^2 + 1)y^3$

$\dfrac{1}{y^3} \dfrac{dy}{dx} = -\dfrac{(x^2 + 1)}{x^2}$ ** .... Form integrals of both sides with respect to x.

Here we are assuming that $y \neq 0$.

$\displaystyle\int y^{-3} \dfrac{dy}{dx} dx = \int (-x^{-2} - 1) dx$

$\displaystyle\int y^{-3} dy = \int (-x^{-2} - 1) dx$ .... Integrate each side.

$-\dfrac{1}{2y^2} = \dfrac{1}{x} - x + C$

Is this the complete solution?

Earlier in the solution (at line **) we assumed that $y \neq 0$. However, when $y = 0$, the original differential equation becomes $\dfrac{dy}{dx} = \dfrac{0}{x^2} = 0$. Since $y = 0$ is the x-axis, it is true that its gradient is 0 at all points and so $y = 0$ is also a solution of the differential equation and must be included for the general solution to be complete.

The complete general solution is, therefore, $-\dfrac{1}{2y^2} = \dfrac{1}{x} - x + C$ or $y = 0$.

In Worked example 10.2, we saw that, when rearranging to the form $f(y)\dfrac{dy}{dx} = g(x)$, there are cases where $f(y)$ is a function such as, for example, $\dfrac{1}{y^3}$. In these cases, a value (or values) of y must be excluded for the separation of the variables to take place.

As a final step, when finding the general solution in these cases, we must check if the value (or values) we have excluded when separating the variables gives (or give) a valid solution of the original differential equation. When they do, we should include them.

WORKED EXAMPLE 10.3

It is given that, for $0 < x < 300$, $\dfrac{dx}{dt} = \dfrac{x(300 - x)}{600}$ and $x = 50$ when $t = 0$. Find the particular solution of the differential equation, giving your answer as an expression for x in terms of t.

Answer

$\dfrac{dx}{dt} = \dfrac{x(300 - x)}{600}$ .... Separate the variables. The derivative is with respect to t and so the expression in x needs to be moved to the left-hand side.

$$\frac{1}{x(300-x)}\frac{dx}{dt} = \frac{1}{600}$$

$$\int \frac{1}{x(300-x)}\frac{dx}{dt}\,dt = \frac{1}{600}\int dt$$

Form integrals of both sides with respect to t.

Note that it is simpler to leave the 600 on the right-hand side here.

$$\int \frac{1}{x(300-x)}\,dx = \frac{1}{600}\int dt$$

Write the expression on the left in terms of its partial fractions.

$$\frac{1}{300}\int\left(\frac{1}{x} + \frac{1}{300-x}\right)dx = \frac{1}{600}\int dt$$

Integrate each side.

$$\frac{1}{300}(\ln x - \ln(300-x)) = \frac{1}{600}t + C$$

The modulus signs are not required as we know that $0 < x < 300$.

$$\ln x - \ln(300-x) = \frac{1}{2}t + C$$

$$\ln\frac{x}{300-x} = \frac{1}{2}t + C$$

$x = 50$ when $t = 0$.

$$C = \ln\frac{50}{300-50},\ C = \ln 0.2$$

$$\ln\frac{x}{300-x} = \frac{1}{2}t + \ln 0.2$$

$$\frac{x}{300-x} = e^{\frac{1}{2}t + \ln 0.2}$$

$$\frac{x}{300-x} = e^{\frac{1}{2}t} \times e^{\ln 0.2}$$

$$x = 0.2e^{\frac{1}{2}t}(300-x)$$

$$x + 0.2e^{\frac{1}{2}t}x = 60e^{\frac{1}{2}t}$$

$$x(1 + 0.2e^{\frac{1}{2}t}) = 60e^{\frac{1}{2}t}$$

$$x = \frac{60e^{\frac{1}{2}t}}{1 + 0.2e^{\frac{1}{2}t}}$$

EXERCISE 10A

1 Find the general solution of each of the following differential equations.

   a $\ \dfrac{dy}{dx} = x^3 - 6$ b $\ \dfrac{dA}{dr} = 2\pi r$ c $\ s^2\dfrac{ds}{dt} = \cos(t+5)$

   d $\ V^{-1}\dfrac{dV}{dt} = 2$ e $\ \dfrac{dy}{dx} = \dfrac{y+1}{x}$ f $\ x + \sec x \sin y \dfrac{dy}{dx} = 1$

2 Find the particular solution for each of the following differential equations, using the values given.

   a $\ \cos^2 x \dfrac{dy}{dx} = y$ given $y = 5$ when $x = 0$

   b $\ xy\dfrac{dy}{dx} = \dfrac{x^3 - x}{1 - \sqrt{y}}$ given $y = 4$ when $x = 3$

3 Given that $\dfrac{dy}{dx} = e^{x-y}$ and $y = \ln 2$ when $x = 0$, obtain an expression for y in terms of x.

4 The gradient of a curve at the point (x, y) is $5y^3 x$.

   a Write down the differential equation satisfied by y.

   b Given that the curve passes through the point $(1, 1)$, find y^2 as a function of x.

5 a Express $\dfrac{1}{(2-x)(x+1)}$ in the form $\dfrac{A}{2-x} + \dfrac{B}{x+1}$.

> **TIP**
>
> $\dot{x}$ is an alternative notation for $\dfrac{dx}{dt}$.

   b The variables x and t satisfy the differential equation

   $$\frac{dx}{dt} = (2-x)(x+1).$$

   It is given that $t = 0$ when $x = 0$.

   i Solve this differential equation and obtain an expression for x in terms of t.

   ii State what happens to the value of x when t becomes large.

PS 6 The variables w and t satisfy the differential equation

$$\frac{dw}{dt} = 0.001(100 - w)^2.$$

Given that when $t = 0$ $w = 0$, show that the solution to the differential equation can be written in the form $w = 100 - \dfrac{1000}{t + 10}$.

7 Solve the differential equation $4\dfrac{dx}{dt} = (3x^2 - x)\cos t$ given that $x = 0.4$ when $t = 0$. Give your answer as an expression for x in terms of t.

8 Find the general solution of the differential equation $xy \dfrac{dy}{dx} = \dfrac{x^2 - 1}{e^{y+1}}$.

9 Find the equation of the curve that satisfies the equation $\dfrac{dy}{dx} = \dfrac{1 + y^2}{4 + x^2}$ and passes through the point $(0, 1)$. Give your answer in the form $y = f(x)$.

10 Find the general solution of the equation $x^3 \dfrac{dy}{dx} + 2y = 1$.

11 Solve the differential equation $\dfrac{dx}{dt} = 4x \cos^2 t$ given that $x = 1$ when $t = 0$. Give your answer as an expression for x in terms of t.

12 a Use the substitution $u = t^2$ to find $\displaystyle\int t e^{t^2} \, dt$.

   b Solve the differential equation $\dfrac{dx}{dt} = \dfrac{t(e^{t^2} + 5)}{x^2}$ given that $x = 1$ when $t = 0$.

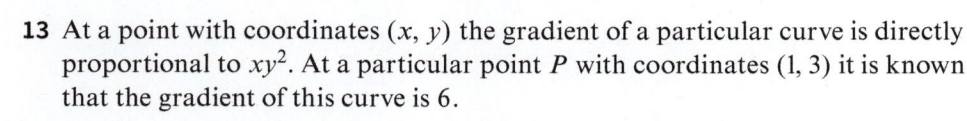

PS 13 At a point with coordinates (x, y) the gradient of a particular curve is directly proportional to xy^2. At a particular point P with coordinates $(1, 3)$ it is known that the gradient of this curve is 6.

Find and simplify an expression for y in terms of x.

14 The variables x and t satisfy the differential equation

$$(3x + 2x^3)\frac{dx}{dt} = k(3x^2 + x^4)$$

for $x > 0$ where k is a constant. When $t = 0$, $x = 1$ and when $t = 0.5$, $x = 2$.

Solve the differential equation, finding the exact value of k, and show that $7^{2t} = \dfrac{3x^2 + x^4}{4}$.

15 a Use the substitution $u = \ln x$ to find $\int \dfrac{1}{x \ln x} \, dx$.

  b Given that $x > 0$, $y > 0$, find the general solution of the differential equation $\ln x \dfrac{dy}{dx} = \dfrac{y}{x}$.

M **16** A type of medicine that reacts to strong light can be used to treat some skin problems.

It is important that, after treatment with this type of medicine, the person stays out of bright light until the amount of medicine remaining in the skin is at a safe level.

Jonty is treated with a medicine of this type.

After the treatment, 65 mg remains in the treated area of skin.

The rate at which the medicine disappears from the treated area of Jonty's skin is modelled by the differential equation $\dfrac{dm}{dt} = -0.05\,m$, where m is the amount of medicine, in milligrams, in the skin t hours after the treatment.

Jonty is safe to go out in bright light when no more than 6 mg of the medicine remains in his skin.

How long must Jonty wait before he goes out in bright light?

> **TIP**
>
> Notice that the rate at which the medicine disappears is with respect to time but the words 'with respect to time' are usually not included.

EXPLORE 10.1

Newton's law of cooling or heating

Many differential equations arise from modelling natural laws and relationships. One example is Newton's law of cooling or heating:

> The rate of change of temperature of an object is proportional to the difference in temperature of the object and its surroundings.

We define T as the temperature of the object

  t as time

  T_0 as the temperature of the object when $t = 0$

  s as the constant temperature of the surroundings

  k as the constant of proportionality.

Discuss the following.

- Why it is necessary to define the variables and constants.
- Why the differential equation $\dfrac{dT}{dt} = k(T - s)$, $k > 0$ is satisfied by T.
- Other possible correct ways to represent Newton's law of cooling or heating as a differential equation with this set of defined variables and constants.
- Why the general solution of the given form of the differential equation may be written as:

  $|T - s| = A e^{kt}$ where $A = e^C$.

- How many arbitrary constants there are in the general solution of **any** first order differential equation.
- Why the modulus in the general solution is required.
- How the general solution can be used to represent the temperature of an object cooling down or warming up.

10.2 Forming a differential equation from a problem

The key to forming a differential equation from a problem is the successful interpretation of the language used in the question.

For example, when you are given a problem such as:

The rate of increase of **the number of bacteria in a dish** is k times **the number of bacteria in the dish**. The number of bacteria present in the dish at time t is x...

Replacing the words **the number of bacteria in a dish** by x the problem becomes:

The rate of increase of x is k times x.

This rate of increase of x is with respect to time, t, and is much easier to translate into a differential equation: $\dfrac{dx}{dt} = kx$.

WORKED EXAMPLE 10.4

Velocity is defined to be the rate of change of displacement with respect to time.

The velocity of a particle at time t seconds is inversely proportional to its displacement, s metres, from an origin, O.

a Form a differential equation to represent this data.

b Initially, the particle has a displacement of 2 metres from O.
After 1 second, the particle has a displacement of 5 metres from O.
Solve the differential equation giving your answer in the form $t = f(s)$.

c Find t when the displacement of the particle from O is 10 metres.

Answer

a The **velocity** of a particle at time t seconds **is inversely proportional to** its **displacement**, s metres, from the origin, O.

$v = \dfrac{ds}{dt}$ and so $\dfrac{ds}{dt} \propto \dfrac{1}{s}$, which means that $\dfrac{ds}{dt} = \dfrac{k}{s}$ ····· At the moment, you do not have enough information to find the constant of proportionality, k

b $\quad \dfrac{ds}{dt} = \dfrac{k}{s}$ .... Separate the variables s and t.

$s \dfrac{ds}{dt} = k$ .... Form integrals of both sides with respect to t.

$\displaystyle\int s \dfrac{ds}{dt}\, dt = \int k\, dt$

$\displaystyle\int s\, ds = \int k\, dt$ .... Integrate each side.

$\dfrac{s^2}{2} = kt + C$ .... $s = 2$ when $t = 0$.

$C = 2, \dfrac{s^2}{2} = kt + 2$ .... $s = 5$ when $t = 1$.

$\dfrac{5^2}{2} = k + 2, \quad k = 10.5$

$\dfrac{s^2}{2} = \dfrac{21}{2} t + 2$ .... Write in the form $t = f(s)$.

$t = \dfrac{s^2 - 4}{21}$

c Find t when $s = 10$.

$t = \dfrac{100 - 4}{21} = 4.5714\ldots = 4.57$ seconds correct to 3 significant figures.

WORKED EXAMPLE 10.5

The population density, P, of a town can be modelled as a continuous variable.

In the year when records began:
- the population density, P, of the town was increasing at the rate of 0.5 people per hectare per year
- the population density of the town was 2 people per hectare.

It is suggested that, t years after records began, the rate of increase of the population density is inversely proportional to the square of the population density.

a Form a differential equation to model this information.

b Solve the differential equation, giving your answer in the form $P = f(t)$.

c The table shows the actual population density of the town, every 5 years for the first 20 years after records began.

t	5	10	15	20
P	3.5	4	5.5	7

Comment on how good a fit the model is to the actual data in the longer term.

Answer

a The rate of increase of P is inversely proportional to P^2:

$$\frac{dP}{dt} \propto \frac{1}{P^2} \qquad \frac{dP}{dt} = \frac{k}{P^2}$$

When $t = 0$, $\frac{dP}{dt} = 0.5$ and $P = 2$, therefore:

$$\frac{dP}{dt} = \frac{k}{P^2} \qquad 0.5 = \frac{k}{4}, k = 2$$

$$\frac{dP}{dt} = \frac{2}{P^2}$$

b $\frac{dP}{dt} = \frac{2}{P^2}$ .... Separate the variables P and t.

$P^2 \frac{dP}{dt} = 2$ .... Form integrals of both sides with respect to t.

$$\int P^2 \frac{dP}{dt} dt = \int 2 \, dt$$

$$\int P^2 \, dP = 2 \int dt$$ .... Integrate each side. $t = 0$, $P = 2$

$$\frac{P^3}{3} = 2t + C$$

$$\frac{8}{3} = C$$

$\frac{P^3}{3} = 2t + \frac{8}{3}$ .... Write in the form $P = f(t)$.

$P^3 = 6t + 8$

$P = \sqrt[3]{6t + 8}$

c

t	5	10	15	20
P	3.5	4	5.5	7
Model prediction (to 2 significant figures)	3.4	4.1	4.6	5.0

In the longer term, the model is not a particularly good fit as the values it generates when time increases are under-estimates of the known data.

Chapter 10: Differential equations

WORKED EXAMPLE 10.6

The value, V, of a motorbike as it gets older can be modelled as a continuous variable.

A model is suggested to estimate the rate of decrease of the value of the motorbike. This model is that the value V when the motorbike is t months old decreases at a rate proportional to V.

a Write down a differential equation that is satisfied by V.

b The motorbike had a value of \$10 000 when it was new.

Solve the differential equation, showing that $V = 10\,000e^{-kt}$, where k is a positive constant.

c After 3 years the motorbike has lost half of its value when new.

Find the value of the motorbike when it was 2 years old.

Answer

a The rate of decrease of V is proportional to V. ········ This describes exponential decay. The rate is decreasing and so the constant of proportionality is negative.

$$\frac{dV}{dt} \propto V \text{ and so } \frac{dV}{dt} = -kV, \text{ where } k \text{ is a positive constant.}$$

b $\quad \dfrac{dV}{dt} = -kV$ ········ Separate the variables V and t.

$\quad \dfrac{1}{V}\dfrac{dV}{dt} = -k$ ········ Form integrals of both sides with respect to t.

$\quad \displaystyle\int \dfrac{1}{V}\dfrac{dV}{dt}\,dt = \int -k\,dt$

$\quad \displaystyle\int \dfrac{1}{V}\,dV = \int -k\,dt$ ········ Integrate each side.

$\quad \ln|V| = -kt + C$ ········ $t = 0$, $V = 10\,000$

$\quad \ln 10\,000 = C$ ········ For this particular solution, V is positive and so the modulus is not required.

$\quad \ln V = -kt + \ln 10\,000$ ········ Write in the form $V = f(t)$.

$\quad V = e^{-kt + \ln 10\,000}$

$\quad V = e^{-kt} \times e^{\ln 10\,000}$

$\quad V = 10\,000e^{-kt}$

c $\quad V = 10\,000e^{-kt}$ ········ $t = 3$, $V = 5000$

$\quad 5000 = 10\,000e^{-3k}$

$\quad \dfrac{5000}{10\,000} = e^{-3k}$

$\quad \ln 0.5 = -3k$

$\quad k = \dfrac{-\ln 0.5}{3}$

$\quad V = 10\,000e^{-2\left(\frac{-\ln 0.5}{3}\right)}$

$\quad V = \$6299.605\ldots = \$6300$

correct to 3 significant figures

EXERCISE 10B

M **1** Write the following as differential equations. Do **not** solve the equations.

   a For height h and time t:

      The rate of decrease of the height of water in a tank is proportional to the square of the height.

   b For number of infected cells n and time t:

      The rate of increase of the number of infected cells in a body is proportional to the number of infected cells in the body at that time.

   c For velocity v and time t:

      The rate of decrease of the velocity of a particle moving in a fluid is proportional to the product of v and $v+1$.

   d For volume V and time t:

      The rate at which oil is leaking from an engine is proportional to the volume of oil in the engine at that time.

   e For cost C and time t:

      The rate at which the cost of an item increases is proportional to the cube of the cost at that time.

   f For speed v and time t:

      The rate of change of the speed of a skydiver just after he jumps out of a plane is inversely proportional to $e^{\frac{t}{2}}$.

M **2** A large tank is full of water. The tank starts to leak.

   t seconds after the tank starts to leak the depth of water in the tank is x metres.

   The rate of decrease of the depth of water is proportional to the square root of the depth of water.

   a Write down a differential equation relating x, t and a constant of proportionality.

   b At the instant when water starts to leak from the tank, the depth of water is 6.25 metres. Given that 120 seconds after the tank starts to leak, the depth of water is 4 metres, solve the differential equation, expressing t in terms of x.

   c Find the time it takes for the depth of the water to halve.

M **3** The cost of designing an aircraft, $\$A$ per kilogram, at time t years after 1950 can be modelled as a continuous variable. The rate of increase of A is directly proportional to A.

   a Write down a differential equation that is satisfied by A.

   b In 1950, the cost of designing an aircraft was $12 per kilogram. Show that $A = 12e^{kt}$, where k is a positive constant.

   c After 25 years, the cost of designing an aircraft was $48 per kilogram. Find the cost of designing an aircraft after 60 years.

4 A liquid is heated so that its temperature is $x\,°\text{C}$ after t seconds. It is given that the rate of increase of x is proportional to $(100 - x)$. The initial temperature of the liquid is $25\,°\text{C}$.

   a Form a differential equation relating x, t and a constant of proportionality, k, to model this information.

   b Solve the differential equation and obtain an expression for x in terms of t and k.

   c After 180 seconds the temperature of the liquid is $85\,°\text{C}$. Find the temperature of the liquid after 195 seconds.

   d The model predicts that x cannot exceed a certain temperature. Write down this maximum temperature.

5 The diagram shows an inverted cone filled with liquid paint.

An artist cuts a small hole in the bottom of the cone and the liquid paint drips out at a rate of $16\,\text{cm}^3$ per second. At time t seconds after the hole is cut, the paint in the cone is an inverted cone of depth $h\,\text{cm}$.

   a Show that $\dfrac{dV}{dh} = \dfrac{4}{9}\pi h^2$.

   b Hence find an expression for $\dfrac{dh}{dt}$.

   c Solve the differential equation in **part b**, giving t in terms of h.

   d Find the length of time it takes for the depth of the paint to fall to $7.5\,\text{cm}$.

6 The size of a population, P, at time t minutes is to be modelled as a continuous variable such that the rate of increase of P is directly proportional to P.

   a Write down a differential equation that is satisfied by P.

   b The initial size of the population is P_0.
   Show that $P = P_0 e^{kt}$, where k is a positive constant.

   c The size of the population is $1.5P_0$ after 2 minutes.
   Find when the population will be $3P_0$.

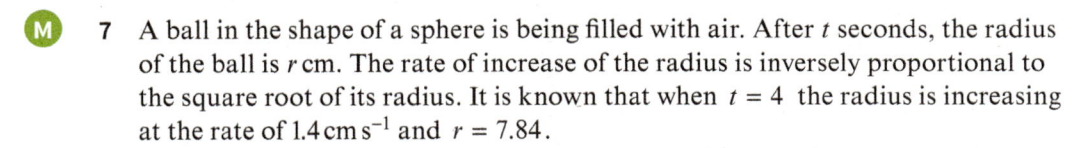

7 A ball in the shape of a sphere is being filled with air. After t seconds, the radius of the ball is $r\,\text{cm}$. The rate of increase of the radius is inversely proportional to the square root of its radius. It is known that when $t = 4$ the radius is increasing at the rate of $1.4\,\text{cm}\,\text{s}^{-1}$ and $r = 7.84$.

   a Form a differential equation relating r and t to model this information.

   b Solve the differential equation and obtain an expression for r in terms of t.

   c How much air was in the ball at the start?

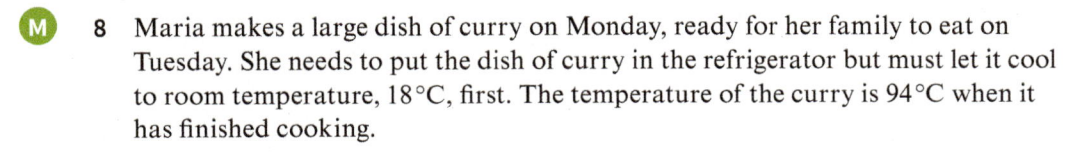

8 Maria makes a large dish of curry on Monday, ready for her family to eat on Tuesday. She needs to put the dish of curry in the refrigerator but must let it cool to room temperature, $18\,°\text{C}$, first. The temperature of the curry is $94\,°\text{C}$ when it has finished cooking.

At 6 pm, Maria places the hot dish in a sink full of cold water and keeps the water at a constant temperature of $7\,°\text{C}$ by running the cold tap and stirring the dish from time to time. After 10 minutes, the temperature of the curry is $54\,°\text{C}$.

> **TIP**
> You must define your own variables in this question as they have not been defined for you.

It is given that the rate at which the curry cools down is proportional to the difference between the temperature of the curry and the temperature of the water in the sink.

At what time can Maria put her dish of curry in the refrigerator?

M   **9**   The half-life of a radioactive isotope is the amount of time it takes for half of the isotope in a sample to decay to its stable form.

Carbon-14 is a radioactive isotope that has a half-life of 5700 years.

It is given that the rate of decrease of the mass, m, of the carbon-14 in a sample is proportional to its mass.

A sample of carbon-14 has initial mass m_0. What fraction of the original amount of carbon-14 would be present in this sample after 2500 years?

M   **10**   Anya carried out an experiment and discovered that the rate of growth of her hair was constant. At the start of her experiment, her hair was 20 cm long. After 20 weeks, her hair was 26 cm long. Form and solve a differential equation to find a direct relationship between time, t, and the length of Anya's hair, L.

M   **11**   A bottle of water is taken out of a refrigerator. The temperature of the water in the bottle is 4°C. The bottle of water is taken outside to drink. The air temperature outside is constant at 24°C.

It is given that the rate at which the water in the bottle warms up is proportional to the difference in the air temperature outside and the temperature of the water in the bottle.

After 2 minutes the temperature of the water in the bottle is 11°C.

   a How long does it take for the water to warm 20°C ?

   b According to the model, what temperature will the water in the bottle eventually reach if the air temperature remains constant and the water is not drunk?

M   **12**   The number of customers, n, of a food shop t months after it opens for the first time can be modelled as a continuous variable. It is suggested that the number of customers is increasing at a rate that is proportional to the square root of n.

   a Form and solve a differential equation relating n and t to model this information.

   b Initially, $n = 0$, and after 6 months the food shop has 3600 customers. Find how many complete months it takes for the number of customers to reach 6800.

   c The food shop has a capacity of serving 14 000 customers per month. Show that the model predicts the shop will have reached its capacity sometime in the 12th month.

M   **13**   Doubling time is the length of time it takes for a quantity to double in size or value.

The number of bacteria in a liquid culture can be modelled as a continuous variable and grows at a rate proportional to the number of bacteria present. Initially, there are 5000 bacteria in the culture. After 3.5 hours there are a million bacteria.

   a What is the doubling time of the bacteria in the culture?

   b What assumption does the model make about the growth of the bacteria and how realistic is this assumption?

Chapter 10: Differential equations

EXPLORE 10.2

Modelling growth

Many of the models we have been using have assumed exponential growth. This type of growth is unlimited as the rate of growth of a population is proportional to the size of the population at that time.

This means that $\dfrac{dP}{dt} = kP$.

Is an exponential growth model appropriate for modelling the population of the planet Earth in the long term?

It is thought that the maximum capacity of the Earth is about 10 billion people.

In 1987, the population of the Earth was about 5 billion people and the growth rate of the population was about 1.84%.

Some people think that the problem of the Earth's increasing population can be solved by sending some people to live on other planets, such as Mars. In 2031, the first people to live on Mars will launch into space. The environment on Mars is different to Earth and many challenges must be overcome. This will all take time.

When a population is modelled as growing **exponentially** at a **constant** rate of r % per year, it takes about $\dfrac{70}{r}$ years for the population to double.

- Use the data given so far to investigate whether going to live on Mars is a good solution to the population problem.
- Comment on any assumptions you have made in your investigation.

There is statistical evidence that the population of the Earth is not growing at a constant rate each year. In 2017, the growth rate had decreased to 1.11% per year. The population needs resources to survive and so the growth rate has decreased as resources have decreased.

A different model, logistic growth, can be used when resources are limited. This model takes into account the changing growth rates of the population.

Here are sketch graphs of exponential and logistic growth models:

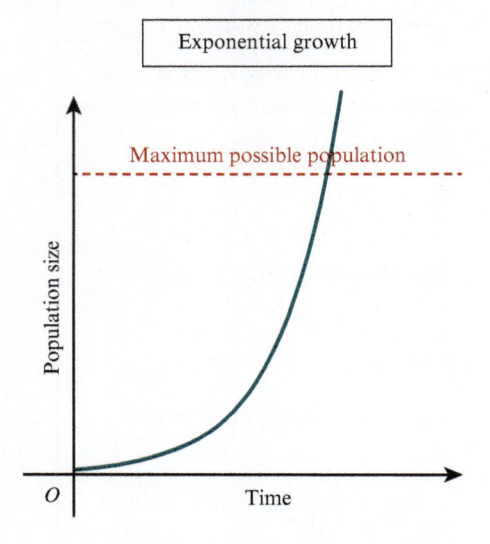

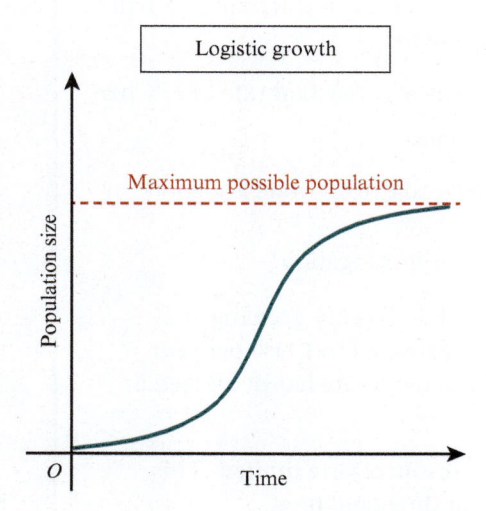

Exponential growth

Logistic growth

For example, for a population of size P,

$$\frac{dP}{dt} = \frac{rP(k-P)}{k}$$

where the constant r is the maximum natural growth rate (birth rate – death rate) and the constant k is the maximum capacity of the limited environment.

- Show that as time increases this model suggests that P tends to k.
- Comment on the limitations of this model.

Checklist of learning and understanding

General procedure to solve a differential equation using the technique of separating the variables

- Separate the variables $f(y)\dfrac{dy}{dx} = g(x)$.
- Integrate both sides $\displaystyle\int f(y)\dfrac{dy}{dx}\,dx = \int g(x)\,dx$

$$\int f(y)\,dy = \int g(x)\,dx.$$

- Solve to find a general solution.
- Substitute any given conditions to find a particular solution.
- Check your answer by differentiating.

Other points to remember

- Take care with modulus signs in expressions such as $\ln|f(x)|$.
- When finding a general solution, check to see if any values excluded when the variables are separated are possible solutions.

END-OF-CHAPTER REVIEW EXERCISE 10

1. The variables x and y satisfy the differential equation
   $$\frac{dy}{dx} = x^2 e^{y+2x}$$
   and it is given that $y = 0$ when $x = 0$.

   Solve the differential equation and obtain an expression for y in terms of x. [9]

2. The number of birds, x, in a particular area of land is recorded every year for t years. x is to be modelled as a continuous variable. The rate of change of the number of birds over time is modelled by
   $$\frac{dx}{dt} = \frac{x(2500 - x)}{5000}.$$
   It is given that $x = 500$ when $t = 0$.

   a Find an expression for x in terms of t. [9]

   b How many birds does the model suggest there will be in the long term? [1]

3. Given that $y = 2$ when $x = 0$, solve the differential equation
   $$y^3 \frac{dy}{dx} = 1 + y^4$$
   obtaining an expression for y^4 in terms of x. [6]

4. The gradient of a curve is such that, at the point (x, y), the gradient of the curve is proportional to $x\sqrt{y}$. At the point $(3, 4)$ the gradient of this curve is -5.

   a Form and solve a differential equation to find the equation of this curve. [7]

   b Find the gradient of the curve at the point $(-3, 4)$. [1]

5. a Given that $x < 5$, find $\displaystyle\int \frac{50}{(5-x)(10-x)} dx$. [3]

   b A chemical reaction takes place between two substances A and B. When this happens, a third substance, C, is produced. After t hours there are $5 - x$ grams of A, there are $10 - x$ grams of B and there are x grams of C present. The rate of increase of x is proportional to the product of $5 - x$ and $10 - x$. Initially, $x = 0$ and the rate of increase of x is 1 gram per hour.

   i Show that x and t satisfy the differential equation $50 \dfrac{dx}{dt} = (5-x)(10-x)$. [1]

   ii Solve this equation, giving x in terms of t. [6]

   iii According to the model, approximately how many grams of chemical C are produced by the reaction? [1]

6. The variables x and y are related by the differential equation
   $$\frac{dy}{dx} = \frac{xe^x}{5y^4}.$$
   It is given that $y = 4$ when $x = 0$. Find a particular solution of the differential equation and hence find the value of y when $x = 3.5$. [8]

7. a Use the substitution $u = \sqrt{y^4 - 1}$ to find $\displaystyle\int \frac{y^3}{\sqrt{y^4 - 1}} dy$. [1]

   b Given that $\dfrac{dy}{dx} = \dfrac{(2x+1)\sqrt{y^4 - 1}}{xy^3}$, $y = \sqrt{3}$ when $x = 1$, find a relationship between x and y. [6]

8 The height, h metres, of a cherry tree is recorded every year for t years after it is planted. It is thought that the height of the tree is increasing at a rate proportional to $8 - h$. When the tree is planted it is 0.5 metres tall and after 5 years it is 2 metres tall.

    a Form and solve a differential equation to model this information. Give your answer in the form $h = f(t)$. [6]

    b According to the model, what will the height of the cherry tree be when it is fully grown? [1]

9 The variables x and y satisfy the differential equation

$$\frac{dy}{dx} = e^{3(x+y)} \text{ and } y = 0 \text{ when } x = 0.$$

Solve the differential equation and obtain an expression for y in terms of x. [6]

10 At the start of a reaction, there are x grams of chemical X present. At time t seconds after the start, the rate of decrease of x is proportional to xt.

    a Using k as constant of proportionality, where $k > 0$, form a differential equation to model this reaction rate. [1]

    b Solve the differential equation obtaining a relation between x, t and k. [5]

    c Initially, there are 150 grams of chemical X present and after 10 seconds this has decreased to 120 grams. Find the time after the start of the reaction when the amount of chemical X has decreased to 1 gram. [3]

11 a Using partial fractions find $\int \frac{1}{P(5-P)} dP$. [4]

    b Given that $P = 3$ when $t = 0$, solve the differential equation $\frac{dP}{dt} = P(5-P)$, obtaining an expression for P in terms of t. [4]

    c Describe what happens to P when $t \to \infty$. [1]

12 The population of a country was 50 million in 2000 and 60 million in 2010. The rate of increase of the population is modelled by

$$\frac{dP}{dt} = \frac{kP(100-P)}{100}$$

Use the model to predict the population of the country in 2025. [7]

13 a Use the substitution $u = x^2$ to find $\int x \sin x^2 \, dx$. [3]

    b Given that $\frac{dy}{dx} = \frac{x \sin x^2}{y}$, $y = -1$ when $x = 0$, find a relationship between x and y. [5]

14 Solve the differential equation

$$\frac{dy}{dx} = \frac{\tan x}{e^{3y}}$$

given that $x = 0$ when $y = 0$. Give your answer in the form $y = f(x)$. [6]

15 There are 2000 mice in a field. At time t hours, x mice are infected with a disease. The rate of increase of the number of mice infected is proportional to the product of the number of mice infected and the number of mice not infected. Initially 500 mice are infected and the disease is spreading at a rate of 50 mice per hour.

    a Form and solve a differential equation to model this data. Give your answer in the form $t = f(x)$. [7]

    b Find how long it takes for 1900 mice to be infected. [1]

16 The variables x and y are related by the differential equation

$$x\frac{dy}{dx} = 1 - y^2.$$

When $x = 2$, $y = 0$. Solve the differential equation, obtaining an expression for y in terms of x. [8]

Cambridge International A Level Mathematics 9709 Paper 31 Q6 November 2012

17 The variables x and θ satisfy the differential equation

$$\frac{dx}{d\theta} = (x + 2)\sin^2 2\theta,$$

and it is given that $x = 0$ when $\theta = 0$. Solve the differential equation and calculate the value of x when $\theta = \frac{1}{4}\pi$, giving your answer correct to 3 significant figures. [9]

Cambridge International A Level Mathematics 9709 Paper 31 Q8 November 2015

Chapter 11
Complex numbers

P3 This chapter is for Pure Mathematics 3 students only.

In this chapter you will learn how to:

- understand the idea of a complex number, recall the meaning of the terms real part, imaginary part, modulus, argument and conjugate and use the fact that two complex numbers are equal if and only if both real and imaginary parts are equal
- carry out operations of addition, subtraction, multiplication and division of two complex numbers expressed in Cartesian form $x + iy$
- use the result that, for a polynomial equation with real coefficients, any non-real roots occur in conjugate pairs
- represent complex numbers geometrically by means of an Argand diagram
- carry out operations of multiplication and division of two complex numbers expressed in polar form $r(\cos\theta + i\sin\theta) = re^{i\theta}$
- find the two square roots of a complex number
- understand in simple terms the geometrical effects of conjugating a complex number and of adding, subtracting, multiplying and dividing two complex numbers
- illustrate simple equations and inequalities involving complex numbers by means of loci in an Argand diagram.

PREREQUISITE KNOWLEDGE

Where it comes from	What you should be able to do	Check your skills

IGCSE / O Level Mathematics	Add, subtract and multiply pairs of binomial expressions.	1 Simplify each of the following.
		  a $(a + bx) + (2a - 3bx)$
		  b $(a + bx)(2a - 3bx)$

Assumed prior knowledge (not in the IGCSE / O Level course)	Manipulate expressions involving surds and calculate with surds.	2 Simplify.
		  a $\sqrt{32} - 3\sqrt{8}$
		  b $(2 - \sqrt{5})(2 + \sqrt{5})$
		  c $\dfrac{4 - \sqrt{3}}{5 + 2\sqrt{3}}$

IGCSE / O Level Mathematics	Draw simple loci. Understand, for example, that the locus of a point which moves so that • it is always the same distance away from a fixed point is a circle • it is always equidistant from two fixed points is a perpendicular bisector.	3 The diagram shows a triangle ABC. Copy the diagram and shade the region that is less than 4 cm from A and nearer to B than to C.
Pure Mathematics 1 Coursebook, Chapter 5	Understand the convention for positive and negative angles and use trigonometry to find the size of an angle that is not acute.	4 a Solve $\tan x = -1$ for $-\pi < x < \pi$ radians. b Find $\pi - \sin^{-1}\left(\dfrac{3}{4}\right)$.

Chapter 9	• Find the magnitude and direction of a vector.
	• Know and use the relationship $\overrightarrow{AB} = \mathbf{b} - \mathbf{a}$ when the position vector of A is $\mathbf{a}$ and the position vector of B is $\mathbf{b}$.
		5 The point A has position vector $\mathbf{a} = 3\mathbf{i} + 4\mathbf{j}$ and the point B has position vector $\mathbf{b} = 5\mathbf{i} - 7\mathbf{j}$. Find:
		  a the magnitude of $\mathbf{a}$
		  b the angle made between $\mathbf{a}$ and the positive x-axis
		  c $\overrightarrow{BA}$

What are complex numbers?

To understand the structure of a **complex number** we must first introduce a new set of numbers, the set of **imaginary numbers**.

The number system we use today has developed over time as new sets of numbers have been needed to model new situations. Early mathematicians used natural numbers for counting. Later, zero and negative numbers were accepted as valid numbers and the set of integers was established. In Pythagoras' time, mathematicians understood that rational numbers existed as they were the values in between integers. Pythagorean mathematicians thought that all numbers could be written as rational numbers. These same mathematicians were shocked by the discovery of irrational numbers such as $\sqrt{2}$ and they did not trust them as valid quantities at the time. Even as late as the 17th and 18th centuries, some famous mathematicians still doubted that negative numbers were of any use and thought that subtracting something from nothing was an impossible operation!

We will start with some simple facts.

- In the set of natural or counting numbers ($\mathbb{N}$) there are no negatives and no zero.
- In the set of integers ($\mathbb{Z}$) there are no part or fractional values.
- In the set of rational numbers ($\mathbb{Q}$) there are no irrational quantities such as π.
- In the set of real numbers ($\mathbb{R}$) there are no numbers that when squared result in a negative value.

Until now, the numbers with which we have been working have all been **real numbers**. That is, they belong to the set of real numbers, which includes the rational and irrational numbers. Real numbers can be represented on a 1-dimensional diagram called a number line.

In the 17th century, the set of imaginary numbers (I) was accepted as a valid number set. René Descartes described the collection of number sets that had been accepted as valid before that point as the set of real numbers. The new set, which he did not like very much, he described as imaginary. He intended this to be an insult as he thought they were not of great importance. Many other famous mathematicians agreed with him. They were, however, quite wrong about that. This name has often confused students who are new to this area of study.

So what is an imaginary number and how is it related to a complex number?

In the new set of imaginary numbers there **are** numbers that when squared result in a negative value.

> **KEY POINT 11.1**
>
> The set of complex numbers combines the real numbers and the imaginary numbers into a 2-dimensional rather than a 1-dimensional number system. Descartes described the numbers in the set of complex numbers as the sum of a real part and an imaginary part. Complex numbers are represented as points on a plane rather than points on a number line because of their 2-dimensional nature.

Clearly, accepting new sets of numbers as being valid is something mathematicians have struggled with over time. You might not feel comfortable with this new idea. The set of real, imaginary and complex numbers does have application in the real world. Without imaginary and, therefore, complex numbers there would be no mobile telephones, for example, as they are used to find audio signals. Complex numbers are used to generate fractals, which are used in the study of weather systems and earthquakes. Fractals are also used to make realistic computer-generated images for movies.

11.1 Imaginary numbers

In the set of imaginary numbers we introduce a new number i and we define $i^2 = -1$.

This means that $i = \sqrt{-1}$.

Note that:

- Engineers call this quantity j as i is used in engineering for a different quantity.
- Just as with some other important numbers, i is described as a letter rather than a digit.

Consider the equation $x^2 + 1 = 0$.

How many solutions does this equation have in the set of real numbers?

$x^2 + 1 = 0 \quad x^2 = -1 \quad x = \pm\sqrt{-1}$ No solutions in the set of real numbers.

How many solutions does this equation have in the set of imaginary numbers?

$x^2 + 1 = 0 \quad x^2 = -1 \quad x = \pm\sqrt{-1} = \pm i$ Two solutions in the set of imaginary numbers.

We can use laws of surds to extend the use of i to find other imaginary numbers.

WORKED EXAMPLE 11.1

Write the following numbers in their simplest form.

   a $\sqrt{-9}$ b $\sqrt{-5}$ c $\sqrt{-18}$

Answer

a $\sqrt{-9} = \sqrt{9 \times -1}$ When solving $x^2 + 1 = 0$ we needed ± in the
       $= \sqrt{9} \times \sqrt{-1} = 3i$ solution. Why do we not need it here?

b $\sqrt{-5} = \sqrt{5 \times -1}$ When the multiplier is a simple surd, it is better
       $= \sqrt{5} \times \sqrt{-1} = i\sqrt{5}$ to write the i before the surd to avoid any
                                               confusion with $\sqrt{5}i$ or we can write $(\sqrt{5})i$.

c $\sqrt{-18} = \sqrt{2 \times 9 \times -1}$
          $= \sqrt{2} \times \sqrt{9} \times \sqrt{-1}$
          $= \sqrt{2} \times 3 \times i = (3\sqrt{2})i$

> **TIP**
>
> Simplest form for an imaginary number means using simple surd form and i.

The imaginary number i is called the unit imaginary number as it has a coefficient of 1.

All other imaginary numbers are multiples of i.

Imaginary numbers are therefore of the form bi or iy, where b is a real number or y is a real number.

We calculate with imaginary numbers in the same way that we calculate with real numbers.

> **TIP**
>
> Many calculators are able to calculate in the set of complex numbers. You must practise without a calculator but can use one to check your answer.
>
> You need to demonstrate your ability to work with these numbers without a calculator. It is therefore important to practise writing down your method steps. Working without a calculator will also help to improve your understanding and recall of this subject (and do not forget that not all calculators are able to calculate with complex numbers).

Chapter 11: Complex numbers

WORKED EXAMPLE 11.2

Without using a calculator, find:

a $(3i^2) + (3i)^2$ b $-8i + (-4i)^3$ c $\sqrt{\dfrac{9i + 16i}{4i}}$ d i^{-6}.

Answer

a $(3i^2) + (3i)^2 = 3i^2 + 9i^2 = 12i^2 = -12$

b $-8i + (-4i)^3 = -8i + 64i = 56i$

c $\sqrt{\dfrac{9i+16i}{4i}} = \sqrt{\dfrac{25i}{4i}} = \sqrt{\dfrac{25}{4}} = \dfrac{5}{2}$

d $i^{-6} = \dfrac{1}{i^6} = \dfrac{1}{(i^2)^3} = \dfrac{1}{(-1)^3} = -1$

EXERCISE 11A

Do not use a calculator in this exercise.

1 Write the following numbers in their simplest form.

 a $\sqrt{-144}$ b $\sqrt{-\dfrac{36}{81}}$ c $\sqrt{-90}$ d $\sqrt{-16} + \sqrt{-81}$

2 Simplify.

 a $-5i^2 + 3i^8$ b $9i - (i\sqrt{2})^3$ c $\sqrt{\dfrac{100i^4 - 16i^2}{4}}$ d $\dfrac{-5}{6i^2}$

3 Solve.

 a $x^2 + \dfrac{64}{25} = 0$ b $4x^2 + 7 = 0$ c $12x^2 + 3 = 0$

EXPLORE 11.1

Without using a calculator, find i^{200}.

11.2 Complex numbers

A complex number is defined as a number of the form $x + iy$, where x and y are real values and i is the unit imaginary number.

Note: Other general cases are common too, such as $a + bi$, where a and b are real.

A complex number is often denoted by the letter z so $z = x + iy$:

- the real number x is called the real part of z, Re z
- the real number y is called the imaginary part of z, Im z.

When $x = 0$, then $z = iy$ and z is completely imaginary.

When $y = 0$, then $z = x$ and z is completely real.

TIP

Sometimes iy is considered to be Im z.

The relationship between the set of complex numbers (ℂ) and other sets of numbers is often represented using nested rings.

Equal complex numbers

Two complex numbers are equal if and only if their real parts are equal and their imaginary parts are equal.

So when $z_1 = x + iy$ and $z_2 = a + bi$ then $z_1 = z_2$ means that $x = a$ and $y = b$.

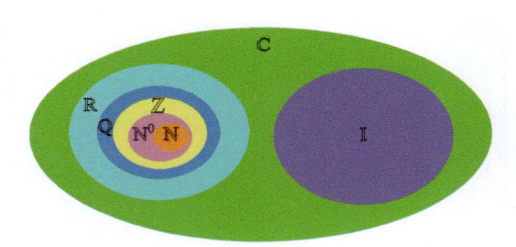

Complex conjugate

The complex conjugate of $z = x + iy$ is defined as $z^* = x - iy$.

> **TIP**
> Sometimes z^* is written as $\bar{z}$.

WORKED EXAMPLE 11.3

Solve the equation $5z^2 + 14z + 13 = 0$.

Answer

Using the quadratic formula $a = 5, b = 14, c = 13$.

$$z = \frac{-14 \pm \sqrt{14^2 - 4(5)(13)}}{10}$$

$$z = \frac{-14 \pm \sqrt{-64}}{10}$$

$$z = \frac{-14 \pm \sqrt{64 \times -1}}{10}$$

$$z = \frac{-14 \pm 8i}{10}$$

$$z = -\frac{7}{5} + \frac{4i}{5} \text{ or } -\frac{7}{5} - \frac{4i}{5}$$ .... Note that the two roots are complex conjugates of each other.

WORKED EXAMPLE 11.4

$(2x + y) + i(y - 5) = 0$

Find the value of x and the value of y.

Answer

Two complex numbers are equal if and only if their real parts are equal and their imaginary parts are equal.

$(2x + y) + i(y - 5) = 0 + 0i$

Therefore

$2x + y = 0$ ------(1)

and

$y - 5 = 0$ -------(2)

From (2): $y = 5$

Substituting into (1): $2x + 5 = 0$, $x = -\frac{5}{2}$

Chapter 11: Complex numbers

Calculating with complex numbers

The processes we need to use are the same as those we use when manipulating expressions involving surds.

For example, when $z_1 = 2 + 3i$ and $z_2 = 8 - i$ we have:

Addition
$z_1 + z_2 = 2 + 3i + 8 - i$ — Collect like terms.
$z_1 + z_2 = 10 + 2i$

Subtraction
$z_1 - z_2 = 2 + 3i - (8 - i)$ — Collect like terms.
$z_1 - z_2 = -6 + 4i$

Multiplication
$z_1 z_2 = (2 + 3i)(8 - i)$ — Expand the brackets and simplify.
$z_1 z_2 = 16 - 2i + 24i - 3i^2$ — $i^2 = -1$
$z_1 z_2 = 16 + 22i + 3$ — It is important to show all your working out clearly.
$z_1 z_2 = 19 + 22i$

$z_1 z_1^* = (2 + 3i)(2 - 3i)$ — Expand the brackets and simplify.
$z_1 z_1^* = 4 - 6i + 6i - 9i^2$ — $i^2 = -1$
$z_1 z_1^* = 4 + 9$
$z_1 z_1^* = 13$

Division
$\dfrac{z_2}{z_1} = \dfrac{8 - i}{2 + 3i}$ — Multiply the numerator and denominator by the complex conjugate of z_1.

$\dfrac{z_2}{z_1} = \dfrac{(8 - i)(2 - 3i)}{(2 + 3i)(2 - 3i)}$ — Expand the brackets and simplify.

$\dfrac{z_2}{z_1} = \dfrac{16 - 24i - 2i + 3i^2}{13}$ — $i^2 = -1$

$\dfrac{z_2}{z_1} = \dfrac{16 - 24i - 2i - 3}{13}$ — We have already found 13 so we did not show full working. You should usually show all your working to demonstrate you understand the method you are using.

$\dfrac{z_2}{z_1} = \dfrac{13 - 26i}{13} = 1 - 2i$

> **KEY POINT 11.2**
>
> When we multiply a complex number by its conjugate the result is always a real number as, if we let $z = x + iy$, and so $z^* = x - iy$, then $zz^* = (x + iy)(x - iy) = x^2 + y^2$. Since x and y are real, $x^2 + y^2$ is also real.

EXERCISE 11B

Do not use a calculator in this exercise.

1. $z_1 = 5 - 3i$ and $z_2 = 1 + 2i$

   Write each of the following in the form $x + iy$, where x and y are real values.

   a $z_1 + z_2^*$ **b** $z_1^* - z_2$ **c** $z_1 z_2$ **d** $\dfrac{z_1}{z_2}$

2. Solve these equations.

   a $z^2 + 2z + 13 = 0$ **b** $z^2 + 4z + 5 = 0$ **c** $2z^2 - 2z + 5 = 0$

   d $z^2 - 6z + 15 = 0$ **e** $3z^2 + 8z + 10 = 0$ **f** $2z^2 + 5z + 4 = 0$

3. Find the value of x and the value of y.

   a $(x + 2y) + i(3x - y) = 1 + 10i$

   b $(x + y - 4) + 2xi = (5 - y)i$

   c $(x - y) + (2x - y)i = -1$

4. Write each of the following in the form $x + iy$, where x and y are real values.

   a $(1 + 3i)(2 - i)$ **b** $(4 - 5i)(4 + 5i)$ **c** $(7 - 3i)^2$

   d $(3 - i)^4$ **e** $\dfrac{17 - i}{3 + i}$ **f** $\dfrac{-1 + 11i}{6 - 5i}$

   g $\dfrac{13(1 + i)}{2 + 3i}$ **h** $\dfrac{(3 - 2i)^2}{5 + i}$

5. **a** Show that $z = \dfrac{1}{5} + \dfrac{2\sqrt{6}}{5}i$ is a solution of the equation $5z^2 - 2z + 5 = 0$.

   b Write down the other solution of the equation.

PS 6. Find the quadratic equations that have the following roots.

   a $\alpha = -7i$ $\beta = 7i$

   b $\alpha = 1 + 5i$ $\beta = 1 - 5i$

   c $\alpha = 2 - 3i$ $\beta = 2 + 3i$

   d $\alpha = -\dfrac{5}{2} - \dfrac{\sqrt{31}}{2}i$ $\beta = -\dfrac{5}{2} + \dfrac{\sqrt{31}}{2}i$

> **TIP**
>
> $(z - \alpha)(z - \beta) = 0$
> $z^2 - (\alpha + \beta)z + \alpha\beta = 0$

PS 7. Find the complex number z satisfying the equation $z + 4 = 3iz^*$.

   Give your answer in the form $x + iy$, where x and y are real.

PS 8. Find the complex number u, given that $u(3 - 5i) = 13 + i$.

PS 9. $z = 5 + i\sqrt{3}$ is a root of a quadratic equation. Find this quadratic equation.

M 10. In an electrical circuit, the voltage (volts), current (amperes) and impedance (ohms) are related by the equation

   voltage = current × impedance

   The voltage in a particular circuit is 240 V and the impedance is $48 + 36i$ ohms.

   Find the current.

11.3 The complex plane

As complex numbers have two dimensions, one real and one imaginary, a diagrammatical representation must have two dimensions. This means it has to be a plane and not a line. Jean-Robert Argand (1768–1822) is credited for the diagram used to do this.

Argand diagrams

The complex number $x + iy$ can be represented by the Cartesian coordinates (x, y).

The horizontal axis is called the real axis and the vertical axis is called the imaginary axis.

For example, we can represent the numbers $z_1 = 2 + 3i$ and $z_2 = 5 - 3i$ by the points $A(2, 3)$ and $B(5, -3)$ on an Argand diagram.

We can also represent the numbers $z_1 = 2 + 3i$ and $z_2 = 5 - 3i$ by the position vectors $\overrightarrow{OA} = \begin{pmatrix} 2 \\ 3 \end{pmatrix}$ and $\overrightarrow{OB} = \begin{pmatrix} 5 \\ -3 \end{pmatrix}$.

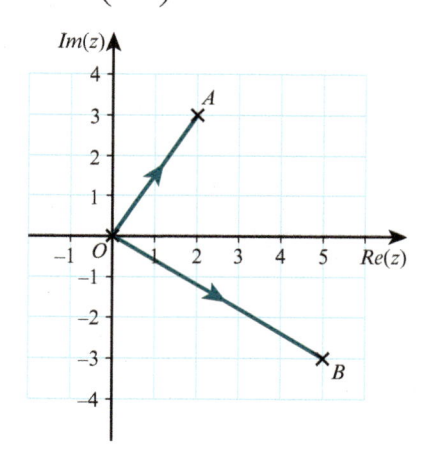

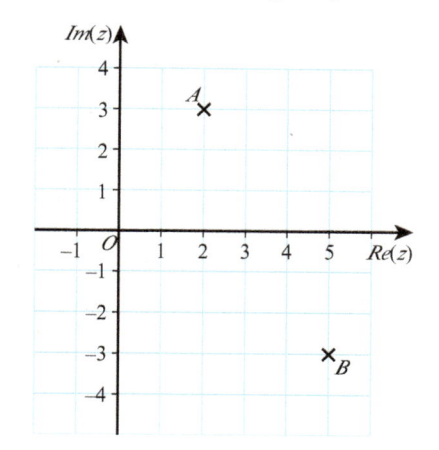

> **DID YOU KNOW?**
>
> Real numbers can be ordered along a number line. Complex numbers cannot be ordered as they are represented by points in a plane rather than on a line.

> **TIP**
>
> When drawing an Argand diagram, make sure you choose equal scales for the real and imaginary axes. This will give you a good picture and not stretch any shape you draw.

Modulus-argument form

When a complex number is written as $x + iy$ where x and y are real values, we say it is in **Cartesian form**. There are other ways of writing complex numbers.

The vector representation of a complex number on an Argand diagram helps us to understand a different representation called the **modulus-argument form**.

Before we do this, we need to define the **modulus** and the **argument** of a complex number.

The **modulus of a complex number** $x + iy$ is the magnitude of the position vector $\begin{pmatrix} x \\ y \end{pmatrix}$.

Therefore, the modulus of $x + iy$ is defined to be $\sqrt{x^2 + y^2}$.

Notation: The modulus of $z = x + iy$ is $|z|$.

Also note that the plural of modulus is moduli.

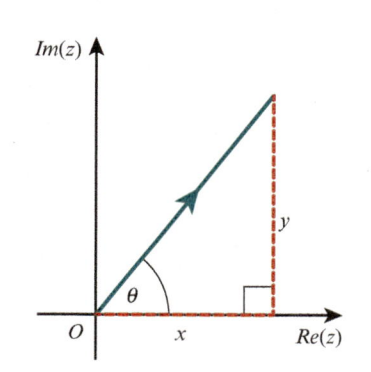

Cambridge International AS & A Level Mathematics: Pure Mathematics 2 & 3

The **argument of a complex number** $x + iy$ is the direction of the position vector $\begin{pmatrix} x \\ y \end{pmatrix}$.

More precisely, it is the angle made between the positive real axis and the position vector. The **principal argument**, θ, of a complex number is an angle such that $-\pi < \theta \leqslant \pi$. Sometimes it might be more convenient to give θ as an angle such that $0 \leqslant \theta < 2\pi$. Usually, the argument is given in radians.

Notation: The argument of $z = x + iy$ is arg z.

We can use trigonometry to find the argument. For example, when θ is acute, $\tan \theta = \dfrac{y}{x}$.

A diagram **must** be drawn first to check the position of the complex number.

WORKED EXAMPLE 11.5

Find the modulus and argument of each of the following complex numbers.

   a $5 + 12i$ **b** $-3 + 4i$ **c** $12 - 5i$ **d** $-4 - 3i$

Answer

A simple sketch of the Argand diagram is always necessary when finding the argument.

a $|5 + 12i| = \sqrt{5^2 + 12^2} = \sqrt{169} = 13$ *From the definition of the modulus.*

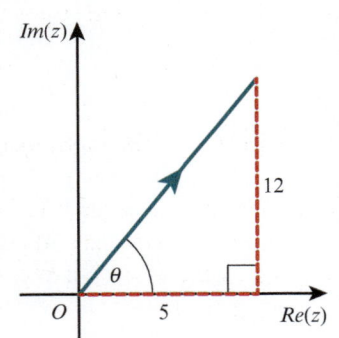

The diagram shows that the angle is in the first quadrant and so the angle is acute and positive.

$\arg(5 + 12i) = \tan^{-1}\left(\dfrac{12}{5}\right) = 1.1760...$ *From the definition of the argument.*

$= 1.18$ radians correct to 3 significant figures

b $|-3 + 4i| = \sqrt{(-3)^2 + 4^2} = \sqrt{25} = 5$ *From the definition of the modulus.*

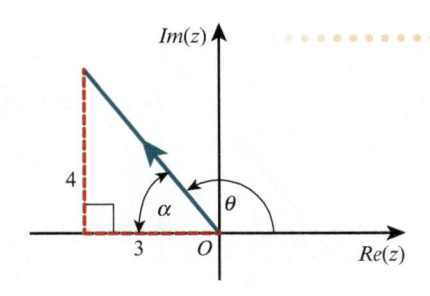

The diagram shows that the angle is in the second quadrant and is obtuse and positive. This is calculated by subtracting the value of the acute angle α from π. Remember that the principal argument is $-\pi < \theta \leqslant \pi$.

$\theta = \pi - \alpha$

$$\arg(-3 + 4i) = \pi - \tan^{-1}\left(\frac{4}{3}\right) = 2.214\ldots$$ *From the definition of the argument.*

$= 2.21$ radians correct to 3 significant figures

c $|12 - 5i| = \sqrt{12^2 + (-5)^2} = \sqrt{169} = 13$ *From the definition of the modulus.*

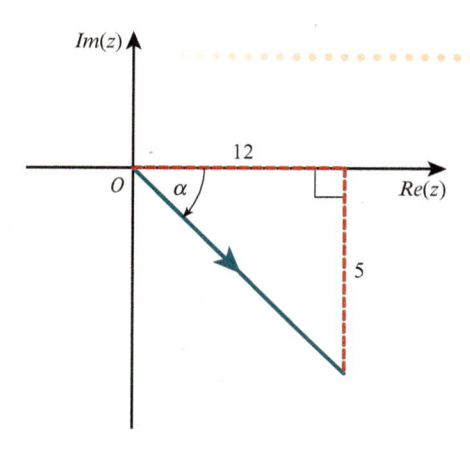

The diagram shows that the angle is in the fourth quadrant and is acute and negative. Again, recall that the principal argument is $-\pi < \theta \leq \pi$.

$\theta = -\alpha$

$$\arg(12 - 5i) = -\tan^{-1}\left(\frac{5}{12}\right) = -0.3947\ldots$$ *From the definition of the argument.*

$= -0.395$ radians correct to 3 significant figures

d $|-4 - 3i| = \sqrt{(-4)^2 + (-3)^2} = \sqrt{25} = 5$ *From the definition of the modulus.*

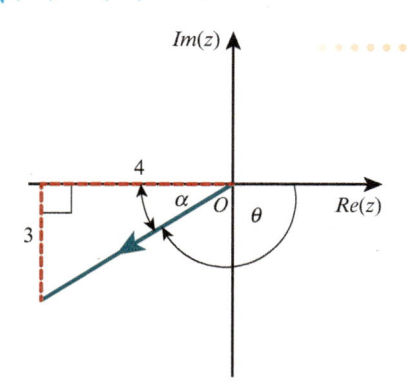

The diagram shows the angle is in the third quadrant and is obtuse and negative. Again recall that the principal argument is $-\pi < \theta \leq \pi$.

$\theta = -\pi + \alpha$

$$\arg(-4 - 3i) = -\pi + \tan^{-1}\left(\frac{3}{4}\right) = -2.498\ldots$$ *From the definition of the argument.*

$= -2.50$ radians correct to 3 significant figures

Using trigonometry, we can now write a complex number in modulus-argument form.

$\cos\theta = \dfrac{x}{r}$ and $\sin\theta = \dfrac{x}{r}$.

Therefore, $x = r\cos\theta$ and $y = r\sin\theta$.

Substituting these expressions into the Cartesian form of a complex number, we have

$z = x + iy$

$z = r\cos\theta + i(r\sin\theta)$

$z = r(\cos\theta + i\sin\theta)$

where r is $|z|$ and θ is $\arg z$.

When a complex number is written as $z = r(\cos\theta + i\sin\theta)$ where r is $|z|$ and θ is $\arg z$, we say it is in modulus-argument form.

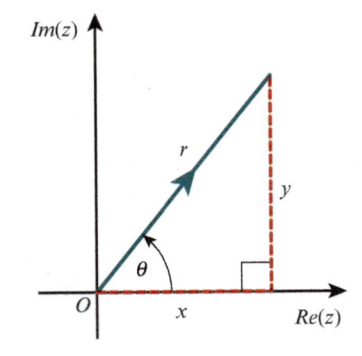

WORKED EXAMPLE 11.6

$u = 6 - 3i$ and $w = -7 + 5i$

Write each of u and w in modulus-argument form.

Answer

$u = 6 - 3i$

$r = |u| = \sqrt{6^2 + (-3)^2}$ .... From the definition of the modulus.

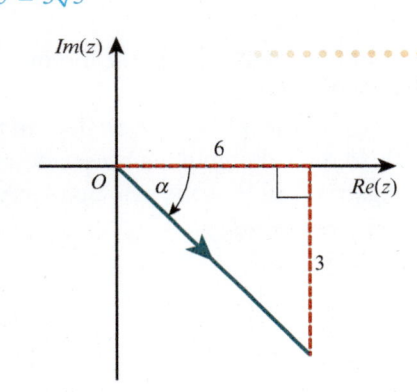

$\quad\quad = \sqrt{45} = 3\sqrt{5}$

.... The diagram shows the angle to be acute and negative.

$\theta = -\alpha$

$\arg(6 - 3i) = -\tan^{-1}\left(\dfrac{3}{6}\right)$

$\quad\quad\quad\quad = -0.4636...$ .... From the definition of the argument.

$\quad\quad\quad\quad = -0.464$ radians correct to 3 significant figures

$u = 3\sqrt{5}(\cos(-0.464) + i\sin(-0.464))$

$w = -7 + 5i$

$r = |w| = \sqrt{(-7)^2 + 5^2}$ .... From the definition of the modulus.

$\quad\quad = \sqrt{74}$

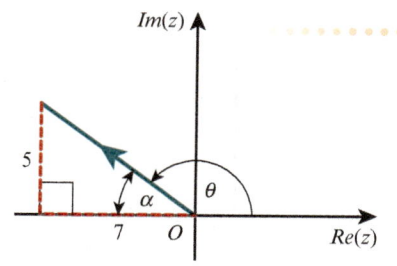

$\theta = \pi - \alpha$

$\arg(-7 + 5i) = \pi - \tan^{-1}\left(\dfrac{5}{7}\right)$

$= 2.5213...$

$= 2.52$ radians correct to 3 significant figures

$w = \sqrt{74}(\cos(2.52) + i\sin(2.52))$

> The diagram shows the angle to be obtuse and positive.

> From the definition of the argument.

Exponential form

The mathematician Leonhard Euler (1707–1783) discovered the relationship $\cos\theta + i\sin\theta = e^{i\theta}$.

This is the basis for another representation, called the **exponential form**.

$z = r(\cos\theta + i\sin\theta)$

$z = re^{i\theta}$

where r is $|z|$ and θ is $\arg z$.

The coordinates (r, θ), where r is $|z|$ and θ is $\arg z$, are called the **polar coordinates**.

When a complex number is written as $z = re^{i\theta}$, where r is $|z|$ and θ is $\arg z$, we say it is in exponential form.

The modulus-argument and exponential forms of a complex number are **polar forms**. A polar form uses polar coordinates; a **Cartesian form** uses Cartesian coordinates.

WORKED EXAMPLE 11.7

$z_1 = 1 + i$ $z_2 = 5e^{\frac{i\pi}{3}}$ $z_3 = 2\left(\cos\dfrac{\pi}{12} + i\sin\dfrac{\pi}{12}\right)$

a Write z_1 in
- modulus-argument form
- exponential form.

b Write z_2 in
- modulus-argument form
- Cartesian form.

c Write z_3 in
- exponential form
- Cartesian form.

Answer

a

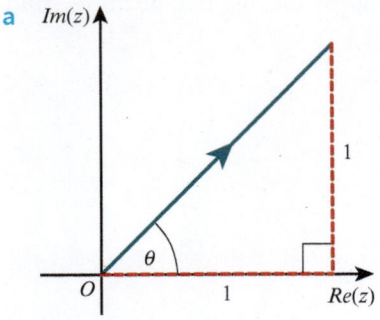

$|z_1| = \sqrt{1^2 + 1^2} = \sqrt{2}$ — From the definition of the modulus.

The diagram shows the angle to be acute and positive.

$\arg(1+i) = \tan^{-1}(1) = \dfrac{\pi}{4}$ — From the definition of the argument.

Modulus-argument form:

$$z_1 = \sqrt{2}\left(\cos\dfrac{\pi}{4} + i\sin\dfrac{\pi}{4}\right)$$

Exponential form:

$$z_1 = \sqrt{2}\,e^{\frac{i\pi}{4}}$$

b $|z_2| = 5 \quad \theta = \dfrac{\pi}{3}$

Modulus-argument form:

$$z_2 = 5\left(\cos\dfrac{\pi}{3} + i\sin\dfrac{\pi}{3}\right)$$

Comparing the modulus-argument form with $x + iy$.

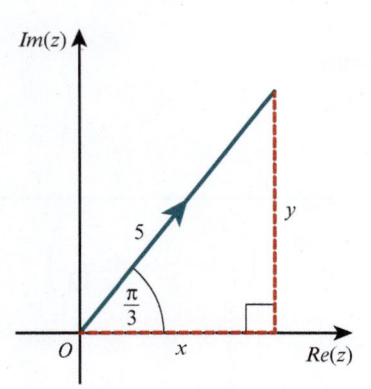

$x = 5\cos\dfrac{\pi}{3} \quad y = 5\sin\dfrac{\pi}{3}$ — Evaluate.

$x = \dfrac{5}{2} \quad y = \dfrac{5\sqrt{3}}{2}$ — Write in Cartesian form, $x + iy$.

Cartesian form:

$$z_2 = \dfrac{5}{2} + i\left(\dfrac{5\sqrt{3}}{2}\right)$$

c $|z_3| = 2$ $\theta = \dfrac{\pi}{12}$

Exponential form: $z_3 = 2e^{\frac{i\pi}{12}}$ .... Comparing the modulus-argument form with $x + iy$.

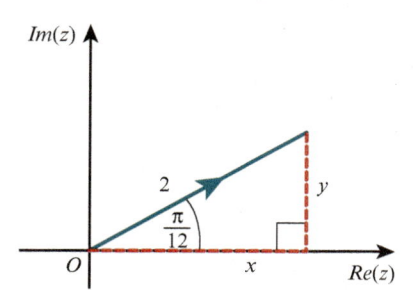

$x = 2\cos\dfrac{\pi}{12}$ $y = 2\sin\dfrac{\pi}{12}$ .... Evaluate.

$x = \dfrac{\sqrt{6} + \sqrt{2}}{2}$ $y = \dfrac{\sqrt{6} - \sqrt{2}}{2}$ .... Write in Cartesian form, $x + iy$.

Cartesian form: $z_3 = \dfrac{\sqrt{6} + \sqrt{2}}{2} + i\left(\dfrac{\sqrt{6} - \sqrt{2}}{2}\right)$

WORKED EXAMPLE 11.8

$u = 5e^{2i}$ and $w = 10e^{-3i}$

a Find the modulus and argument of:

  i uw ii $\dfrac{w}{u}$

b Show u, w, uw and $\dfrac{w}{u}$ on a single Argand diagram.

Answer

a i $uw = 5 \times e^{2i} \times 10 \times e^{-3i}$ .... Using laws of indices.

   $uw = 50 \times e^{2i-3i}$

   $uw = 50e^{-i}$

 ii $\dfrac{w}{u} = \dfrac{10 \times e^{-3i}}{5 \times e^{2i}}$ .... Using laws of indices.

   $\dfrac{w}{u} = 2 \times e^{-3i-2i}$

   $\dfrac{w}{u} = 2e^{-5i}$

   -5 is less than $-\pi$ and so this argument is not a principal argument.

   The equivalent principal argument, in this example, is $\theta = 2\pi - 5$

   $\dfrac{w}{u} = 2e^{(2\pi-5)i}$

TIP

The exponential form of a complex number is very useful when multiplying or dividing two complex numbers.

b
$$u = 5(\cos 2 + i \sin 2) = -2.08 + 4.55i \qquad A \approx (-2, 4.5)$$
$$w = 10(\cos(-3) + i \sin(-3)) = -9.90 - 1.41i \qquad B \approx (-10, -1.5)$$
$$uw = 50(\cos(-1) + i \sin(-1)) = 27.0 - 42.1i \qquad C \approx (27, -42)$$
$$\frac{w}{u} = 2(\cos(2\pi - 5) + i \sin(2\pi - 5)) = 0.567 + 1.92i \qquad D \approx (0.5, 2)$$

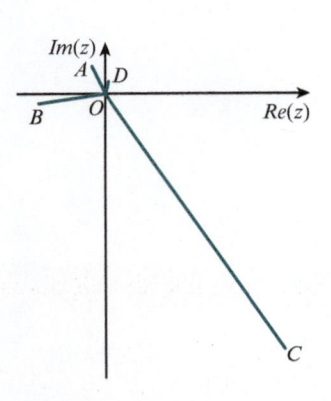

> **TIP**
>
> It is not essential to find the Cartesian form, but you might find it easier to plot in the correct quadrant if you do.

Using laws of indices and the exponential form of a complex number, we can derive some very important and useful results.

Let the complex number z_1 be such that $|z_1| = r_1$ and $\arg z_1 = \theta_1$ and let the complex number z_2 be such that $|z_2| = r_2$ and $\arg z_2 = \theta_2$ then:

$$z_1 = r_1 e^{i\theta_1}, \ z_2 = r_2 e^{i\theta_2}, \ z_1 z_2 = r_1 e^{i\theta_1} \times r_2 e^{i\theta_2} = r_1 r_2 e^{i(\theta_1 + \theta_2)} \text{ and } \frac{z_1}{z_2} = \frac{r_1 e^{i\theta_1}}{r_2 e^{i\theta_2}} = \frac{r_1}{r_2} e^{i(\theta_1 - \theta_2)}$$

Therefore, the complex number $z_1 z_2$ is such that:

$$|z_1 z_2| = r_1 r_2 = |z_1||z_2| \text{ and } \arg(z_1 z_2) = \theta_1 + \theta_2 = \arg z_1 + \arg z_2,$$

and the complex number $\frac{z_1}{z_2}$ is such that:

$$\left|\frac{z_1}{z_2}\right| = \frac{r_1}{r_2} = \frac{|z_1|}{|z_2|} \text{ and } \arg\left(\frac{z_1}{z_2}\right) = \theta_1 - \theta_2 = \arg z_1 - \arg z_2.$$

These results greatly simplify the multiplication and division of complex numbers.

We can see that, when in exponential form,

- to multiply two complex numbers we multiply the moduli and add the arguments
- to divide two complex numbers we divide the moduli and subtract the arguments.

WORKED EXAMPLE 11.9

The complex numbers u and w are given by $u = 9 + 12i$ and $w = -5 - 3i$.

Find the modulus and argument of:

a uw **b** $\dfrac{u}{w}$

Answer

$$|u| = \sqrt{9^2 + 12^2} = \sqrt{225} = 15 \qquad \text{By definition for the moduli of } u \text{ and } w.$$
$$|w| = \sqrt{(-5)^2 + (-3)^2} = \sqrt{34}$$

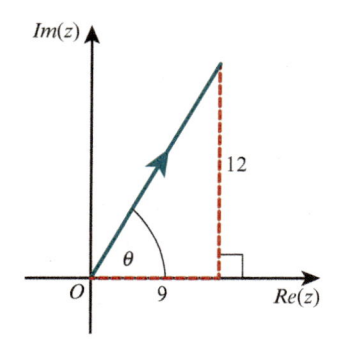

From the diagram, the angle for u is acute and positive.

$$\arg u = \tan^{-1}\left(\frac{12}{9}\right) = 0.9272...$$

$= 0.927$ radians correct to 3 significant figures

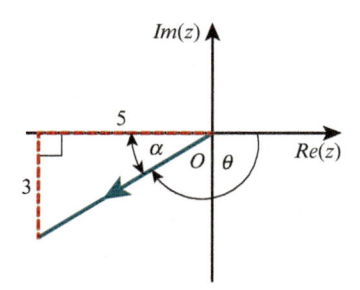

From the diagram, the angle for w is obtuse and negative.

$\theta = -\pi + \alpha$

$$\arg w = -\pi + \tan^{-1}\left(\frac{3}{5}\right) = -2.601$$

$= -2.60$ radians correct to 3 significant figures

a $|uw| = |u||w| = 15\sqrt{34}$

Using $|z_1 z_2| = r_1 r_2 = |z_1||z_2|$

$\arg(uw) = \arg u + \arg w = 0.9272... + (-2.601...)$
$= -1.67$ radians correct to 3 significant figures

Using $\arg(z_1 z_2) = \theta_1 + \theta_2 = \arg z_1 + \arg z_2$

b $\left|\dfrac{u}{w}\right| = \dfrac{|u|}{|w|} = \dfrac{15}{\sqrt{34}} = \dfrac{15\sqrt{34}}{34}$

Using $\left|\dfrac{z_1}{z_2}\right| = \dfrac{r_1}{r_2} = \dfrac{|z_1|}{|z_2|}$

$\arg\left(\dfrac{u}{w}\right) = \arg u - \arg w = 0.9272... - (-2.601...)$
$= 3.528$ radians

Using $\arg\left(\dfrac{z_1}{z_2}\right) = \theta_1 - \theta_2 = \arg z_1 - \arg z_2$

However, $3.528...$ is more than π and so this argument is not a principal argument. The equivalent principal argument, in this example, is

$\theta = -(2\pi - 3.528...)$
$= -2.754... = -2.75$ radians correct to 3 significant figures

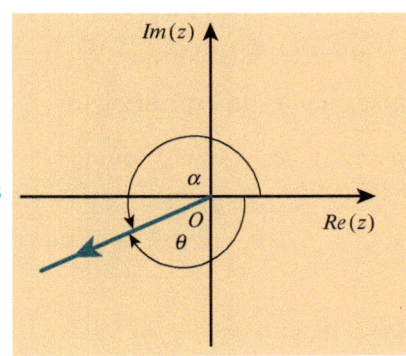

Cambridge International AS & A Level Mathematics: Pure Mathematics 2 & 3

EXERCISE 11C

1. $u = 5 - 2i$

   a On an Argand diagram, show the points A, B and C representing the complex numbers u, u^* and $-u$, respectively.

   b The points $ABCD$ form a rectangle. Write down the complex number represented by the point D.

2. $z_1 = 1 + 5i$ and $z_2 = -7 - i$

   a On an Argand diagram, show the points P and Q representing the complex numbers z_1 and z_2, respectively.

   b Write down the complex number represented by the point R, the midpoint of PQ.

3. Find the modulus and argument of each of the following complex numbers.

   a $-12 + 5i$ b $5i$ c $8 + 15i$

   d $60 - 11i$ e $-9 - 40i$ f $-1 - i\sqrt{3}$

   g $\sqrt{5} + 2i$ h $-24 - 7i$ i $k(1 - i)$, where $k > 0$

4. [Argand diagram showing points A at approximately $(-1, 3)$, B at approximately $(2, 1)$, and C at approximately $(0, -3)$ with $-3 \le \text{Re}(z) \le 3$ and $-4 \le \text{Im}(z) \le 4$]

   a Write each of the complex numbers shown in the Argand diagram in modulus-argument form.

   b Show that ABC is a right-angled, isosceles triangle.

5. Write each of these complex numbers in Cartesian form.

   a $3\left(\cos\dfrac{\pi}{3} + i\sin\dfrac{\pi}{3}\right)$ b $5\left(\cos\dfrac{3\pi}{8} + i\sin\dfrac{3\pi}{8}\right)$

   c $\dfrac{e^{i\pi}}{2}$ d $3e^{-\frac{i\pi}{4}}$

6. Given that $z(4 - 9i) = 8 + 3i$, find the value of $|z^2|$ and the value of $\arg z^2$.

7. $w = 5\left(\cos\dfrac{\pi}{6} + i\sin\dfrac{\pi}{6}\right)$ and $z = \dfrac{3 - 7i}{5 - 2i}$

   a Write w in exponential form.

   b Write z in exponential form.

   c Find and simplify an expression for $\dfrac{z}{w}$.

8. **PS** The complex number $a + bi$ is denoted by z. Given that $|z| = 5$ and $\arg z = \dfrac{\pi}{6}$, find the value of a and the value of b.

9. Given that $z = r(\cos\theta + i\sin\theta)$, find and simplify an expression for

   a zz^* b $\dfrac{z}{z^*}$

10. **P** $z = r(\cos\theta + i\sin\theta)$

    a Show that $z^2 = r^2(\cos 2\theta + i\sin 2\theta)$.

    b Show that $z - \dfrac{1}{z} = i(2r\sin\theta)$.

Chapter 11: Complex numbers

EXPLORE 11.2

Using Argand diagrams and position vectors, investigate the transformations of the plane that are:

- the geometrical relationship between a complex number and its conjugate
- the geometrical effects of combining two complex numbers by addition, subtraction, multiplication and division.

11.4 Solving equations

We have already solved quadratic equations with complex roots.

From the Pure Mathematics 1 Coursebook, Chapter 1, we know that when solving quadratic equations for solutions that are real:

$b^2 - 4ac$	Nature of roots	Roots
> 0	2 real and distinct roots	α and β
$= 0$	2 real and equal roots (repeated roots)	α and β, $\alpha = \beta$
< 0	0 real roots	

We can now update this for solving quadratic equations for solutions that are complex:

$b^2 - 4ac$	Nature of roots	Roots
> 0	2 real and distinct roots	α and β
$= 0$	2 real and equal roots (repeated roots)	α and β, $\alpha = \beta$
< 0	2 complex roots of the form $x \pm iy$ with $y \neq 0$ (a complex conjugate pair). This means that one root is the conjugate of the other root.	α and α^*

Using the set of complex numbers to find solutions, a quadratic equation always has two roots.

Higher order polynomial equations

We solved cubic and quartic equations for real roots in Chapter 1.

We will now consider solving these equations for complex roots.

The Fundamental Theorem of Algebra states that a polynomial equation of degree n has n complex roots.

TIP

The degree of a quadratic equation is 2, the degree of a cubic equation is 3... the degree of a polynomial equation is the same as its largest power.

Cambridge International AS & A Level Mathematics: Pure Mathematics 2 & 3

> **WORKED EXAMPLE 11.10**
>
> **a** Show that $z = 4$ is a root of the equation $z^3 - 3z^2 - 3z - 4 = 0$.
>
> **b** Find all complex solutions of $z^3 - 3z^2 - 3z - 4 = 0$.
>
> **Answer**
>
> **a** Let $f(z) = z^3 - 3z^2 - 3z - 4$ then
>
> $$f(4) = 4^3 - 3(4^2) - 3(4) - 4$$
> $$= 64 - 48 - 12 - 4 = 64 - 64 = 0 \checkmark$$
>
> **b** As $z = 4$ is a root then $z - 4$ is a factor, therefore
>
> $z^3 - 3z^2 - 3z - 4 = (z - 4)(z^2 + az + 1)$ .... Multiply out and compare.
>
> $$z^3 - 3z^2 - 3z - 4 = z^3 + az^2 + z$$
> $$ - 4z^2 - 4az - 4$$
> $$\overline{ z^3 + (a-4)z^2 + (1-4a)z - 4}$$
>
> .... This can also be done by synthetic or long division.
>
> $z^3 - 3z^2 - 3z - 4 = z^3 + (a-4)z^2 + (1-4a)z - 4$ .... Compare the coefficients of z^2.
>
> $-3 = a - 4$, therefore $a = 1$ .... Check, using the coefficients of z.
>
> $-3 = 1 - 4(1) \checkmark$
>
> $(z - 4)(z^2 + z + 1) = 0$
>
> $z = 4$ .... Given.
>
> $z^2 + z + 1 = 0$
>
> $z = \dfrac{-1 \pm \sqrt{1^2 - 4(1)(1)}}{2}$
>
> $z = \dfrac{-1 \pm \sqrt{-3}}{2}$
>
> $z = \dfrac{-1 \pm i\sqrt{3}}{2}$

> **KEY POINT 11.3**
>
> When a polynomial equation with real coefficients has complex roots, they occur in complex conjugate **pairs**.

A **cubic** equation has three roots (by the Fundamental Theorem of Algebra).

A cubic equation can always be written as the product of a linear factor and a quadratic factor.

This means that a cubic equation must always have **at least one real root**.

Cubic equations either have

- three real roots or
- one real root and two complex roots of the form $x \pm iy$ with $y \neq 0$ (a complex conjugate pair).

> **TIP**
> To understand this, think about the shape of the graph of a cubic equation.

A **quartic** equation has four roots (by the Fundamental Theorem of Algebra).

Quartic equations either have

- four real roots or
- two real and two complex roots of the form $x \pm iy$ with $y \neq 0$ (a complex conjugate pair) or
- four complex roots of the form $x \pm iy$ with $y \neq 0$ (two complex conjugate pairs).

A quartic equation can **never** have one real and three complex roots. Make sure you understand why this is true.

WORKED EXAMPLE 11.11

It is given that $z - 2i$ is a factor of the equation $z^4 + 13z^2 + 36 = 0$. Solve this equation.

Answer

As $z - 2i$ is a factor $z + 2i$ must also be a factor .... Complex roots of the form $x \pm iy$ with $y \neq 0$ occur in pairs.

$(z - 2i)(z + 2i) = z^2 + 2iz - 2iz - 4i^2$
$= z^2 + 4$

$z^4 + 13z^2 + 36 = (z^2 + 4)(z^2 + kz + 9)$ .... Multiply out and compare.

$$z^4 + 13z^2 + 36 = z^4 + kz^3 + 9z^2$$
$$ + 4z^2 + 4kz + 36$$
$$\overline{ z^4 + kz^3 + 13z^2 + 4kz + 36}$$

This can also be done by synthetic or long division.

$z^4 + 13z^2 + 36 = z^4 + kz^3 + 13z^2 + 4kz + 36$ .... Compare the coefficients of z^3.

$k = 0$ .... Check, using the coefficients of z.
$0 = 4(0)$ ✓
$z^4 + 13z^2 + 36 = (z^2 + 4)(z^2 + 9) = 0$

$z = \pm 2i$ .... From factors given.
$z^2 + 9 = 0$ so $z = \pm\sqrt{-9}$
$z = \pm 3i$

$z = \pm 2i, z = \pm 3i$ .... 4 complex roots of the form $x \pm iy$ with $y \neq 0$.

Finding the square roots of a complex number

We can find the square roots of a complex number.

WORKED EXAMPLE 11.12

Find the square roots of $-2 - (2\sqrt{3})i$ and show the position of these roots on an Argand diagram.

Answer

Find $x + iy$, where x and y are real, such that $x + iy = \pm\sqrt{-2 + (2\sqrt{3})i}$.

$x + iy = \pm\sqrt{-2 + (2\sqrt{3})i}$ .... Square both sides.

$(x + iy)^2 = -2 + (2\sqrt{3})i$ .... Multiply out the left-hand side.

$x^2 + 2xyi + i^2 y^2 = -2 + (2\sqrt{3})i$ .... Collect real and imaginary parts.

$(x^2 - y^2) + (2xy)i = -2 + (2\sqrt{3})i$ .... Equate real and imaginary parts.

$x^2 - y^2 = -2$ ------(1)

$2xy = 2\sqrt{3}$ --------(2) .... Solve simultaneously.

From (2): $y = \dfrac{\sqrt{3}}{x}$ and so $x^2 - \dfrac{3}{x^2} = -2$ .... Multiply by x^2.

$x^4 - 3 = -2x^2$ .... Rearrange and let $u = x^2$.

$u^2 + 2u - 3 = 0$ .... Solve.

$(u + 3)(u - 1) = 0$

$u = -3, \ u = 1$

$x^2 = -3, \ x^2 = 1$ .... x is real, so disregard $x^2 = -3$.

$x = \pm 1$ .... Substitute into $y = \dfrac{\sqrt{3}}{x}$.

$y = \dfrac{\sqrt{3}}{\pm 1} = \pm\sqrt{3}$

Therefore, the square roots of $-2 - (2\sqrt{3})i$ are $1 + i\sqrt{3}$ and $-1 - i\sqrt{3}$.

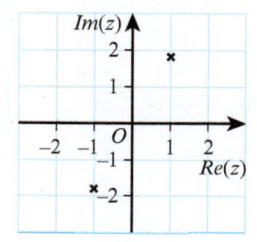

> Notice that the modulus of the two square roots is the same and the difference between the arguments of the square roots is π radians.

> **TIP**
> When we want to work in exponential form, we square $re^{i\theta}$ to obtain $r^2 e^{i(2\theta)}$. We then equate this to the complex number whose square roots we are finding.

Chapter 11: Complex numbers

The cube roots of one (unity)

By the Fundamental Theorem of Algebra, $z^3 - 1 = 0$ has three roots.

We know that one of these roots is 1, since $z = \sqrt[3]{1} = 1$ is the real solution we are familiar with.

To find the two complex roots of the form $x \pm iy$ with $y \neq 0$, we can use this method:

As $z = 1$ is a root then $z - 1$ is a factor and so

$z^3 - 1 = (z-1)(z^2 + kz + 1)$ Multiply out and compare.

$z^3 - 1 = z^3 + kz^2 + z$
$ - z^2 - kz - 1$
$\overline{ z^3 + (k-1)z^2 + (1-k)z - 1}$

This can also be done by synthetic or long division.

$z^3 - 1 = z^3 + (k-1)z^2 + (1-k)z - 1$ Compare the coefficients of z^2.

$k - 1 = 0, k = 1$ Check, using the coefficients of z.

$0 = 1 - 1$ ✓

$z^3 - 1 = (z-1)(z^2 + z + 1)$ Solve.

$z = 1, z = \dfrac{-1 \pm \sqrt{1^2 - 4(1)(1)}}{2}, z = \dfrac{-1 \pm \sqrt{-3}}{2}$

$z = 1, z = \dfrac{-1 + i\sqrt{3}}{2}, z = \dfrac{-1 - i\sqrt{3}}{2}$

These roots can be very useful when solving related equations.

WORKED EXAMPLE 11.13

Solve $z^3 = 8$.

Answer

$z = \sqrt[3]{8} = 2\sqrt[3]{1}$ As z is $2 \times \sqrt[3]{1}$ replace $\sqrt[3]{1}$ by each of the cube roots of 1 we have already found.

$z = 2 \times 1, z = 2 \times \dfrac{-1 + i\sqrt{3}}{2}, z = 2 \times \dfrac{-1 - i\sqrt{3}}{2}$ Simplify.

$z = 2, z = -1 + i\sqrt{3}, z = -1 - i\sqrt{3}$

EXERCISE 11D

1 a Given that $z_1 = -\text{i}$ is a root of the equation $z^3 + z^2 + z + k = 0$, find the value of the constant k.

   b Write down the other complex root of the equation and find the third root, stating whether it is real or complex.

2 Solve:

   a $(z-5)^3 = 8$

   b $(2z+3)^3 = \dfrac{1}{64}$.

3 Find the roots of the equation $2z^2 + z + 3 = 0$, giving your answer in modulus-argument form.

4 It is given that $z - 3$ is a factor of $z^3 - 3z^2 + 25z - 75$.

   Solve $z^3 - 3z^2 + 25z - 75 = 0$.

5 $(x + \text{i}y)^2 = 55 + 48\text{i}$

   Find the value of x and the value of y, when x and y are real and positive.

6 It is given that $2z + 1$ is a factor of $2z^3 - 11z^2 + 14z + 10$.

   Solve $2z^3 - 11z^2 + 14z + 10 = 0$.

7 Given that $z = 3\text{i}$ is a root, solve $z^4 - 2z^3 + 14z^2 - 18z + 45 = 0$.

8 Find the square roots of:

   a $24 - 10\text{i}$

   b $7 + (6\sqrt{2})\text{i}$

   c $\dfrac{5}{4} - \text{i}\sqrt{6}$

   d $7 - 24\text{i}$

   e $-4 + (2\sqrt{5})\text{i}$

   f $\dfrac{1}{2} e^{\frac{\pi}{2}\text{i}}$

9 $z^3 - 12z^2 + pz + q = 0$ where p and q are real constants.

   a Given that $z = 2 + \text{i}$ is a root of this equation, write down another root of the equation.

   b Find the value of p and the value of q.

   c Represent the roots of this equation on an Argand diagram.

PS 10 $\text{f}(z) = z^4 - z^3 + 20z^2 - 16z + 64 = (z^2 + a)(z^2 + bz + 4)$

   Solve the equation $\text{f}(z) = 0$.

11.5 Loci

A **locus** is a path traced out by a point as it moves following a particular rule. The rule is expressed as an inequality or an equation.

In order to draw the correct locus on an Argand diagram, we need to interpret the inequality or equation we have been given.

Circles with centre (0, 0)

- $|z| = r$, $|z| < r$, $|z| \leq r$ where r is a constant.

In this case, the length of the position vector representing the complex number, z, is the constant value r. The argument of z can vary.

A point that moves so that is it always the same distance from a fixed point is a circle. The fixed point is the centre of the circle. The fixed distance is the radius of the circle.

Recall that a position vector is simply a means of locating a point in space relative to an origin, usually O.

For the complex number $z = x + iy$, where x and y are real, the locus of the point $P(x, y)$ that satisfies:

- $|z| = r$ is the circumference of a circle with centre (0, 0) and radius r
- $|z| < r$ is all points within a circle with centre (0, 0) and radius r, but the circumference is not included
- $|z| \leq r$ is all points within a circle with centre (0, 0) and radius r or the circumference of the circle.

WORKED EXAMPLE 11.14

On separate Argand diagrams, sketch these loci.

a $|z| = 5$ b $|z| < 3$ c $|z| \leq 4$

Answer

a $|z| = 5$ is the circumference of a circle, centre (0, 0), radius 5.

It is **only** the circle, not the points within the circle.

The circle is drawn as a solid line as it is included.

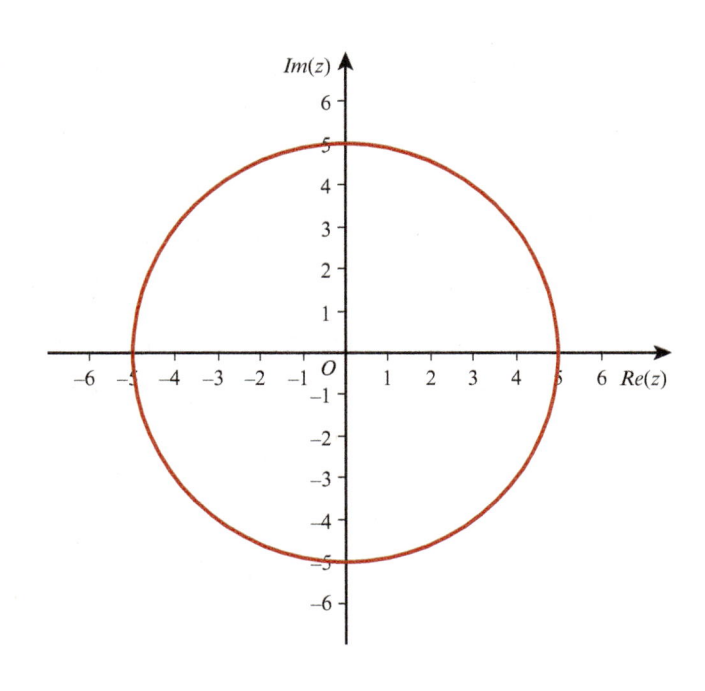

b $|z| < 3$ is a circle, centre (0, 0), radius 3.

It is only the points **within** the circle.

The circumference of the circle is drawn as a dotted line as it is **not** included.

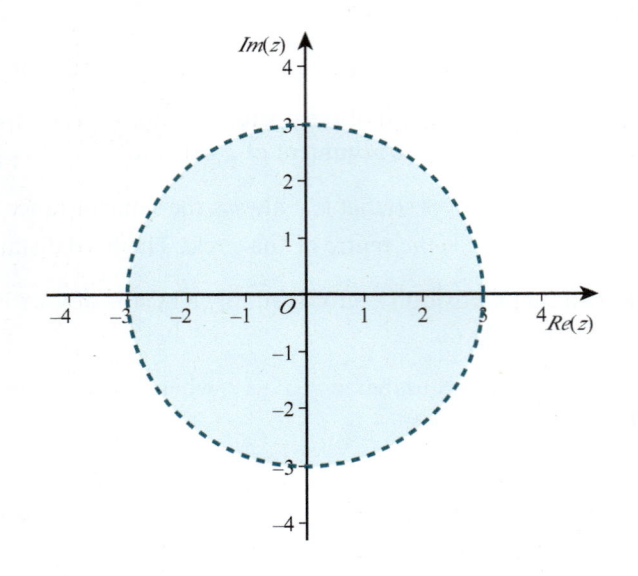

c $|z| \leq 4$ is a circle, centre (0, 0), radius 4.

It is **all** points within the circle and on the circumference.

The circumference of the circle is drawn as a solid line as it is included.

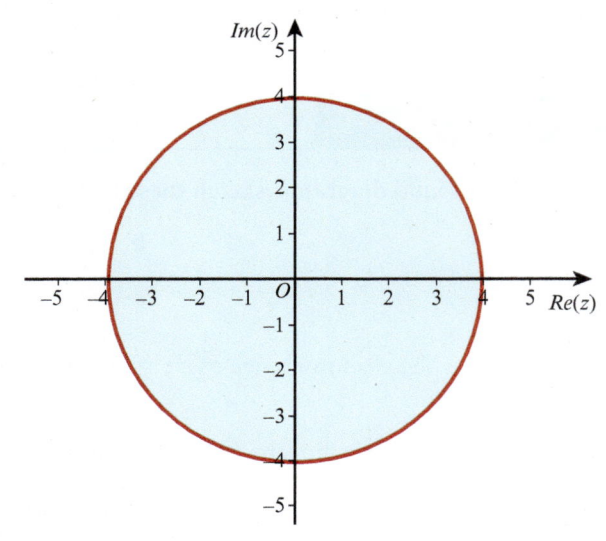

Circles with centre (a, b)

- $|z - z_1| = r$, $|z - z_1| < r$, $|z - z_1| \leq r$ where r is a constant and z_1 is the fixed point $a + b\text{i}$.

We need to understand the vector representation of the complex number $z - z_1$.

Let $P(x, y)$ be the moving point that represents $z = x + \text{i}y$ on an Argand diagram.

Let $Q(a, b)$ be the fixed point that represents $z_1 = a + b\text{i}$ on the Argand diagram.

$z = \overrightarrow{OP}$ and $z_1 = \overrightarrow{OQ}$

Therefore, $\overrightarrow{QP} = -z_1 + z$

$\overrightarrow{QP} = z - z_1$

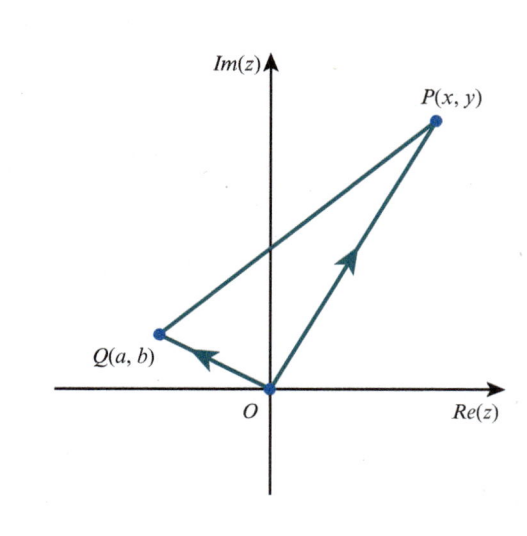

We now interpret $|z - z_1| = r$ as:

the length of the vector $\overrightarrow{QP}$ is r.

For the complex number $z = x + iy$, where x and y are real, and the fixed point $z_1 = a + bi$, where a and b are real, the locus of the point $P(x, y)$ that satisfies

- $|z - z_1| = r$ is the circumference of a circle with centre (a, b) and radius r
- $|z - z_1| < r$ is all points within a circle with centre (a, b) and radius r, but the circumference is not included
- $|z - z_1| \leqslant r$ is all points within a circle with centre (a, b) and radius r or the circumference of the circle.

WORKED EXAMPLE 11.15

On separate Argand diagrams, sketch these loci.

a $|z - (2 + 4i)| = 3$ **b** $|z - 3i| < 4$ **c** $|z + 5| \leqslant 5$

Answer

a $z_1 = 2 + 4i$

$|z - (2 + 4i)| = 3$ is the circumference of a circle, centre $(2, 4)$, radius 3.

It is **only** the circle, not the points within the circle.

The circle is drawn as a solid line as it is included.

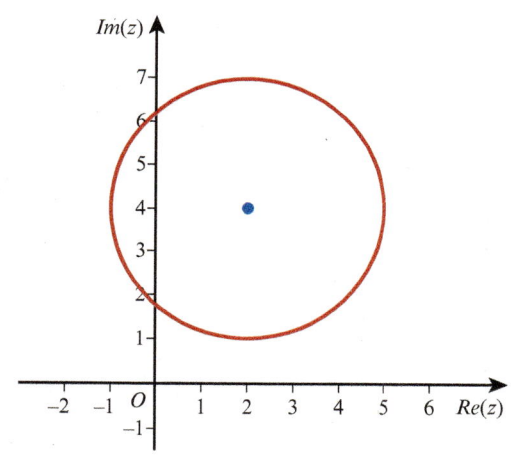

b $z_1 = 3i$

$|z - 3i| < 4$ is a circle, centre $(0, 3)$, radius 4.

It is only the points **within** the circle.

The circumference of the circle is drawn as a dotted line as it is not included.

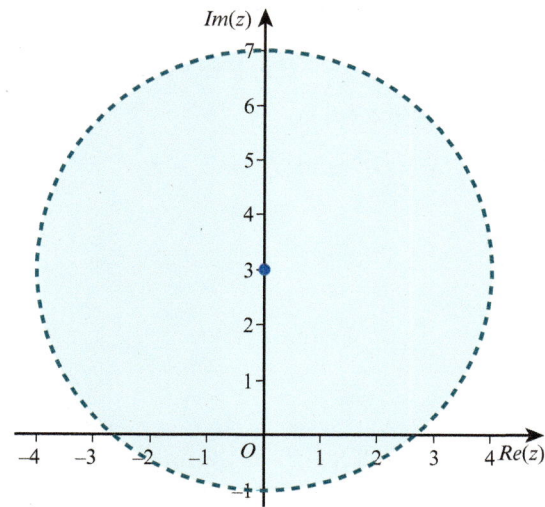

c $z - z_1 = z + 5 = z - (-5)$ .... Take care with signs.

$z_1 = -5$

$|z + 5| \leq 2$ is a circle, centre $(-5, 0)$, radius 2.

It is **all** points within the circle and on the circumference.

The circumference of the circle is drawn as a solid line as it is included.

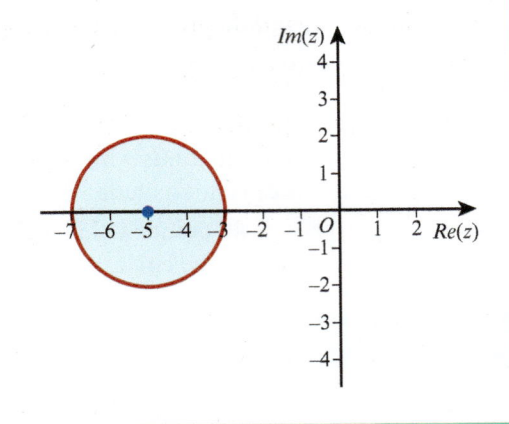

Half-lines and part-lines

- $\arg(z - z_1) = \alpha$, where α is a fixed angle and z_1 is the fixed point $a + b\text{i}$.

    With $z - z_1$ defined as previously

    $z = \overrightarrow{OP}$, $z_1 = \overrightarrow{OQ}$ and $\overrightarrow{QP} = z - z_1$

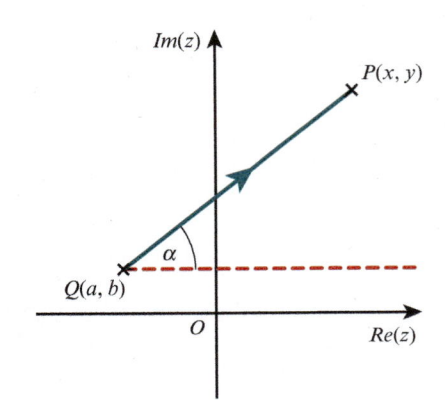

    We now define $\arg(z - z_1)$ as the angle made between the direction parallel to the real axis and the direction of the vector $\overrightarrow{QP}$.

    Points below Q are not part of the locus as the angle made between these points and the relevant horizontal line is $\pi - \alpha$, not α.

- $\arg(z_1 - z) = \beta$, where β is a fixed angle and z_1 is the fixed point $a + b\text{i}$.

    With $z_1 - z = -(z - z_1)$,

    $z = \overrightarrow{OP}$, $z_1 = \overrightarrow{OQ}$ and $\overrightarrow{PQ} = z_1 - z$

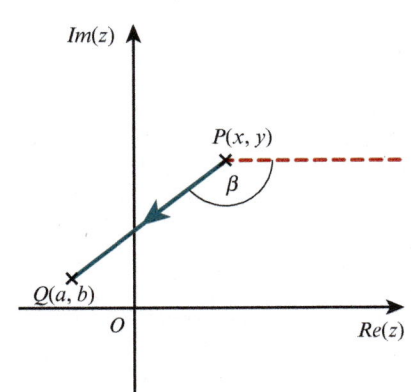

    We now define $\arg(z_1 - z)$ as the angle made between the direction parallel to the real axis and the direction of the vector $\overrightarrow{PQ}$.

    Points above P are not part of the locus as the angle made between these points and the relevant horizontal line is $\pi - \beta$, not β.

Chapter 11: Complex numbers

WORKED EXAMPLE 11.16

On a single Argand diagram, sketch the loci $|z| = 4$ and $\arg(z + 2 + 3i) = \dfrac{\pi}{4}$.

Show that there is only one complex number, z, that satisfies both loci.

Label this point as P on your diagram.

Answer

$|z| = 4$ .... Circle, centre (0, 0), radius 4

$\arg(z + 2 + 3i) = \dfrac{\pi}{4}$

$\quad z - z_1 = z + 2 + 3i$

$\qquad\quad\; = z - (-2 - 3i)$

$\quad\; z_1 = -2 - 3i$

Half-line from $(-2, -3)$ at an angle of $\dfrac{\pi}{4}$ radians.

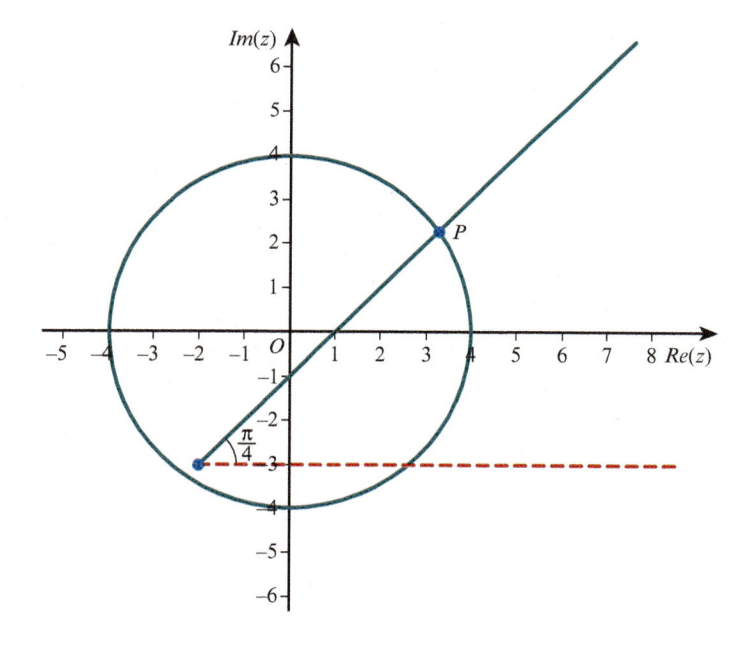

Perpendicular bisectors

- $|z - z_1| = |z - z_2|$, $|z - z_1| < |z - z_2|$, $|z - z_1| \leqslant |z - z_2|$ where z_1 is the fixed point $a + bi$ and z_2 is the fixed point $c + di$.

  Let $P(x, y)$ be the moving point that represents $z = x + iy$ on an Argand diagram,

  $\quad Q(a, b)$ be the fixed point that represents $z_1 = a + bi$ on the Argand diagram,

  $\quad R(c, d)$ be the fixed point that represents $z_2 = c + di$ on the Argand diagram,

  then

- when $|z - z_1| = |z - z_2|$, the point $P(x, y)$ moves so that it is always the same distance from Q as it is from R. The locus is, therefore, the perpendicular bisector of the line QR.
- $|z - z_1| < |z - z_2|$ is all points such that P is nearer to Q than to R. The perpendicular bisector marks the boundary of this region but is not included.
- $|z - z_1| \leqslant |z - z_2|$ is all points such that P is nearer to Q than to R or equidistant from Q and R. The perpendicular bisector marks the boundary of this region and is included.

WORKED EXAMPLE 11.17

On an Argand diagram, sketch the locus $|z - 4 + i| = |z + 5i|$.

Find the Cartesian equation of this locus.

Answer

$|z - 4 + i| = |z - (4 - i)|$ .... Write as $|z - z_1|$ and find z_1

$z_1 = 4 - i$

$|z + 5i| = |z - (-5i)|$ .... Write as $|z - z_2|$ and find z_2.

$z_2 = -5i$

Midpoint of QP, $(4, -1)$ and $(0, -5)$, ... The locus is the perpendicular bisector of $(4, -1)$ and $(0, -5)$.

is $\left(\dfrac{4+0}{2}, \dfrac{-1+(-5)}{2} \right) = (2, -3)$

Gradient of $QP = 1$ and so the gradient of perpendicular bisector $= -1$.

Cartesian equation is $y + 3 = -1(x - 2)$

$y = -x - 1$

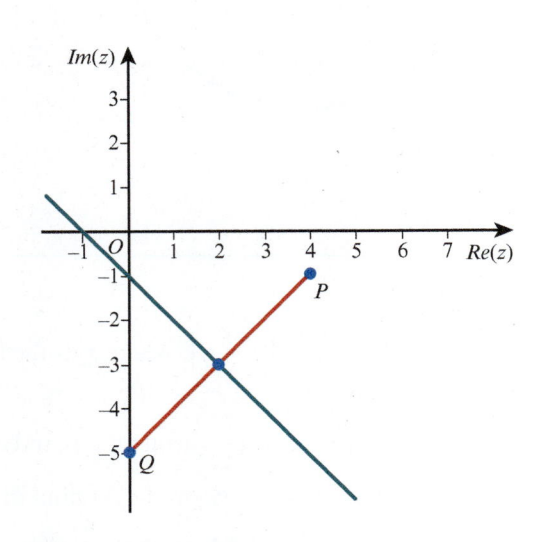

EXERCISE 11E

1 Describe, in words, the locus represented by each of the following.

   a $\arg(z - 2 + 3i) = \dfrac{\pi}{12}$

   b $|z - 10| < |z - 6i|$

   c $|z + 6 - i| = 7$

   d $\arg(z) = \dfrac{5\pi}{12}$

2 On different Argand diagrams, sketch the following loci. Shade any regions where needed.

   a $|z - (2 + i)| = |z - (4 + 3i)|$

   b $\arg(z + 3i) = -\dfrac{\pi}{3}$

   c $|z + 4 - 5i| \leqslant 2$

   d $|z - 2i| \leqslant |z - 4|$

3 On a single Argand diagram, sketch the loci $|z| = 4$ and $|z + 2| = |z - 4|$. Hence determine complex numbers that satisfy both loci, giving your answers in Cartesian form.

4 On an Argand diagram, sketch the locus $|z| = |z + 8i|$. Find the Cartesian equation of this locus.

5 Sketch the locus $|z - (3 - 6i)| = 3$ on an Argand diagram. Write down the Cartesian equation of this locus.

6 Sketch the loci $\arg(z + 4 + 2i) = \dfrac{5\pi}{6}$ and $|z - 5i| = 4$ on an Argand diagram. Determine whether or not there is a complex number, z, that satisfies both loci.

7 For complex numbers z satisfying $|z - 8 - 16i| = 2\sqrt{5}$, find the least possible value of $|z|$ and the greatest possible value of $|z|$.

8 By sketching the appropriate loci on an Argand diagram, find the value of z that satisfies $|z| = 13$ and $\arg(z - 12) = \dfrac{\pi}{2}$.

9 On a single Argand diagram, sketch the loci $|z + 3 - 2i| = 4$ and $\arg(z + 1) = \dfrac{\pi}{4}$. Hence determine the value of z that satisfies both loci, giving your answer in Cartesian form.

10 The complex number z is represented by the point $P(x, y)$ on an Argand diagram.

   It is given that the locus of P is $|z - 5 - 5i| = 5$.

   a Write down the Cartesian equation of the locus of P.

   b Sketch an Argand diagram to show the locus of P.

   c Write down the greatest possible value of $\arg z$ and the least possible value of $\arg z$.

Checklist of learning and understanding

$i^2 = -1$

Cartesian form: $x + iy$ where x and y are real values.

Arithmetic operations on $z_1 = a + bi$ and $z_2 = c + di$:

- Addition: $\quad z_1 + z_2 = (a + c) + (b + d)i$

- Subtraction: $\quad z_1 - z_2 = (a - c) + (b - d)i$

- Multiplication: $\quad z_1 z_2 = (ac - bd) + (ad + bc)i$

- Division: $\quad \dfrac{z_1}{z_2} = \dfrac{(ac + bd) + (bc - ad)i}{c^2 - d^2}$

- Modulus: $\quad |z| = \sqrt{x^2 + y^2}$

- Argument: $\quad$ Found using a diagram with $\tan \theta = \dfrac{y}{x}$

$|z_1 z_2| = r_1 r_2 = |z_1||z_2|$ and $\arg(z_1 z_2) = \theta_1 + \theta_2 = \arg z_1 + \arg z_2$,

$\left|\dfrac{z_1}{z_2}\right| = \dfrac{r_1}{r_2} = \dfrac{|z_1|}{|z_2|}$ and $\arg\left(\dfrac{z_1}{z_2}\right) = \theta_1 - \theta_2 = \arg z_1 - \arg z_2$

Polar forms

- Modulus-argument form: $r(\cos \theta + i \sin \theta)$

- Exponential form: $\quad re^{i\theta}$

Roots of equations occur in complex conjugate pairs

- Quadratic equations have $\quad$ 2 real $\quad$ or $\quad$ 2 complex roots of the form $x \pm iy$ with $y \neq 0$.

- Cubic equations have $\quad$ 3 real $\quad$ or $\quad$ 1 real and 2 complex roots of the form $x \pm iy$ with $y \neq 0$.

- Quartic equations have $\quad$ 4 real $\quad$ or $\quad$ 2 real and 2 complex roots of the form $x \pm iy$ with $y \neq 0$,

$\quad\quad\quad\quad\quad\quad\quad\quad\quad\quad\quad\quad\quad\quad\quad$ or $\quad$ 4 complex roots of the form $x \pm iy$ with $y \neq 0$.

The cube roots of one are: $z = 1$, $z = \dfrac{-1 + i\sqrt{3}}{2}$, $z = \dfrac{-1 - i\sqrt{3}}{2}$

END-OF-CHAPTER REVIEW EXERCISE 11

1. **a** Express $\dfrac{5-2i}{1+3i}$ in the form $x+iy$ where x and y are real numbers. [3]

   b Solve $w^2 - 2w + 26 = 0$. [2]

   c On a sketch of an Argand diagram, shade the region whose points represent complex numbers satisfying the inequality $|z+1-5i| \leq 2$. [3]

2. **a** The complex number z is defined as $z = k - 6i$, where k is a real value.

   Find and simplify expressions, in terms of k, for zz^* and $\dfrac{z}{z^*}$, giving your answers in the form $x+iy$ where x and y are real. [3]

   b The complex numbers u and w are defined as $u = 4\left(\cos\dfrac{5\pi}{12} + i\sin\dfrac{5\pi}{12}\right)$ and $w = 2e^{i\pi}$.

   Find and simplify expressions for uw and $\dfrac{u}{w}$, giving your answers in the form $re^{i\theta}$, where $r > 0$ and $-\pi < \theta \leq \pi$. [5]

3. The complex number $w = 1 + 2i$.

   a Represent w and w^* by points P and Q on an Argand diagram with origin O and describe the polygon OPQ. [2]

   b Given also that $u = -3 - i$, write the complex number $\dfrac{u}{w}$ in the form $r(\cos\theta + i\sin\theta)$, where $r > 0$ and $-\pi < \theta \leq \pi$. [4]

4. $z = 2 - 5i$

   a Find the real values x and y such that $z^* = (2x+1) + (4x+y)i$. [2]

   b On an Argand diagram, show the points A, B and C representing the complex numbers z, z^* and $-z$. What type of triangle is ABC? [3]

   c Without using a calculator, express $\dfrac{z^*}{-z}$

   i in the form $x+iy$ where x and y are real [3]

   ii in the form $r(\cos\theta + i\sin\theta)$, where $r > 0$ and $-\pi < \theta \leq \pi$. [3]

5. $z^2 + (4\sqrt{3})z + 13 = 0$

   a Find the roots of this equation, giving your answers in the form $x+iy$ where x and y are real. [2]

   b On an Argand diagram with origin O, show the position vectors $\overrightarrow{OA}$ and $\overrightarrow{OB}$ representing the roots of the equation. Describe the geometrical relationship between $\overrightarrow{OA}$ and $\overrightarrow{OB}$. [2]

   c Find the modulus and argument of each root. [3]

6. $z = 4\sqrt{3} - 4i$

   a Find the exact values of the modulus and argument of z. [3]

   b Given that $w = 2\sqrt{2}\left(\cos\dfrac{\pi}{12} + i\sin\dfrac{\pi}{12}\right)$, write $\dfrac{z}{w}$ in the form $re^{i\theta}$, where $r > 0$ and $-\pi < \theta \leq \pi$. [3]

7 a Find the complex number w satisfying the equation $w^* - 2 - 2i = 3iw$. Give your answer in the form $x + iy$ where x and y are real. [5]

   b i On a single Argand diagram, sketch the loci $|z - 3 - 3i| = 2$ and $\arg(z - 3 - 3i) = \dfrac{\pi}{3}$. [4]

      ii Hence determine the value of z that satisfies both loci, giving your answer in the form $x + iy$, where x and y are real. [2]

8 a $(x + iy)^2 = 7 - (6\sqrt{2})i$

   Given that x and y are real numbers, find the values of x and the values of y. [5]

   b i Show that $z - 3$ is a factor of $2z^3 - 4z^2 - 5z - 3$. [1]

      ii Solve $2z^3 - 4z^2 - 5z - 3 = 0$. [4]

9 a Given that $z_1 = 5 - 3i$ and $z_1 z_2 = 21 + i$, find z_2, giving your answer in the form $x + iy$, where x and y are real. [2]

   b Solve $(3z + 1)^3 = -27$. [3]

10 a It is given that $w = 1$ is a root of the equation $f(w) = 2w^4 + 5w^3 - 2w^2 + w - 6 = 0$.

      i Show that $w + 3$ is a factor of $2w^4 + 5w^3 - 2w^2 + w - 6$. [1]

      ii Solve the equation. [4]

   b i On an Argand diagram, sketch the locus $|z + 1 - i\sqrt{3}| = 1$. [2]

      ii Write down the minimum value of $\arg z$. [1]

      iii Find the maximum value of $\arg z$. [2]

11 a i Given that $z_1 = -\dfrac{3}{2} + \dfrac{\sqrt{7}}{2}i$ is a root of the equation $z^2 + pz + q = 0$, where p and q are real constants, find the value of p and the value of q. [3]

      ii Find $|z_1|$. [1]

   b i Find the roots of the equation $z^3 + 1 = 0$. [3]

      ii On an Argand diagram, show the points A, B and C representing the roots of the equation. What type of triangle is ABC? [2]

12 $z = \sqrt{5} - i$

   a Show that $\dfrac{z}{z^*} = \dfrac{2}{3} - \dfrac{\sqrt{5}}{3}i$. [3]

   b Find the value of $\left|\dfrac{z}{z^*}\right|$ and $\arg\left(\dfrac{z}{z^*}\right)$. [2]

   c Find and simplify a quadratic equation with integer coefficients that has roots $\dfrac{z}{z^*}$ and its conjugate. [2]

13 The complex number z is defined by $z = \dfrac{k - 4i}{2k - i}$ where k is an integer.

   a The imaginary part of z is $\operatorname{Im} z = \dfrac{7}{5}$. Find the value of k. [4]

   b Find the argument of z. [3]

14 i Without using a calculator, solve the equation
$$3w + 2iw^* = 17 + 8i,$$
where w^* denotes the complex conjugate of w. Give your answer in the form $a + bi$. [4]

ii In an Argand diagram, the loci
$$\arg(z - 2i) = \frac{1}{6}\pi \text{ and } |z - 3| = |z - 3i|$$
intersect at the point P. Express the complex number represented by P in the form $re^{i\theta}$, giving the exact value of θ and the value of r correct to 3 significant figures. [5]

Cambridge International A Level Mathematics 9709 Paper 31 Q7 June 2013

15 Throughout this question the use of a calculator is not permitted.

i The complex numbers u and v satisfy the equations
$$u + 2v = 2i \text{ and } iu + v = 3.$$
Solve the equations for u and v, giving both answers in the form $x + iy$, where x and y are real. [5]

ii On an Argand diagram, sketch the locus representing complex numbers z satisfying $|z + i| = 1$ and the locus representing complex numbers w satisfying $\arg(w - 2) = \frac{3}{4}\pi$. Find the least value of $|z - w|$ for points on these loci. [5]

Cambridge International A Level Mathematics 9709 Paper 31 Q8 November 2013

16 Throughout this question the use of a calculator is not permitted.

The complex numbers w and z satisfy the relation
$$w = \frac{z + i}{iz + 2}.$$

i Given that $z = 1 + i$, find w, giving your answer in the form $x + iy$, where x and y are real. [4]

ii Given instead that $w = z$ and the real part of z is negative, find z, giving your answer in the form $x + iy$, where x and y are real. [4]

Cambridge International A Level Mathematics 9709 Paper 31 Q5 November 2014

Cambridge International AS & A Level Mathematics: Pure Mathematics 2 & 3

CROSS-TOPIC REVIEW EXERCISE 4

P3 This exercise is for Pure Mathematics 3 students only.

1 Relative to the origin O, the position vectors of the points A and B are given by

$$\overrightarrow{OA} = \begin{pmatrix} -5 \\ 0 \\ 3 \end{pmatrix} \text{ and } \overrightarrow{OB} = \begin{pmatrix} 1 \\ 7 \\ 2 \end{pmatrix}.$$

  a Find a vector equation of the line AB. [3]

  b The line AB is perpendicular to the line L with vector equation:

$$\mathbf{r} = \begin{pmatrix} 4 \\ 2 \\ -3 \end{pmatrix} + \mu \begin{pmatrix} m \\ 3 \\ 9 \end{pmatrix}.$$

   i   Find the value of m. [3]

   ii Show that the line AB and the line L do not intersect. [4]

2 Given that $y = 0$ when $x = 1$, solve the differential equation

$$xy \frac{dy}{dx} = y^2 + 4,$$

obtaining an expression for y^2 in terms of x. [6]

Cambridge International A Level Mathematics 9709 Paper 31 Q5 June 2010

3 The point $P(p, q, -1)$ lies on the line L with vector equation

$\mathbf{r} = \mathbf{i} - \mathbf{j} + 2\mathbf{k} + \lambda(\mathbf{i} + 8\mathbf{j} + \mathbf{k}).$

  a Find the value of each of the constants p and q. [2]

  b The position vector of Q, relative to the origin O, is $\overrightarrow{OQ} = -10\mathbf{i} + \mathbf{j} - 5\mathbf{k}$.

   i   Find the unit vector in the direction $\overrightarrow{OQ}$. [2]

   ii Find angle POQ and hence find the exact area of triangle POQ. [5]

4 The complex number z is given by

$z = (\sqrt{3}) + \mathrm{i}.$

   i   Find the modulus and argument of z. [2]

   ii The complex conjugate of z is denoted by z^*. Showing your working, express in the form $x + \mathrm{i}y$, where x and y are real,

       a $2z + z^*$,   b $\dfrac{\mathrm{i}z^*}{z}$. [4]

   iii On a sketch of an Argand diagram with origin O, show the points A and B representing the complex numbers z and $\mathrm{i}z^*$ respectively. Prove that angle $AOB = \dfrac{1}{6}\pi$. [3]

Cambridge International A Level Mathematics 9709 Paper 31 Q6 November 2010

5   The variables x and θ are related by the differential equation

$$\sin 2\theta \frac{dx}{d\theta} = (x+1)\cos 2\theta,$$

where $0 < \theta < \frac{1}{2}\pi$. When $\theta = \frac{1}{12}\pi$, $x = 0$. Solve the differential equation, obtaining an expression for x in terms of θ, and simplifying your answer as far as possible. [7]

Cambridge International A Level Mathematics 9709 Paper 31 Q4 November 2011

6   $PQRS$ is a parallelogram. The vertices, P, Q and R have position vectors, relative to an origin O,

$$\overrightarrow{OP} = \begin{pmatrix} 1 \\ 4 \\ 5 \end{pmatrix}, \overrightarrow{OQ} = \begin{pmatrix} 3 \\ 2 \\ 0 \end{pmatrix} \text{ and } \overrightarrow{OR} = \begin{pmatrix} 2 \\ -1 \\ 5 \end{pmatrix}.$$

a   Find $\overrightarrow{OS}$. [2]

b   Find the lengths of the sides of the parallelogram. [3]

c   Find the interior angles of the parallelogram. [3]

7   The complex number u is defined by $u = \dfrac{(1+2i)^2}{2+i}$.

i   Without using a calculator and showing your working, express u in the form $x + iy$, where x and y are real. [4]

ii Sketch an Argand diagram showing the locus of the complex number z such that $|z - u| = |u|$. [3]

Cambridge International A Level Mathematics 9709 Paper 31 Q4 June 2012

8   The complex number $2 + 2i$ is denoted by u.

i   Find the modulus and argument of u. [2]

ii Sketch an Argand diagram showing the points representing the complex numbers 1, i and u. Shade the region whose points represent the complex numbers z which satisfy both the inequalities $|z - 1| \leq |z - i|$ and $|z - u| \leq 1$. [4]

iii Using your diagram, calculate the value of $|z|$ for the point in this region for which $\arg z$ is least. [3]

Cambridge International A Level Mathematics 9709 Paper 31 Q7 June 2010

9   A certain substance is formed in a chemical reaction. The mass of substance formed t seconds after the start of the reaction is x grams. At any time the rate of formation of the substance is proportional to $(20 - x)$. When $t = 0$, $x = 0$ and $\dfrac{dx}{dt} = 1$.

i   Show that x and t satisfy the differential equation

$$\frac{dx}{dt} = 0.05(20 - x).$$ [2]

ii Find, in any form, the solution of this differential equation. [5]

iii Find x when $t = 10$, giving your answer correct to 1 decimal place. [2]

iv State what happens to the value of x as t becomes very large. [1]

Cambridge International A Level Mathematics 9709 Paper 31 Q10 November 2010

10 The line L_1 has vector equation

$$\mathbf{r} = -2\mathbf{j} + \mathbf{k} + \lambda(\mathbf{i} + 3\mathbf{k}).$$

The line L_2 passes through the point $P(-2, 1, -1)$ and is parallel to L_1.

   a Write down a vector equation of the line L_2. [2]

   b Find the shortest distance from P to the line L_1. [5]

11 In a certain country the government charges tax on each litre of petrol sold to motorists. The revenue per year is R million dollars when the rate of tax is x dollars per litre. The variation of R with x is modelled by the differential equation

$$\frac{dR}{dx} = R\left(\frac{1}{x} - 0.57\right),$$

where R and x are taken to be continuous variables. When $x = 0.5$, $R = 16.8$.

   i Solve the differential equation and obtain an expression for R in terms of x. [6]

   ii This model predicts that R cannot exceed a certain amount. Find this maximum value of R. [3]

Cambridge International A Level Mathematics 9709 Paper 31 Q7 November 2014

12 The complex number z is defined by $z = \dfrac{9\sqrt{3} + 9i}{\sqrt{3} - i}$. Find, showing all your working,

   i an expression for z in the form $re^{i\theta}$, where $r > 0$ and $-\pi < \theta \leq \pi$, [5]

   ii the two square roots of z, giving your answers in the form $re^{i\theta}$, where $r > 0$ and $-\pi < \theta \leq \pi$. [3]

Cambridge International A Level Mathematics 9709 Paper 31 Q5 June 2014

13 The variables x and y are related by the differential equation

$$\frac{dy}{dx} = \frac{6ye^{3x}}{2 + e^{3x}}.$$

Given that $y = 36$ when $x = 0$, find an expression for y in terms of x. [6]

Cambridge International A Level Mathematics 9709 Paper 31 Q4 June 2014

14 a Show that the straight line L with vector equation

$$\mathbf{r} = \begin{pmatrix} 1 \\ 2 \\ -10 \end{pmatrix} + \lambda \begin{pmatrix} 8 \\ -1 \\ 3 \end{pmatrix}$$

intersects with the line through the points A and B with coordinates $(0, 2, 7)$ and $(7, 1, 27)$, respectively and find the position vector of this point of intersection. [5]

   b Find the acute angle between these two lines. [4]

15 The variables x and y satisfy the differential equation $x\dfrac{dy}{dx} = y(1 - 2x^2)$, and it is given that $y = 2$ when $x = 1$. Solve the differential equation and obtain an expression for y in terms of x in a form not involving logarithms. [6]

Cambridge International A Level Mathematics 9709 Paper 31 Q4 June 2016

16 The complex number u is defined by $u = \dfrac{6 - 3i}{1 + 2i}$.

   i Showing all your working, find the modulus of u and show that the argument of u is $-\dfrac{1}{2}\pi$. [4]

   ii For complex numbers z satisfying $\arg(z - u) = \dfrac{1}{4}\pi$, find the least possible value of $|z|$. [3]

   iii For complex numbers z satisfying $|z - (1 + i)u| = 1$, find the greatest possible value of $|z|$. [3]

Cambridge International A Level Mathematics 9709 Paper 31 Q8 June 2011

17 With respect to the origin O, the position vectors of the points A, B, C and D are given by

$$\overrightarrow{OA} = \begin{pmatrix} 2 \\ 1 \\ 5 \end{pmatrix}, \overrightarrow{OB} = \begin{pmatrix} 3 \\ m \\ 4 \end{pmatrix}, \overrightarrow{OC} = \begin{pmatrix} 1 \\ -5 \\ m+11 \end{pmatrix} \text{ and } \overrightarrow{OD} = \begin{pmatrix} 2 \\ -4 \\ -2m \end{pmatrix}.$$

   a In the case where ABC is a right angle, find the possible values of the constant m. [3]

   b In the case where D is the midpoint of the line BC, find the value of the constant m. [2]

   c In the case where $m = -5$, find whether the lines AB and CD intersect. [5]

18 The complex number w is defined by $w = \dfrac{22 + 4i}{(2 - i)^2}$.

   i Without using a calculator, show that $w = 2 + 4i$. [3]

   ii It is given that p is a real number such that $\dfrac{1}{4}\pi \leq \arg(w + p) \leq \dfrac{3}{4}\pi$. Find the set of possible values of p. [3]

   iii The complex conjugate of w is denoted by w^*. The complex numbers w and w^* are represented in an Argand diagram by the points S and T respectively. Find, in the form $|z - a| = k$, the equation of the circle passing through S, T and the origin. [3]

Cambridge International A Level Mathematics 9709 Paper 31 Q8 June 2015

19 Given that $y = 1$ when $x = 0$, solve the differential equation

$$\dfrac{dy}{dx} = 4x(3y^2 + 10y + 3),$$

obtaining an expression for y in terms of x. [9]

Cambridge International A Level Mathematics 9709 Paper 31 Q7 June 2015

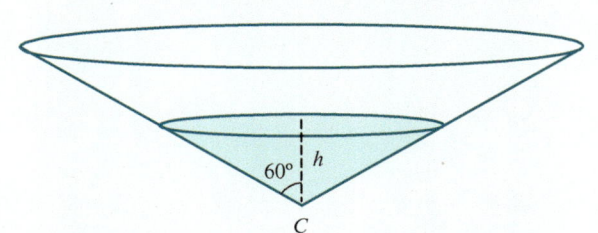

20

A tank containing water is in the form of a cone with vertex C. The axis is vertical and the semi-vertical angle is $60°$, as shown in the diagram. At time $t = 0$, the tank is full and the depth of water is H. At this instant, a tap at C is opened and water begins to flow out. The volume of water in the tank decreases at a rate proportional to $\sqrt{h}$ where h is the depth of water at time t. The tank becomes empty when $t = 60$.

i Show that h and t satisfy a differential equation of the form

$$\frac{dh}{dt} = -Ah^{-\frac{3}{2}},$$

where A is a positive constant. [4]

ii Solve the differential equation given in **part i** and obtain an expression for t in terms of h and H. [6]

iii Find the time at which the depth reaches $\frac{1}{2}H$.

[The volume V of a cone of vertical height h and base radius r is given by $V = \frac{1}{3}\pi r^2 h$.] [1]

Cambridge International A Level Mathematics 9709 Paper 31 Q10 November 2013

21 a Find, in the form $\mathbf{r} = \mathbf{a} + \lambda\mathbf{b}$, a vector equation of the line AB where the points have coordinates $A(2, 5, 7)$ and $B(9, -1, -2)$. [3]

b Find the obtuse angle between the line AB and a line in the direction of $\mathbf{i} + 3\mathbf{j} + 2\mathbf{k}$. [4]

22 The complex number $3 - i$ is denoted by u. Its complex conjugate is denoted by u^*.

i On an Argand diagram with origin O, show the points A, B and C representing the complex numbers u, u^* and $u^* - u$ respectively. What type of quadrilateral is $OABC$? [4]

ii Showing your working and without using a calculator, express $\dfrac{u^*}{u}$ in the form $x + iy$, where x and y are real. [3]

iii By considering the argument of $\dfrac{u^*}{u}$, prove that

$$\tan^{-1}\left(\frac{3}{4}\right) = 2\tan^{-1}\left(\frac{1}{3}\right).$$ [3]

Cambridge International A Level Mathematics 9709 Paper 31 Q9 November 2015

23 The complex number $1 + (\sqrt{2})i$ is denoted by u. The polynomial $x^4 + x^2 + 2x + 6$ is denoted by $p(x)$.

i Showing your working, verify that u is a root of the equation $p(x) = 0$, and write down a second complex root of the equation. [4]

ii Find the other two roots of the equation $p(x) = 0$. [6]

Cambridge International A Level Mathematics 9709 Paper 31 Q9 November 2012

24. The diagram shows a cuboid $OABCDEFG$ with a horizontal base $OABC$. The cuboid has a length OA of $10\,\text{cm}$, a width AB of $4\,\text{cm}$ and a height BF of $h\,\text{cm}$. The point M is the midpoint of CB and the point N is the point on DG such that $\overrightarrow{ON} = \mathbf{j} + 3\mathbf{k}$. The unit vectors $\mathbf{i}$, $\mathbf{j}$ and $\mathbf{k}$ are parallel to OA, OC and OD, respectively.

    a Write down the value of h.   [1]

    b Find the unit vector in the direction $\overrightarrow{ON}$.   [2]

    c Find angle NME.   [4]

    d Find, in the form $\mathbf{r} = \mathbf{a} + \lambda \mathbf{b}$, a vector equation of the line MF.   [3]

25. The variables x and y are related by the differential equation
$$\frac{dy}{dx} = \frac{6xe^{3x}}{y^2}.$$

It is given that $y = 2$ when $x = 0$. Solve the differential equation and hence find the value of y when $x = 0.5$, giving your answer correct to 2 decimal places.   [8]

Cambridge International A Level Mathematics 9709 Paper 31 Q7 June 2012

26. i Showing all your working and without the use of a calculator, find the square roots of the complex number $7 - (6\sqrt{2})i$. Give your answers in the form $x + iy$, where x and y are real and exact.   [5]

  ii a On an Argand diagram, sketch the loci of points representing complex numbers w and z such that $|w - 1 - 2i| = 1$ and $\arg(z - 1) = \dfrac{3}{4}\pi$.   [4]

      b Calculate the least value of $|w - z|$ for points on these loci.   [2]

Cambridge International A Level Mathematics 9709 Paper 31 Q10 June 2016

27. The line L_1 has vector equation $\mathbf{r} = \begin{pmatrix} -2 \\ 3 \\ 0 \end{pmatrix} + \lambda \begin{pmatrix} 1 \\ -1 \\ 2 \end{pmatrix}$.

The line L_2 has vector equation $\mathbf{r} = \begin{pmatrix} -4 \\ 5 \\ m \end{pmatrix} + \mu \begin{pmatrix} -1 \\ 0 \\ 1 \end{pmatrix}$.

    a In the case where $m = 2$, show that L_1 and L_2 do not intersect.   [4]

    b Find the value of m in the case where L_1 and L_2 intersect.   [2]

    c For your value of m from part **b**, find the acute angle between L_1 and L_2.   [4]

28 The number of birds of a certain species in a forested region is recorded over several years. At time t years, the number of birds is N, where N is treated as a continuous variable.

The variation in the number of birds is modelled by
$$\frac{dN}{dt} = \frac{N(1800 - N)}{3600}.$$
It is given that $N = 300$ when $t = 0$.

   i   Find an expression for N in terms of t. [9]

   ii   According to the model, how many birds will there be after a long time? [1]

Cambridge International A Level Mathematics 9709 Paper 31 Q10 June 2011

29 **i** Showing your working, find the square roots of the complex number $1 - (2\sqrt{6})i$. Give your answers in the form $x + iy$, where x and y are exact. [5]

   ii   On a sketch of an Argand diagram, shade the region whose points represent the complex numbers z which satisfy the inequality $|z - 3i| \leq 2$. Find the greatest value of $\arg z$ for points in this region. [4]

Cambridge International A Level Mathematics 9709 Paper 31 Q10 November 2011

30 Liquid is flowing into a small tank which has a leak. Initially the tank is empty and t minutes later, the volume of liquid in the tank is V cm^3. The liquid is flowing into the tank at a constant rate of 80 cm^3 per minute. Because of the leak, liquid is being lost from the tank at a rate which, at any instant, is equal to kV cm^3 per minute where k is a positive constant.

   i   Write down a differential equation describing this situation and solve it to show that
$$V = \frac{1}{k}(80 - 80e^{-kt}).$$
[7]

   ii   It is observed that $V = 500$ when $t = 15$, so that k satisfies the equation $k = \frac{4 - 4e^{-15k}}{25}$.

Use an iterative formula, based on this equation, to find the value of k correct to 2 significant figures. Use an initial value of $k = 0.1$ and show the result of each iteration to 4 significant figures. [3]

   iii   Determine how much liquid there is in the tank 20 minutes after the liquid started flowing, and state what happens to the volume of liquid in the tank after a long time. [2]

Cambridge International A Level Mathematics 9709 Paper 31 Q10 June 2013

PURE MATHEMATICS 2 PRACTICE EXAM-STYLE PAPER

Time allowed is 1 hour 15 minutes (50 marks)

1. By writing $\cot x$ as $\dfrac{\cos x}{\sin x}$ show that $\dfrac{d}{dx}(\cot x) = -\operatorname{cosec}^2 x$. [3]

2. Solve the inequality $|3x-1| \geq |2x|$. [4]

3. Solve the equation $3^{2x} - 3^{x+1} = 10$ giving the value of x correct to 3 significant figures. [5]

4. The polynomial $x^3 + 8x^2 + px - 25$ leaves a remainder of R when divided by $x-1$ and a remainder of $-R$ when divided by $x+2$.

   a Find the value of p. [4]

   b Hence, find the remainder when the polynomial is divided by $x+3$. [2]

5. The sequence of values given by the formula
$$x_{n+1} = \sqrt{\dfrac{8x_n^2}{3\sec x_n}},$$
with initial value $x_1 = 1$, converges to α.

   a Use this formula to calculate α correct to 2 decimal places, showing the result of each iteration to 4 decimal places. [3]

   b State an equation satisfied by α and hence find the value of α correct to 7 decimal places. [3]

6. The parametric equations of a curve are $x = \ln(2t+1)$, $y = t - e^{2t}$.

   a Find an expression for $\dfrac{dy}{dx}$ in terms of t. [4]

   b Find the equation of the normal to the curve at the point where $t = 0$. [4]

7. a Express $8\sin\theta + 6\cos\theta$ in the form $R\sin(\theta + \alpha)$, where $R > 0$ and $0° < \alpha < 90°$, giving the value of α correct to 2 decimal places. [3]

   b Hence solve the equation $8\sin\theta + 6\cos\theta = 7$ giving all solutions in the interval $0° < \theta < 360°$. [4]

   c Write down the greatest value of $8\sin\theta + 6\cos\theta + 3$ as θ varies. [1]

8. a Show that $\cos 3x \equiv 4\cos^3 x - 3\cos x$. [5]

   b Hence show that $\displaystyle\int_0^{\frac{\pi}{2}} \cos^3 x \, dx = \dfrac{2}{3}$. [5]

PURE MATHEMATICS 3 PRACTICE EXAM-STYLE PAPER

Time allowed is 1 hour 50 minutes (75 marks)

1. Use logarithms to solve the equation $e^{2x} = 2^{x+5}$, giving your answers correct to 3 decimal places. [3]

2. Show that $\int_0^5 x \ln x \, dx = \frac{25}{4}(\ln 25 - 1)$. [5]

3. The polynomial $2x^3 - 9x^2 + ax + b$, where a and b are constants, is denoted by $f(x)$. It is given that $x - 4$ is a factor of $f(x)$, and that when $f(x)$ is divided by $(x - 1)$ the remainder is -12. Find the value of a and the value of b. [5]

4. Solve the equation $2\sin(x - 60°) = 3\cos x$ for $-180° \leq x \leq 180°$. [5]

5.

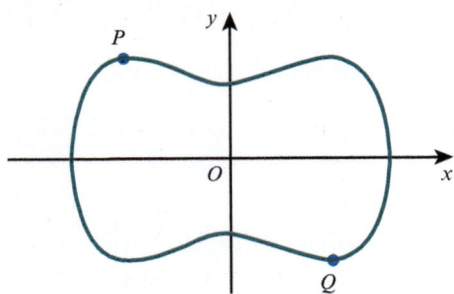

   The diagram shows the curve $y^2 = 4x^2 - x^4 + 5$. The curve is symmetrical about both axes. The point P is one of the curve's maximum points and the point Q is one of the curve's minimum points. Find the exact distance between the points P and Q. [7]

6. $f(x) = \dfrac{7x^2 - 5x + 27}{(x - 2)(x^2 + 5)}$

   a. Express $f(x)$ in partial fractions. [5]

   b. Hence obtain the expansion of $f(x)$ in ascending powers of x, up to and including the term in x^2. [5]

7.

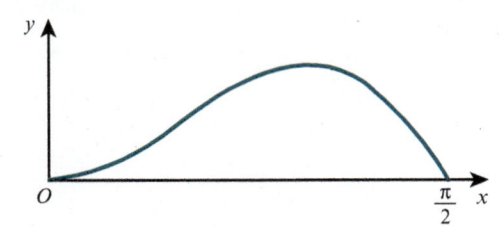

   The diagram shows the curve $y = x \sin 2x$ for $0 \leq x \leq \dfrac{\pi}{2}$.

   a. Find $\dfrac{dy}{dx}$ and show that $x^2 \dfrac{d^2y}{dx^2} - 2x \dfrac{dy}{dx} + 2(1 + 2x^2)y = 0$. [5]

   b. Find the area of the region enclosed by this part of the curve and the x-axis. [5]

8. a. The complex numbers u and w are such that:

   $3u - iw = 15$

   $u + w = 5 + 10i$.

   Without using a calculator, solve these equations to find u and w, giving each answer in the form $x + iy$, where x and y are real. [5]

   b. The complex number z is defined by $z = \sqrt{2}e^{-\frac{\pi}{4}i}$. On an Argand diagram, show the points A, B and C representing the complex numbers z, z^* and z^2, respectively and find the length of the longest side of triangle ABC. [4]

9 a Given the vectors $5\mathbf{i} + 7\mathbf{j} + p\mathbf{k}$ and $\mathbf{i} - 2\mathbf{j} + p\mathbf{k}$ are perpendicular, find the possible values of the constant p. [3]

b The line L_1 passes through the point $(5, 0, 2)$ and is parallel to the vector $\mathbf{i} - 6\mathbf{j} + \mathbf{k}$.

  i Write down a vector equation of the line L_1. [2]

  ii The line L_2 has vector equation:
  $$\mathbf{r} = 4\mathbf{j} + 3\mathbf{k} + \mu(3\mathbf{i} + 9\mathbf{j} + \mathbf{k})$$
  Show that L_1 and L_2 do not intersect. [4]

10 A liquid is heated so that its temperature, $x\,°\text{C}$, at time t seconds satisfies the differential equation
$$\frac{dx}{dt} = \alpha(100 - x)$$
where α is a positive constant. The temperature at $t = 0$ is $25\,°\text{C}$.

a Show that $\ln\left(\dfrac{75}{100-x}\right) = \alpha t$. [6]

b It is given that $x = 500\alpha$ when $t = 2$. Show that $\alpha = 0.2 - 0.15e^{-2\alpha}$. [2]

c Use an iterative formula based on the equation in **part b** to find the value of α correct to 2 significant figures. Use a starting value of 0.1 and show the result of each iteration to 4 significant figures. [3]

d Find the temperature of the liquid when $t = 30$. [1]

Answers

1 Algebra

Prerequisite knowledge

1. **a** 357 **b** 381
   c 133 remainder 27

2. Straight line, gradient 2, crossing axes at $\left(\frac{5}{2}, 0\right)$ and $(0, -5)$.

Exercise 1A

1. **a** $-1, \frac{5}{2}$ **b** $-2, 3$ **c** $-6, \frac{22}{3}$
   d $-15, 3$ **e** $-20, 40$ **f** 7

2. **a** $\frac{9}{7}, \frac{11}{3}$ **b** $-1, 3$ **c** $\frac{7}{4}, \frac{23}{6}$
   d 1, 3 **e** 2 **f** $-7, 3$

3. **a** $-1, -\frac{1}{3}$ **b** $-3, \frac{3}{5}$ **c** 2, 4
   d $-4, -\frac{6}{5}$ **e** $-7, 1$ **f** $0, \frac{12}{13}$

4. **a** ± 3
   b $-1, 2, \frac{1}{2}\left(-1-\sqrt{33}\right), \frac{1}{2}\left(\sqrt{33}-1\right)$
   c $-2, \pm 1$
   d $\sqrt{3}-1, 1+\sqrt{5}$
   e $1, 2, 1-\sqrt{3}, 1+\sqrt{3}$
   f $0, 2, 6$

5. **a** $x = 0, y = 4$ or $x = -\frac{8}{3}, y = \frac{16}{3}$
   b $x = -1, y = 3$ or $x = -\frac{5}{2}, y = \frac{15}{2}$

6. $x = \frac{4}{5}$ or $x = \frac{6}{5}$

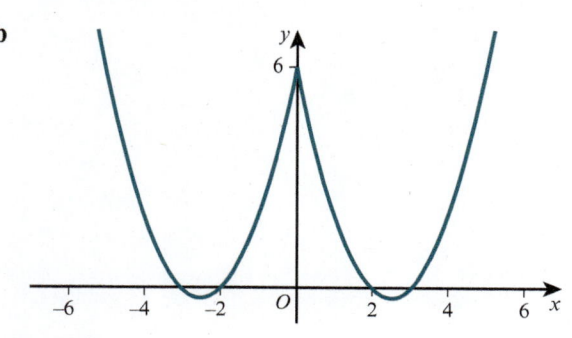

7. **a** $x = \pm 2, x = \pm 3$
   b [graph showing W-shaped curve with y-intercept 6, crossing x-axis at ± 2 and ± 3]
   c $x = 0$

8. $x = \pm \frac{3}{4}$

9. $x = 7, y = 5$

Exercise 1B

1. **a** ∨-shaped graph, vertex = $(-2, 0)$, y-intercept = 2
   $$y = \begin{cases} x + 2 & \text{if } x \geq -2 \\ -(x+2) & \text{if } x < -2 \end{cases}$$
   b ∨-shaped graph, vertex = $(3, 0)$, y-intercept = 3
   $$y = \begin{cases} x - 3 & \text{if } x \geq 3 \\ 3 - x & \text{if } x < 3 \end{cases}$$
   c ∨-shaped graph, vertex = $(10, 0)$, y-intercept = 5
   $$y = \begin{cases} \frac{1}{2}x - 5 & \text{if } x \geq 10 \\ 5 - \frac{1}{2}x & \text{if } x < 10 \end{cases}$$

2. **a**

x	0	1	2	3	4	5	6
y	5	4	3	2	3	4	5

   b ∨-shaped graph, vertex = $(3, 2)$, y-intercept = 5
   c Translation $\begin{pmatrix} 3 \\ 2 \end{pmatrix}$

3. **a** Translation $\begin{pmatrix} -1 \\ 2 \end{pmatrix}$ **b** Translation $\begin{pmatrix} 5 \\ -2 \end{pmatrix}$
   c Reflection in x-axis, translation $\begin{pmatrix} 0 \\ 2 \end{pmatrix}$
   d Stretch, stretch factor 2, with $y = 0$ invariant, translation $\begin{pmatrix} 0 \\ -3 \end{pmatrix}$.
   e Reflection in x-axis, translation $\begin{pmatrix} -2 \\ 1 \end{pmatrix}$.
   f Stretch, stretch factor 2, with $y = 0$ invariant, reflection in x-axis, translation $\begin{pmatrix} 0 \\ 5 \end{pmatrix}$

4. **a** ∨-shaped graph, vertex = $(-1, 2)$
   b ∨-shaped graph, vertex = $(5, -2)$
   c ∧-shaped graph, vertex = $(0, 2)$
   d ∨-shaped graph, vertex = $(0, -3)$
   e ∧-shaped graph, vertex = $(-2, 1)$
   f ∧-shaped graph, vertex = $(0, 5)$

5. $3 \leq f(x) \leq 14$

6 a, b

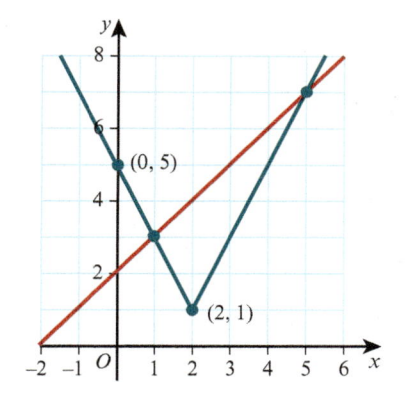

[graph showing V-shape with vertex (2,1), point (0,5), and line crossing]

c $x = 1$ or $x = 5$

7 a, b

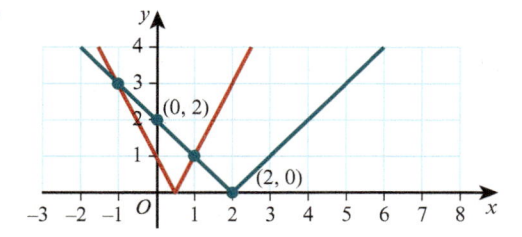

[graph showing V-shapes with points (0,2) and (2,0)]

c $x = \pm 1$

8 a

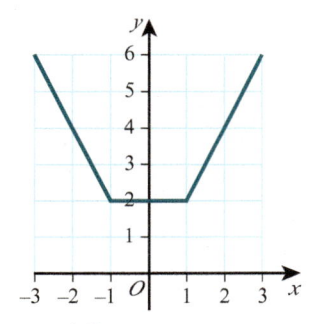

[graph showing trough shape]

b $x = \pm 2$

Exercise 1C

1 $3 < x < 7$

2 a

[graph showing $y = |2x-1|$ and $y = 4 - |x-1|$]

b $x < -\dfrac{2}{3}$ or $x > 2$

3 a $-2 \leqslant x \leqslant 1$ b $-2 < x < 6$
  c $x \leqslant -\dfrac{5}{3}$ or $x \geqslant 3$

4 a $\dfrac{7}{3} \leqslant x \leqslant 3$ b $x > \dfrac{1}{3}$
  c $x \geqslant 1$

5 a $x < -\dfrac{4}{3}$ or $x > 4$ b $x < 2$ or $x > \dfrac{8}{3}$
  c $x \geqslant \dfrac{1}{2}$ d $3 < x < 7$
  e $x \leqslant 2$ or $x \geqslant 8$ f $1 < x < \dfrac{9}{5}$

6 $x < -\dfrac{5}{4}$ or $x > \dfrac{5}{4}$

Exercise 1D

1 a $x^2 + 3x - 1$ b $x^2 - 5x + 7$
  c $3x^2 - 4x + 2$ d $x^2 - 2x + 5$
  e $-5x^2 + 3x - 4$
  f $-6x^3 - 6x^2 - 6x - 19$

2 a Quotient = $x^2 + x + 4$, remainder = -8
  b Quotient = $6x^2 + 19x + 38$, remainder = 70
  c Quotient = $4x^2 + \dfrac{5}{2}x + \dfrac{1}{4}$, remainder = $\dfrac{5}{4}$
  d Quotient = $-2x^2 + 3x + 9$, remainder = -36
  e Quotient = $x + 3$, remainder = $-10x - 4$
  f Quotient = $5x^2 - 7$, remainder = $15 - 13x$

3 a Proof b Proof

4 a Proof
  b $(x-2)(2x-3)(x+7)$

5 a Proof b $\pm\dfrac{1}{2}, -\dfrac{4}{3}$

6 a Proof b Proof

7 $-3, \dfrac{1}{2}, 5$

Exercise 1E

1 Proof

2 Proof

3 $a = -4$

4 $a = \dfrac{1-b}{3}$

5 $a = -2, b = 1$

6 a $p = -1, q = -6$ b Proof

7 $-4, 0, 1$

8 a $p = -7, q = -6$
   b $(x+1)(x+2)(x-3)$ and $(x+1)(x+3)(x+4)$

9 a $p = -19, q = 30$
   b $(x-1)(x+2)(x-5)(x+3)$

10 a $\pm 2, 5$ b $-7, -1, 3$
   c $-\frac{5}{2}, 2, 3$ d $-4, -2, \frac{1}{3}$
   e $-3, \pm 2, 1$ f $-\frac{1}{2}, 1, 4$

11 Proof

12 $0 < k < 5$

Exercise 1F

1 a 6   b 8   c 4   d $-\frac{31}{4}$

2 a $a = 5$ b $b = 8$

3 $a = 2, b = 2$

4 $a = 4, b = 0$

5 a $a = 6, b = -14$
   b $(3x-4)(2x+1)(x+2)$

6 a $p = 14$ b 57

7 a $a = 5, b = -15$
   b $2, \frac{-7+\sqrt{53}}{2}, \frac{-7-\sqrt{53}}{2}$

8 a $k = \frac{5}{2}$ b -36

9 2550

10 $a = 2, b = -4, c = -2$.

End-of-chapter review exercise 1

1 $-\frac{4}{3}, \frac{2}{7}$

2 $x \leqslant -\frac{4}{5}$ or $x \geqslant 2$

3 $1 < x < \frac{5}{3}$

4 $\pm 5, \pm \sqrt{3}$

5 a $a = 6$
   b $(3x+1)(2x+3)(x-4)$

6 a $(3x-1)(2x+3)(x-5)$
   b $\pm \frac{1}{3}, \pm 5$

7 $a = -10, b = 8$

8 a Quotient $= x - 3$, remainder $= 2x - 6$
   b Proof

9 a Quotient $= 4x^2 + 4x - 3$, remainder $= 5$
   b $-\frac{3}{2}, -1, \frac{1}{2}, 1$

10 a $k = 7$
   b $\frac{1}{2}(-7-\sqrt{41}), \frac{1}{2}(-7+\sqrt{41}), \frac{1}{2}(7-\sqrt{53}), \frac{1}{2}(7+\sqrt{53})$

11 a $a = 6$
   b $(2x-1)(x+3)(x-2)(x+1)$

12 a $a = 7$ b $-5, \frac{2}{3}, 2$

13 a 33
   b Quotient $= 2x + 13$, remainder $= 41x - 15$

14 a $a = 3, b = -10$ b $3x - 4$

15 a $a = -21$ b $-\frac{5}{3}, -\frac{1}{2}, 2$

16 a $a = -3, b = -11$
   b $(2x-1)(x+2)(x-3)$

17 a $a = 2, b = -16$ b -18

18 a Quotient $= 5x^2 - 8x + 9$
   b Proof

19 a $k = -29$ b $-2, \frac{1}{4}, 9$
   c $\pm 3, \pm \frac{1}{2}$

20 a $a = 3, b = 50$
   b Quotient $= 2x - 3$, remainder $= 56 - 4x$

21 a $a = -11, b = 30$ b $-2, \frac{3}{2}, 5$

22 i $x + 2, 3x + 4$ ii Proof

23 i −16 ii $-\frac{1}{2}, \frac{3}{2}, 3$

24 i $a = 2, b = -6$
ii $(2x + 3)(x - 1)(x - 3)$

2 Logarithmic and exponential functions

Prerequisite knowledge

1 a $\frac{1}{25}$ b 4 c 1

2 $\frac{1}{5}$

3 a $2x^{-\frac{7}{2}}$ b $\frac{1}{5}x^{\frac{5}{2}}$ c $\frac{8x^{15}}{27}$

Exercise 2A

1 a $2 = \log_{10} 100$ b $x = \log_{10} 200$
c $x = \log_{10} 0.05$

2 a 1.72 b 2.40
c −0.319

3 a $10\,000 = 10^4$ b $x = 10^{1.2}$
c $x = 10^{-0.6}$

4 a 75.9 b 575
c 0.0398

5 a 2 b −4 c 1.5
d $\frac{1}{3}$ e $2\frac{1}{3}$ f 0.5

6 $f^{-1}(x) = \log_{10}(x + 3)$

7 $10\sqrt{10}$

Exercise 2B

1 a $2 = \log_5 25$ b $4 = \log_2 16$
c $-5 = \log_3 \frac{1}{243}$ d $-10 = \log_2 \frac{1}{1024}$
e $x = \log_8 15$ f $y = \log_x 6$
g $b = \log_a c$ h $5y = \log_x 7$

2 a $2^3 = 8$ b $3^4 = 81$ c $8^0 = 1$
d $16^{\frac{1}{2}} = 4$ e $8^{\frac{1}{3}} = 2$ f $2^4 = y$
g $a^0 = 1$ h $x^y = 5$

3 a 8 b 9 c 1 d 7

4 a 4 b 11 c 3

5 a 3 b 2 c $\frac{1}{2}$ d −3

e −2 f $\frac{3}{2}$ g $-\frac{1}{2}$ h $-\frac{4}{3}$

6 a 3 b $\frac{1}{3}$ c $\frac{5}{2}$ d −4

e −6 f $\frac{5}{2}$ g $\frac{2}{3}$ h $\frac{9}{2}$

7 $f^{-1}(x) = 3 + 2^{x-1}$

8 a $\sqrt{3}$ b $-3, \frac{1}{2}$

9 $\log_4 3, \log_2 2, \log_3 4, \log_2 3, \log_3 9, \log_3 20, \log_2 8$

Exercise 2C

1 a $\log_2 77$ b $\log_6 5$ c $\log_5 2$
d $\log_3 2$ e $\log_2 18$ f $\log_4 8$

2 a 3 b $2 \log_6 10$ c $2 + \log_2 3$

3 a $\log_5 4$ or $2 \log_5 2$
b $\log_3 4$ or $2 \log_3 2$

4 $2^3, 2^{-2}, -\frac{3}{2}$

5 a 3 b $\frac{7}{4}$ c −1 d $-\frac{3}{2}$

6 $y = \frac{x - 2}{x^2}$

7 $z = \frac{1}{1 - 3y^3}$

8 a 5^y b $2 + y$
c $\frac{3}{2}y - 3$ d $\frac{3}{y}$

9 a $3 + x$ b $\frac{1}{2} - y$
c 4^{x+y} d $2x - \frac{1}{2}y - 1$

10 a 3 b $-\frac{9}{2}$
c 16 d $\frac{5}{3}$

11 $x = 2 \log_3 2 - 2 \log_2 5$,
$y = \log_2 5 - 2 \log_3 2$,
$z = \log_2 5 - \log_3 2$

Exercise 2D

1 a 10 b 3.5 c $-\frac{46}{9}$ d 54

2 a $\frac{41}{2}$ b $\frac{9}{7}$ c $-\frac{3}{10}$ d $\frac{7}{3}$

3 a 5 b 10 c 3, 6 d $\frac{1}{4}$

4 **a** 8   **b** 12   **c** $\frac{5}{3}$ **d** 4

5 **a** 8, 32   **b** 0.1, 1000

    c $\frac{1}{3125}$, 25   **d** $2\sqrt{2}, \frac{\sqrt{2}}{2}$

6 **a** $x = 3, y = 27$ **b** $x = \frac{1}{2}, y = 1$

    c $x = 4, y = -12$ **d** $x = \frac{625}{4}, y = \frac{25}{2}$

7 $\log_{10} x = 3, \log_{10} y = -2$

Exercise 2E

1 **a** 1.80   **b** 5.13   **c** 0.946

    d 3.64   **e** 3.86   **f** 1.71

    g 0.397   **h** 0.682   **i** −0.756

    j 6.76   **k** −0.443   **l** −15.4

2 **a** Proof   **b** 1.26

3 **a** −0.322   **b** 1.11   **c** 1.83

    d 1.03   **e** 3.21   **f** 0.535

4 2, 3

5 **a** 0.431, 0.683   **b** 0, 2.32

    c 0, 1.77   **d** 0.792, 1.58

6 2, 2.58

7 **a** 2.81   **b** 1.46

    c 0.431, 1.29   **d** 0.792, 0.161

8 **a** 1.58   **b** −0.792, 0.792

    c 1.58, 2.32   **d** 1.37

9 **a** 1.77   **b** 0.510   **c** 1.98

10 0, $20^{0.8}$

11 **a** 1.29, 1.66   **b** 1.63   **c** 1.91, 2.91

    d 0.834   **e** −0.515   **f** 1.87

    g ±1.89   **h** ±2.81

12 **a** 13.5   **b** 1.45

13 0, 1, 2

Exercise 2F

1 **a** $x < \frac{\log 5}{\log 2}$ **b** $x \geq \frac{\log 7}{\log 5}$

    c $x < \frac{\log 3}{\log \frac{2}{3}}$ **d** $x > \frac{\log 0.3}{\log 0.8}$

2 **a** $x > 5 - \frac{1}{\log 8}$

    b $x > \frac{1}{2}\left(\frac{\log 20}{\log 3} - 5\right)$

    c $x \leq \frac{1}{2}\left(\frac{\log \frac{3}{2}}{\log 5} - 1\right)$

    d $x > 3 - \frac{\log \frac{4}{7}}{\log \frac{5}{6}}$

3 $x < 0$ or $x > \frac{\log 2}{\log 5}$

4 Proof

5 **a** 61   **b** 1

Exercise 2G

1 **a** 20.1   **b** 14.9

    c 2.23   **d** 0.135

2 **a** 1.10   **b** 0.336

    c −0.105   **d** −1.90

3 **a** 2   **b** 3   **c** 30   **d** 2

4 **a** 5   **b** 15   **c** 4   **d** $\frac{1}{3}$

5 **a** 2.89   **b** 1.61   **c** 1.08   **d** 2.89

6 **a** $\ln 13$ **b** $\frac{1}{3}\ln 7$

    c $\frac{1}{2}(1 + \ln 6)$ **d** $2[\ln(4) - 3]$

7 **a** $x > \ln 10$

    b $x \leq \frac{1}{5}(2 + \ln 35)$

    c $x < -\frac{1}{2}(3 + \ln 5)$

8 a 148 b 0.0183
   c 405 d −0.432

9 a 1.22 b 5.70 c 1.41
   d 0.690 e 1.16 f 1.08

10 a $y = \dfrac{1}{x-2}$
   b $y = \dfrac{2}{ex^2 - 1}$

11 a $\ln 3$ b $\ln 2, \ln 3$
   c $\ln\left(\dfrac{5}{2}\right)$ d $\ln 7$

12 $f^{-1}(x) = \ln\left(\dfrac{x-2}{5}\right)$

13 0.151

14 a $x = e,\ y = \dfrac{5}{e}$
   b $x = 2\ln 2,\ y = -\dfrac{1}{5}\ln 2$

15 $-\dfrac{1}{2} < x \leqslant 3$

Exercise 2H

1 a $\ln y = ax + b,\ Y = \ln y,\ X = x,\ m = a,\ c = b$
   b $\log y = ax - b,\ Y = \log y,\ X = x,\ m = a,\ c = -b$
   c $\ln y = -b\ln x + \ln a,\ Y = \ln y,\ X = \ln x,\ m = -b,\ c = \ln a$
   d $\ln y = x\ln b + \ln a,\ Y = \ln y,\ X = x,\ m = \ln b,\ c = \ln a$
   e $x^2 = -by + \ln a,\ Y = x^2,\ X = y,\ m = -b,\ c = \ln a$
   f $\ln y = -\dfrac{a}{b}\ln x + \dfrac{\ln 8}{b},\ Y = \ln y,\ X = \ln x,\ m = -\dfrac{a}{b},\ c = \dfrac{\ln 8}{b}$
   g $\ln x = -y\ln a + \ln b,\ Y = \ln x,\ X = y,\ m = -\ln a,\ c = \ln b$
   h $\ln y = -bx + \ln a,\ Y = \ln y,\ X = x,\ m = -b,\ c = \ln a$

2 $a = 66,\ n = -0.53$

3 $k = 9.5,\ n = 0.42$

4 a $\log_{10} y = \dfrac{3}{2}x + 2$
   b $y = 100 \times 10^{\frac{3}{2}x}$

5 a $\ln y = 3\ln x - 2$
   b $y = \dfrac{x^3}{e^2}$

6 Gradient $= \dfrac{\ln 3}{\ln 5}$, y-intercept $= \left(0, \dfrac{\ln 3}{2\ln 5}\right)$

7 a [graph of $\ln m$ vs t showing points at $t = 10, 20, 30, 40, 50$ with $\ln m$ values approximately 3.7, 3.5, 3.3, 3.1, 2.9]

   b $m_0 = 50,\ k = 0.02$
   c 35 days

8 a $\ln(T - 25) = -nt + \ln k$
   b $k = 45,\ n = 0.08$
   c i 70 °C
      ii 34 minutes
      iii 25 °C

End-of-chapter review exercise 2

1 $x > \dfrac{\log 7}{\log 2}$

2 $p = \dfrac{q^2}{3 + q}$

3 $x < -\dfrac{\log\left(\dfrac{3}{2}\right)}{\log 8}$

4 20.1

5 $-1, \dfrac{2\log 2 - \log 3}{\log 2}$

6 $\dfrac{4}{3}$

7 $K = 7.39,\ m = 1.37$

8 i Proof ii 0, 1.58

9 9.83

10 i $(x+2)(4x+3)(3x-2)$

  ii $3^y = \dfrac{2}{3}$, $y = -0.369$

11 3.81

12 22.281

13 0.438

14 a $\dfrac{2\ln 3}{\ln 5}$

  b $\left(1, -\dfrac{2\ln 3}{\ln 5}\right)$

15 $K = 1.73$, $b = 1.65$

3 Trigonometry

Prerequisite knowledge

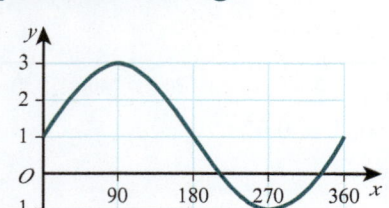

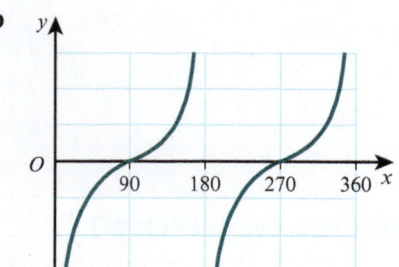

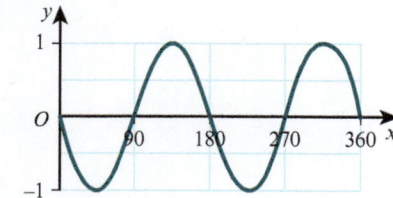

1 a [graph of sine-like curve from 0 to 360°, amplitude 3]

  b [graph of tangent-like curve with asymptotes]

  c [graph of sine-like curve, amplitude 1, period 180°]

2 a $\dfrac{\sqrt{3}}{2}$ b $-\dfrac{\sqrt{2}}{2}$ c $-\dfrac{\sqrt{3}}{3}$

3 a 31.0°, 211.0°

  b 30°, 150°, 270°

Exercise 3A

1 a 2   b $\sqrt{2}$ c $-\dfrac{1}{\sqrt{3}}$

  d 2   e $\sqrt{2}$ f $-\sqrt{3}$

  g $-\dfrac{2}{\sqrt{3}}$ h $-\sqrt{3}$

2 a 2   b $\dfrac{1}{\sqrt{3}}$ c $\sqrt{2}$

  d $-\dfrac{1}{\sqrt{3}}$ e $-\dfrac{2}{\sqrt{3}}$ f $-\sqrt{2}$

  g $\dfrac{1}{\sqrt{3}}$ h $\dfrac{2}{\sqrt{3}}$

3 a 70.5°, 289.5°   b 51.3°, 231.3°
  c 199.5°, 340.5°   d 41.4°, 318.6°

4 a $\dfrac{\pi}{6}, \dfrac{5\pi}{6}$ b π
  c 0.464, 3.61   d 2.76, 5.90

5 a 28.2°, 61.8°   b 37.8°, 142.2°
  c 22.5°, 112.5°   d 24.1°, 155.9°

6 a 60°, 180°   b 35.9°, 84.1°
  c 2.82, 5.96
  d −2.28, −1.44, 0.865, 1.71

7 a −150°, −30°, 30°, 150°
  b −109.5°, −70.5°, 70.5°, 109.5°
  c −112.6°, 112.6°
  d −180°, 0°, 180°
  e −135°, 45°
  f −60°, 60°

8 a 48.2°, 180°, 311.8°
  b 31.0°, 153.4°, 211.0°, 333.4°
  c 19.5°, 160.5°, 203.6°, 336.4°
  d 60°, 180°, 300°
  e 107.6°, 252.4°
  f 27.2°, 152.8°

9 a 41.8°, 138.2°
  b 13.3°, 22.5°, 103.3°, 112.5°
  c 45°, 60°, 120°, 135°
  d 97.2°, 172.8°

10 a $\dfrac{2\pi}{3}, \pi, \dfrac{4\pi}{3}$

   b $\dfrac{7\pi}{6}, \dfrac{11\pi}{6}$

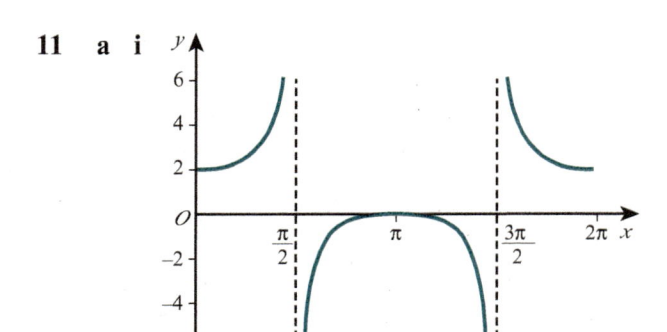

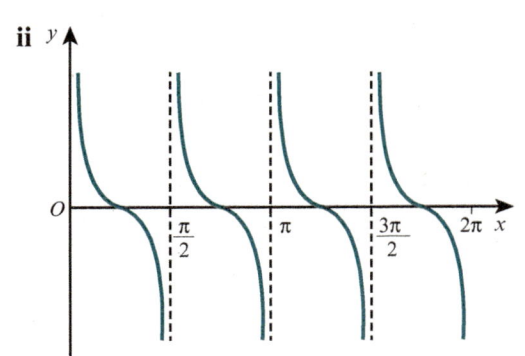

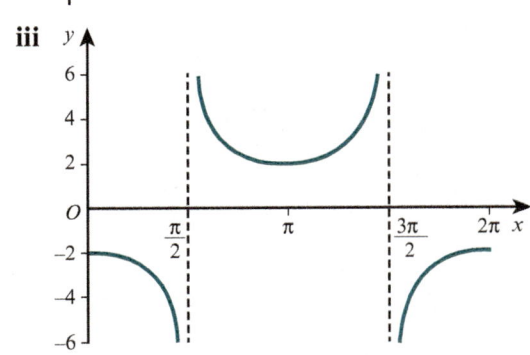

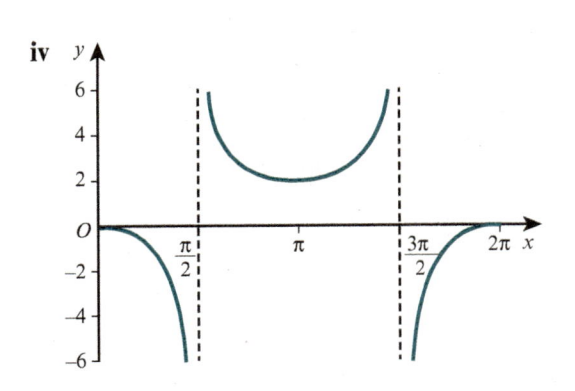

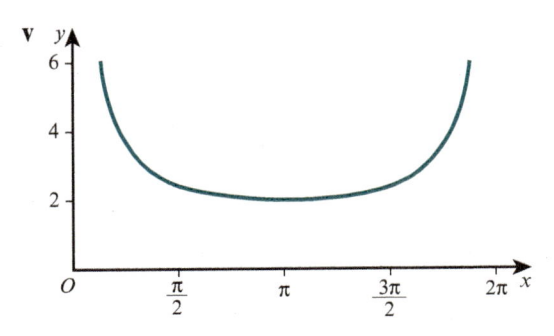

Answers

11 a i *graph*

ii *graph*

iii *graph*

iv *graph*

v *graph*

vi *graph*

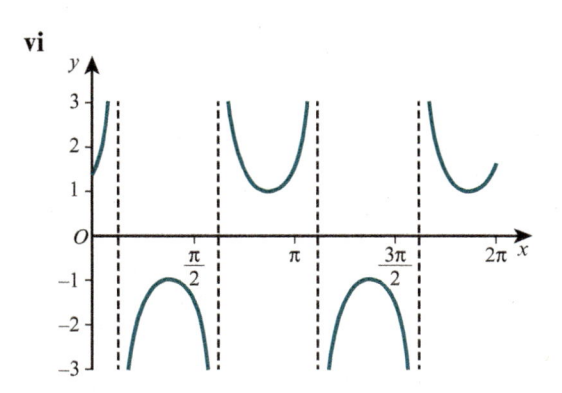

b $x = \dfrac{\pi}{8}$, $x = \dfrac{5\pi}{8}$, $x = \dfrac{9\pi}{8}$, $x = \dfrac{13\pi}{8}$

12 a Proof **b** Proof
c Proof **d** Proof

13 a Proof **b** Proof **c** Proof
d Proof **e** Proof **f** Proof
g Proof **h** Proof

14 a 48.2°, 180° **b** 45°, 63.4°, 161.6°

Exercise 3B

1 $\dfrac{\sqrt{3}}{2}\cos x - \dfrac{1}{2}\sin x$

2 a 1 **b** $\dfrac{\sqrt{2}}{2}$ **c** $\dfrac{1}{2}$
d $\dfrac{\sqrt{3}}{2}$ **e** 1 **f** $\dfrac{\sqrt{3}}{3}$

3 a $\dfrac{\sqrt{6}+\sqrt{2}}{4}$ **b** $2+\sqrt{3}$ **c** $\dfrac{\sqrt{2}-\sqrt{6}}{4}$
d $-2+\sqrt{3}$ **e** $\dfrac{\sqrt{6}-\sqrt{2}}{4}$ **f** $\dfrac{\sqrt{6}-\sqrt{2}}{4}$
g $-2+\sqrt{3}$ **h** $\dfrac{\sqrt{6}+\sqrt{2}}{4}$

4 $\dfrac{4+3\sqrt{3}}{10}$

5 Proof

6 a $\dfrac{33}{65}$ **b** $\dfrac{16}{65}$ **c** $-\dfrac{33}{56}$

7 a $\dfrac{77}{85}$ **b** $\dfrac{36}{85}$ **c** $\dfrac{13}{84}$

8 $\dfrac{t-2}{2t+1}$

9 $\dfrac{1}{2}$

10 a $\dfrac{1}{3}$ b 2

11 a $\dfrac{3}{2}$ b -0.2

12 a Proof b $-30°, 150°$

13 a $19.1°, 199.1°$ b $70.9°, 250.9°$
   c $5.9°, 185.9°$ d $150°, 330°$

14 a $38.4°, 111.6°$ b $18.4°, 116.6°$
   c $16.0°$ d $35.0°$
   e $18.4°, 26.6°$ f $74.1°$

15 $22.5°, 112.5°$

16 Proof

17 $\dfrac{p^2 + q^2 - 2}{2}$

Exercise 3C

1 a $\sin 56°$ b $\cos 68°$ c $\tan 34°$

2 a $\dfrac{24}{25}$ b $-\dfrac{7}{25}$
   c $-\dfrac{24}{7}$ d $-\dfrac{44}{117}$

3 a $\dfrac{336}{625}$ b $-\dfrac{336}{527}$
   c $\dfrac{7}{25}$ d $\dfrac{24}{7}$

4 a $-\dfrac{24}{25}$ b $\dfrac{336}{625}$
   c $\dfrac{24}{7}$ d 2

5 $\dfrac{1}{3}$

6 a $14.5°, 90°, 165.5°, 270°$
   b $60°, 300°$
   c $48.6°, 131.4°, 270°$

7 a $30°, 150°$
   b $33.6°, 180°$
   c $30°, 150°$
   d $33.2°, 90°, 146.8°$
   e $39.2°, 90°, 140.8°$
   f $24.9°, 98.8°$

8 $\dfrac{1}{2}(1 + \cos 4x)$

9 a Proof
  b $\sin 3x \equiv 3\sin x - 4\sin^3 x$

10 $35.3°, 60°, 120°, 144.7°$

11 a Proof b 4

12 a Proof b $-\dfrac{2\pi}{3}, -\dfrac{\pi}{3}, \dfrac{\pi}{3}, \dfrac{2\pi}{3}$

13 a Proof b $-\dfrac{5\pi}{6}, -\dfrac{\pi}{6}, \dfrac{\pi}{6}, \dfrac{5\pi}{6}$

14 a Proof
   b $\sin 3\theta \equiv 3\sin\theta - 4\sin^3\theta$, $\cos 2\theta \equiv 1 - 2\sin^2\theta$
   c Proof
   d $\dfrac{\sqrt{5}-1}{4}$

15 $\dfrac{2\pi}{3} < \theta < \dfrac{4\pi}{3}$

16 $210° \leq \theta \leq 330°$

17 Proof

18 a Proof
   b $0° < x < 45°$ or $120° < x < 135°$

19 $45° < \theta < 135°$ or $225° < \theta < 315°$

Exercise 3D

1 a Proof b Proof c Proof
  d Proof e Proof f Proof
  g Proof h Proof

2 a Proof b Proof c Proof
  d Proof e Proof f Proof
  g Proof h Proof

3 a Proof b Proof

4 Proof

5 Proof

6 Proof

Exercise 3E

1 a $17\sin(\theta - 28.07°)$ b $64.1°, 172°$

2 a $\sqrt{13}\cos(\theta + 56.31°)$ b $12.6°, 234.8°$

3 a $17\sin(\theta - 28.07°)$ b $38.2°, 197.9°$
  c 34

4 a $2\sqrt{13}\sin(\theta-56.31°)$ b 80.9°
   c 49, −3
5 a $5\sin(\theta+53.13°)$ b 103.3°, 330.4°
   c −2
6 a $2\cos\left(\theta-\dfrac{\pi}{3}\right)$ b Proof
7 a $4\sqrt{5}\sin(2\theta+26.57°)$
   b 66.9°, 176.5°, 246.9°, 356.5°
   c $\dfrac{1}{8}$
8 a $\sqrt{3}\cos(\theta+54.74°)$ b 70.5°, 180°
   c $\dfrac{1}{6}$
9 a $\sqrt{2}\cos\left(\theta+\dfrac{\pi}{4}\right)$ b $\dfrac{23\pi}{12}$
   c $-\sqrt{2} \leq k \leq \sqrt{2}$
10 a $\sqrt{10}\sin(\theta-71.57°)$ b 32.3°, 290.8°
   c $1+\sqrt{10}$, 80.8°
11 a $3\cos(\theta-41.81°)$ b 41.8°
   c 302.6°
12 a Proof b $5\sin(\theta+53.13°)$
   c 115.3°, 318.4°
13 a $\sqrt{3}\sin(\theta+60°)$ b 84.7°, 335.3°
14 a $10+\sqrt{5}$, $10-\sqrt{5}$
   b 18.4°, 45°, 198.4°, 225°

End-of-chapter review exercise 3

1 *graph*

2 131.8°
3 $-\dfrac{\sqrt{15}}{8}$
4 0°, 131.8°, 228.2°, 360°

5 41.8°, 138.2°, 194.5°, 345.5°
6 a Proof b 30°, 330°
7 a Proof
   b 61.3°, 118.7°, 241.3°, 298.7°
8 i Proof ii 18.4°, 26.6°
9 a Proof b 35.8°, 125.8°
10 i $R=\sqrt{10}$, $\alpha=18.43°$
   ii 34.6°, 163.8°, 214.6°, 343.8°
11 a Proof
   b i $\dfrac{\sqrt{2}}{2}$
      ii 15°, 75°
12 a Proof
   b $2\sqrt{13}\cos(\theta-56.31°)$
   c 52
13 i $R=2\sqrt{13}$, $\alpha=56.31°$
   ii 80.9°, 211.7°
   iii 60, 8
14 i Proof ii Proof
   iii 0.322, 0.799, −1.12
15 a Proof
   b 60°, 104.5°, 255.5°, 300°
16 a Proof b 21.8°, 161.6°
17 a Proof b $5\sin(2x+36.87°)$
   c 71.6°, 161.6°

Cross-topic review exercise 1

1 $\dfrac{2}{5} < x < 4$
2 0.631, −0.369
3 i $x=\dfrac{7}{6}$ ii 0.222
4 i $-2, \dfrac{2}{5}$ ii −0.569
5 $A=8.5$, $b=1.6$
6 i 1.77 ii ±1.77
7 i $n=1.50$, $C=6.00$
   ii $n\ln x + \ln y = \ln C$ is linear in $\ln y$ and $\ln x$.
8 −0.405, 1.39

9   $\theta = 135°, \phi = 63.4°$ or $\theta = 53.1°, \phi = 161.6°$

10 i $-\dfrac{2}{3}, \dfrac{4}{5}$ \hspace{1em} ii -0.161

11 i $a = -16$ \hspace{1em} ii $\dfrac{3}{2}, -\dfrac{1}{2}, 3$

12 i $a = -4, b = 6$
    ii quotient $= 2x - 4$, remainder $= -2$

13 i $R = \sqrt{10}, \alpha = 71.57°$
    ii $61.2°, 10.4°$

14 i Proof
    ii $(x - 2)(4x + 1)^2$
    iii 2

15 i $a = 6, b = -3$
    ii $(x + 1)(4x + 1)(2x - 1)$

16 i $a = -17, b = 12$
    ii $x = -3, x = \dfrac{4}{3}, x = 1$

17 i $a = 1, b = -10$
    ii quotient $= x - 1, x = 1, x = 2, x = -4$

18 i $a = 19, b = -36$
    ii $(x + 2)(x + 3)(5x - 6), 0.113$

19 i quotient $= x^2 + 2x + 1$, remainder $= 5x + 2$
    ii $p = 7, q = 4$
    iii $x = -1$

20 i $\dfrac{2}{3}$ \hspace{1em} ii $-\dfrac{9}{20}$

21 i $\sqrt{29}\sin(2\theta + 21.80°)$
    ii $13.1°, 55.1°, 193.1°, 235.1°$
    iii $\dfrac{1}{116}$

22 i $a = 2, b = -5$
    ii a Proof
        b $109.5°$

4 Differentiation

Prerequisite knowledge

1 a $15x^2 + \dfrac{6}{x^3} + \dfrac{1}{\sqrt{x}}$

   b $\dfrac{5}{2}x^4 - 4x - \dfrac{1}{2x^2}$

2 a $12(3x - 5)^3$

   b $\dfrac{4}{(1 - 2x)^{\frac{3}{2}}}$

3 $x - 5y = 16$

4 $(0, 2)$ maximum, $(2, -2)$ minimum

Exercise 4A

1 a $(6x - 2)(x - 2)^4$ \hspace{1em} b $5(2x + 1)^2(8x + 1)$

   c $\dfrac{3x + 4}{2\sqrt{x + 2}}$ \hspace{1em} d $\dfrac{3x + 9}{2\sqrt{x + 5}}$

   e $\dfrac{x^2(7x - 3)}{\sqrt{2x - 1}}$

   f $\dfrac{(13x^2 + 2)(x^2 + 2)^2}{2\sqrt{x}}$

   g $(x - 3)(x + 2)^4(7x - 11)$

   h $2(2x - 1)^4(3x + 4)^3(27x + 14)$

   i $2(3x^2 + 1)(15x^2 - 30x + 1)$

2 -1.5

3 $16x + y = 32$

4 5

5 $-1, \dfrac{3}{5}, 3$

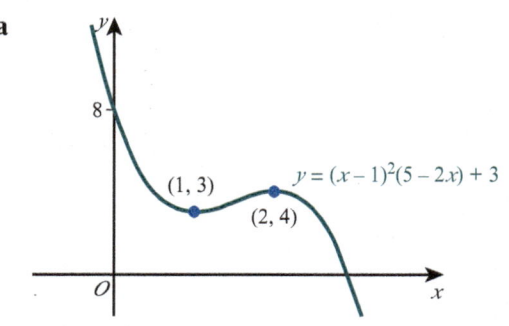

6 $-\dfrac{1}{3}$

7 a

   b 2

Exercise 4B

1 a $-\dfrac{11}{(x - 4)^2}$ \hspace{1em} b $\dfrac{1}{(2 - x)^2}$

   c $\dfrac{2(x^2 - x + 3)}{(2x - 1)^2}$ \hspace{1em} d $\dfrac{11}{(2 - 5x)^2}$

   e $-\dfrac{2(8x + 1)}{(x + 4)^3}$ \hspace{1em} f $-\dfrac{20x^3}{(x^2 - 1)^3}$

g $\dfrac{13x^2 + 30x - 35}{(x^2 + 2x + 5)^2}$

h $-\dfrac{2(x+4)(2x^2 + 12x - 1)}{(x^2 + 1)^4}$

2 $\dfrac{1}{4}$

3 $(-6, -7), (1, 0)$

4 $(2, 1), (8, -5)$

5 $y = 9x - 4$

6 a $\dfrac{-5x - 1}{2\sqrt{x}(5x-1)^2}$ b $\dfrac{x+4}{(2x+3)^{\frac{3}{2}}}$

   c $-\dfrac{x(x^2+1)}{(x^2-1)^{\frac{3}{2}}}$ d $\dfrac{5(x-1)^2(5x+13)}{2(x+2)^{\frac{3}{2}}}$

7 3

8 $3y = x + 7$

9 a $-3, 1, 5$

   b $9, -\dfrac{1}{3}, -\dfrac{5}{3}$

Exercise 4C

1 a $5e^{5x}$ b $-4e^{-4x}$
   c $12e^{6x}$ d $-15e^{-5x}$
   e $2e^{\frac{x}{2}}$ f $2e^{2x-7}$
   g $2xe^{x^2-3}$ h $2 + \dfrac{3e^{\sqrt{x}}}{2\sqrt{x}}$
   i $\dfrac{5\sqrt{e^x}}{2} + 2e^{-2x}$ j $6e^{3x}$
   k $3e^{2x} - e^{-2x}$ l $10xe^{x^2} - 10$

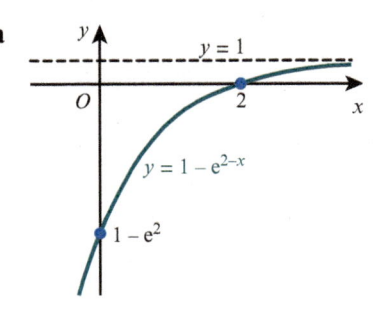

2 a (graph showing $y = 1 - e^{2-x}$, horizontal asymptote $y=1$, x-intercept at 2, y-intercept at $1 - e^2$)

   b $y = -x + 2$

3 0.0283 grams per year

4 a $xe^x + e^x$ b $3x^2 e^{3x} + 2xe^{3x}$
   c $e^{-2x}(5 - 10x)$ d $\dfrac{e^x(2x+1)}{\sqrt{x}}$
   e $\dfrac{e^{6x}(6x-1)}{x^2}$ f $-\dfrac{e^{-2x}(4x+1)}{2x\sqrt{x}}$
   g $\dfrac{3e^x}{(e^x + 2)^2}$ h $3xe^{3x} + 3e^{6x} + e^{3x}$
   i $\dfrac{2x^2 e^x + 5xe^x + 2xe^{2x} - e^x - 2}{(e^x + 2)^2}$

5 $-\dfrac{4}{9}$

6 $\left(-1, -\dfrac{1}{e}\right)$

7 $y = 3x + 3, (-1, 0)$

8 $(3, -e^3)$ minimum

9 $(1, e^2)$ minimum

10 a $x = 0$ minimum, $x = 2$ maximum
   b Proof

11 $x = 1 - \dfrac{1}{\sqrt{2}}, x = 1 + \dfrac{1}{\sqrt{2}}$

12 $\left(\dfrac{1}{2}, 2\right)$

13 Proof

14 $3 + 3\ln 3$

Exercise 4D

1 a $\dfrac{1}{x}$ b $\dfrac{1}{x}$
   c $\dfrac{2}{2x+1}$ d $\dfrac{2x}{x^2+1}$
   e $\dfrac{4}{2x-1}$ f $\dfrac{1}{2(x-3)}$
   g $\dfrac{5}{x+3}$ h $3 - \dfrac{1}{x}$
   i $5 - \dfrac{2}{2x-1}$ j $\dfrac{1}{x \ln x}$
   k $\dfrac{1}{\sqrt{x}(\sqrt{x} - 2)}$ l $\dfrac{1 + 5x}{x(5x + \ln x)}$

2 Possible justification: $\ln 3x = \ln 3 + \ln x$ and $\ln 7x = \ln 7 + \ln x$

$\frac{d}{dx}(\ln 3x) = \frac{d}{dx}(\ln 3 + \ln x) = 0 + \frac{1}{x}$ and

$\frac{d}{dx}(\ln 7x) = \frac{d}{dx}(\ln 7 + \ln x) = 0 + \frac{1}{x}$

3
- **a** $1 + \ln x$
- **b** $2x^2(1 + 3\ln x)$
- **c** $\frac{2x}{2x+1} + \ln(2x+1)$
- **d** $3(1 + \ln 2x)$
- **e** $\frac{1}{\ln x} + \ln(\ln x)$
- **f** $\frac{1 - \ln 5x}{x^2}$
- **g** $-\frac{2}{x(\ln x)^2}$
- **h** $\frac{3x - (3x-2)\ln(3x-2)}{x^2(3x-2)}$
- **i** $\frac{2(4x-1) - 4(2x+1)\ln(2x+1)}{(2x+1)(4x-1)^2}$

4
- **a**

  graph with $x = \frac{3}{2}$, $y = \ln(2x-3)$

- **b** $\frac{2}{7}$

5 -8

6 $2 + 4\ln 10$, $3 + 2\ln 10$

7 $\left(\frac{1}{\sqrt{e}}, -\frac{1}{2e}\right)$, minimum

8 $\left(e, \frac{1}{e}\right)$, maximum

9 $y = 5x - 5$

10
- **a** $\frac{5}{2(5x-1)}$
- **b** $-\frac{3}{3x+2}$
- **c** $\frac{1}{x} + \frac{5}{x+1}$
- **d** $\frac{2}{2x+3} - \frac{1}{x-1}$
- **e** $\frac{3}{3x-1} - \frac{2}{x}$
- **f** $\frac{1}{x} + \frac{1}{x-2} - \frac{1}{x+4}$
- **g** $\frac{1}{x-3} - \frac{1}{x+4} - \frac{1}{x-1}$
- **h** $-\frac{2}{x+1} - \frac{1}{x-2}$
- **i** $\frac{1}{x+2} + \frac{2}{2x-1} - \frac{1}{x} - \frac{1}{x+5}$

11
- **a** $\frac{4x}{2x^2-1}$
- **b** $\frac{9x^2+2}{3x^3+2x}$
- **c** $\frac{2x-4}{(x+1)(x-5)}$

12 -5

Exercise 4E

1
- **a** $\cos x$
- **b** $2\cos x - 3\sin x$
- **c** $-2\sin x - \sec^2 x$
- **d** $6\cos 2x$
- **e** $20\sec^2 5x$
- **f** $-2(3\sin 3x + 2\cos 2x)$
- **g** $3\sec^2(3x+2)$
- **h** $2\cos\left(2x + \frac{\pi}{3}\right)$
- **i** $-6\sin\left(3x - \frac{\pi}{6}\right)$

2
- **a** $3\sin^2 x \cos x$
- **b** $-15\sin 6x$
- **c** $2\sin x(1 + \cos x)$
- **d** $4\sin x(3 - \cos x)^3$
- **e** $12\sin^2\left(2x + \frac{\pi}{6}\right)\cos\left(2x + \frac{\pi}{6}\right)$
- **f** $-12\cos^3 x \sin x + 8\tan\left(2x - \frac{\pi}{4}\right)\sec^2\left(2x - \frac{\pi}{4}\right)$

3
- **a** $x\cos x + \sin x$
- **b** $5(\cos 3x - 3x\sin 3x)$
- **c** $x^2 \sec^2 x + 2x\tan x$
- **d** $\cos^2 2x(\cos 2x - 6x\sin 2x)$
- **e** $15\tan 3x \sec 3x$
- **f** $\sec x(x\tan x + 1)$
- **g** $\frac{x\sec^2 x - \tan x}{x^2}$
- **h** $\frac{1 + 2\cos x}{(2 + \cos x)^2}$
- **i** $\frac{(3x-1)\cos x - 3\sin x}{(3x-1)^2}$
- **j** $-6\cot 2x \csc^3 2x$
- **k** $3(1 - 2x\cot 2x)\csc 2x$
- **l** $\frac{2}{\sin 2x - 1}$

4
- **a** $\cos x \, e^{\sin x}$
- **b** $-2\sin 2x \, e^{\cos 2x}$

c $3\sec^2 3x e^{\tan 3x}$
   d $(\cos x + \sin x)e^{(\sin x - \cos x)}$
   e $(\cos x - \sin x)e^x$
   f $(2\cos 2x + \sin 2x)e^x$
   g $e^x(\cos x - 3\sin x)$
   h $x^2(3 - x\sin x)e^{\cos x}$
   i $-\tan x$
   j $x\cot x + \ln(\sin x)$
   k $-\dfrac{2(\sin 2x + \cos 2x)}{e^{2x+1}}$
   l $\dfrac{(1-2x)\sin 2x + 2x\cos 2x}{e^{2x}}$

5 1

6 $2\sqrt{3} - 6$

7 $\dfrac{\pi}{6}, \dfrac{\pi}{3}$

8 Proof

9 a $\tan x \sec x$ b $-\cot x \csc x$
   c $-\csc^2 x$

10 Proof

11 $y = -13.3x + 12.9$

12 0.464, 2.03

13 $x = \dfrac{\pi}{4}$, maximum

14 $x = \dfrac{\pi}{8}$

15 $x = \dfrac{\pi}{12}$, minimum

16 $x = \dfrac{\pi}{6}$ maximum, $x = \dfrac{5\pi}{6}$ minimum, $x = \dfrac{7\pi}{6}$ maximum, $x = \dfrac{11\pi}{6}$ minimum

17 0.452

Exercise 4F

1 a $5y^4 \dfrac{dy}{dx}$ b $3x^2 + 4y\dfrac{dy}{dx}$
   c $10x + \dfrac{1}{y}\dfrac{dy}{dx}$ d $\cos y \dfrac{dy}{dx}$
   e $18x^2 y^2 \dfrac{dy}{dx} + 12xy^3$ f $2y\dfrac{dy}{dx} + x\dfrac{dy}{dx} + y$
   g $3x^2 - 7x\dfrac{dy}{dx} - 7y + 3y^2 \dfrac{dy}{dx}$
   h $x\cos y \dfrac{dy}{dx} + \sin y - y\sin x + \cos x \dfrac{dy}{dx}$
   i $\dfrac{x^3}{y}\dfrac{dy}{dx} + 3x^2 \ln y$
   j $-2x\sin 2y \dfrac{dy}{dx} + \cos 2y$
   k $5\dfrac{dy}{dx} + e^x \cos y \dfrac{dy}{dx} + e^x \sin y$
   l $-2x\sin y\, e^{\cos y} \dfrac{dy}{dx} + 2e^{\cos y}$

2 a $-\dfrac{3x^2 + 2y}{2x + 3y^2}$ b $\dfrac{5 - 2xy}{x^2 + 2y}$
   c $-\dfrac{4x + 5y}{5x + 2y}$ d $\dfrac{y(2 - \ln y)}{x}$
   e $-\dfrac{2y(e^x y^2 + 1)}{3e^x y^2 + 2}$ f $-\dfrac{y}{2xy^2 + x}$
   g $\dfrac{y^4}{2 - 3xy^3}$ h $\dfrac{y(5xy + 1)}{x(2 - 5xy)}$

3 $-\dfrac{11}{25}$

4 3

5 $y = \dfrac{1}{4}x - \dfrac{9}{4}$

6 $y = \dfrac{3}{8}x + \dfrac{5}{8}$

7 a Proof   b $(-1, 5), (1, -5)$

8 a $(4, 18), (4, -2)$ b Proof
   c $y = 4x + 2$

9 a Proof   b $(1, 2)$

10 $-\dfrac{44}{21}$

11 $(6, -3), (-2, 0)$

12 $5x - 8y = 1$

13 e^{-1}

14 $(-4, -8)$ minimum, $(4, 8)$ maximum

Exercise 4G

1 a $\dfrac{1}{3t}$ b $\dfrac{2 - 2\sin 2\theta}{\cos 2\theta}$

    c $\dfrac{\sin 2\theta}{1-\cos 2\theta}$ d $\dfrac{4\cos 2\theta}{3\sec^2\theta}$

    e $-\dfrac{\sin\theta}{\sec^2\theta}$ f $\dfrac{\sin 2\theta}{\sin\theta - 2\sin 2\theta}$

    g $\dfrac{2\sec^2\theta}{\sin 2\theta}$ h $-e^{2t} - 1$

    i $\dfrac{1}{2}t^2 e^{-t}(t+3)$ j $2e^t(t+3)$

    k $\dfrac{5(1-t)}{t^2}$ l $\dfrac{4}{\sqrt{t}}$

2 $-3, \dfrac{1}{3}$

3 $3\sqrt{3}$

4 $(2, 4)$

5 $x + y = 2$

6   a   Proof   b $\left(e^4, \dfrac{4}{e^2} - 1\right)$

7   a   Proof   b $(1, 2)$

8   a   Proof   b   6, minimum

9 $y = -\dfrac{1}{2}x + 4$

10   a   Proof   b $(1, 2)$ c   Proof

11   a   Proof   b   Proof

End-of-chapter review exercise 4

1   i   Proof   ii $(1, 6)$

2   i   4   ii $\dfrac{2}{25}$

3   i   Proof   ii   Proof

     iii $(e^{-6}, 4e^{-2} + 3)$

4   i   5   ii -3

5   i $5x + 4y - 6 = 0$ ii   Proof

6   i   Proof   ii   Proof   iii $(-3, -2)$

7 $y = 8.66x - 2.53$

8   i   Proof   ii $0.294, 1.865$

     iii $-\dfrac{13}{9}$

9 $5x - 9y + 22 = 0$

10   i $-\dfrac{1}{\sqrt{1-x}\,(x+1)^{\frac{3}{2}}}$ ii $x = \dfrac{1}{2}$

11 $\dfrac{dy}{dx} = -\cos t$

12   i   Proof   ii $(5.47, 0.693)$

13   i   Proof   ii $k = 5, c = 68$

5 Integration

Prerequisite knowledge

1   a $6\cos 2x + 5\sin x$ b $5e^{5x-2}$

    c $\dfrac{2}{2x+1}$

2   Proof

Exercise 5A

1   a $\dfrac{1}{2}e^{2x} + c$ b $-\dfrac{1}{4}e^{-4x} + c$ c $2e^{3x} + c$

    d $8e^{\frac{1}{2}x} + c$ e $-2e^{-x} + c$

    f $\dfrac{1}{2}e^{2x+4} + c$ g $\dfrac{1}{3}e^{3x-1} + c$

    h $-2e^{2-3x} + c$ i $\dfrac{1}{4}e^{8x-3} + c$

2   a $x - e^{-x} + c$ b $\dfrac{5}{4}e^x(e^{3x} + 8) + c$

    c $x - e^{2x} + \dfrac{e^{4x}}{4} + c$ d $x - 2e^{-2x} + c$

    e $\dfrac{1}{2}e^{-2x}(e^x - 4) + c$ f $x - 2e^{-2x} + 4e^{-x} + c$

3   a $\dfrac{1}{3}(e^6 - 1)$ b $\dfrac{1}{4}(e^2 - 1)$

    c $\dfrac{15}{8}$ d $\dfrac{1}{2}e(e^6 - 1)$

    e $4e - \dfrac{4}{e}$ f $\dfrac{1}{2}(-3 + 4e + e^2)$

    g $\dfrac{1}{12}(-17 + 6e^2 + 8e^3 + 3e^4)$

    h $-\dfrac{19}{2} - \dfrac{25}{2e^2} + 2e^2$ i $\dfrac{9}{2} - \dfrac{1}{2e^4} - \dfrac{2}{e^2}$

4 $y = 3e^{2x} - 2e^{-x} + 1$

5 $y = 5e^{-2x} + 2x - 2$

6 $\dfrac{1}{2}\left(4 - \dfrac{1}{e^3} + e\right)$

7 $4e^{\frac{3}{2}} - 4$

8   a   Proof   b $1 + 2e^3$

9   a $7 - 2e^{-2a} - 5e^{-a}$ b   7

10   $6 - 7\ln 2$

11   a 1
     b   Proof

Exercise 5B

1   a $6\ln x + c$
    b $\frac{1}{2}\ln x + c$
    c $\frac{1}{3}\ln(3x+1) + c$
    d $3\ln(2x-5) + c$
    e $-\frac{5}{3}\ln(2-3x) + c$
    f $\frac{3}{10}\ln(5x-1) + c$

2   a $\ln\frac{7}{2}$
    b $\frac{1}{2}\ln 3$
    c $\frac{3}{2}\ln\frac{13}{3}$
    d $\frac{1}{2}\ln\frac{17}{9}$
    e $-\frac{3}{2}\ln\frac{7}{3}$
    f $-\ln 9$

3   a $12 + \frac{5}{3}\ln\frac{14}{5}$
    b $\ln\frac{27}{25}$
    c $4 + \ln 81$

4   a $A = 2$
    b Proof

5   a Quotient $= 3x + 10$, remainder $= 50$
    b Proof

6   $y = x^2 + 3\ln(x+e) - 3\ln 2e$

7   $k = 4e^2 - 3$

8   $(2\ln 2,\ 2\ln 2 - 3)$

Exercise 5C

1   a $-\frac{1}{3}\cos 3x + c$
    b $\frac{1}{4}\sin 4x + c$
    c $-2\cos\frac{x}{2} + c$
    d $-\frac{3}{2}\cos 2x + c$
    e $\frac{5}{3}\sin 3x + c$
    f $\frac{1}{2}\tan 2x + c$
    g $-\frac{2}{5}\sin(1-5x) + c$
    h $-\frac{3}{2}\cos(2x+1) + c$
    i $\frac{2}{5}\tan(5x-2) + c$

2   a $\frac{\sqrt{3}}{8}$
    b 1
    c $\frac{\sqrt{3}}{2}$
    d 2
    e $1 - \frac{\sqrt{3}}{4}$
    f $\frac{5\pi}{2} - 4$
    b $\frac{1}{6}(\sqrt{3}\pi - 3)$

3   a $x\cos x$

4   $y = x + \frac{3}{2}\cos 2x - \frac{\pi}{4}$

5   $y = 3\sin 2x + 2\cos x - 2x + \pi - 3$

6   a $y = 5 - 2\cos\left(2x - \frac{\pi}{2}\right)$
    b $x + 2y = 10 - 2\sqrt{3} + \frac{\pi}{3}$

7   $\frac{\pi}{2} + \sqrt{3}$

8   $\frac{21}{4}$

9   a $\frac{1}{2}(\sqrt{3} - 1)$
    b Proof

10 a $\frac{1}{2}(\sqrt{3} - \sqrt{2})$
    b Proof

Exercise 5D

1   a $\frac{3}{2}x + \frac{3}{4}\sin 2x + c$
    b $2x + 2\sin x + c$
    c $\frac{1}{2}x - \frac{1}{12}\sin 6x + c$
    d $2\tan x - 2x + c$
    e $2\tan 3x - 6x + c$
    f $\frac{3}{8}x + \frac{1}{4}\sin 2x + \frac{1}{32}\sin 4x + c$

2   a $\frac{\pi}{6} - \frac{\sqrt{3}}{8}$
    b $2\sqrt{3} - \frac{\pi}{6}$
    c $\frac{\sqrt{3}}{8} + \frac{\pi}{6}$
    d $\frac{2+\pi}{16}$
    e $\frac{1}{6}(2\pi - 3\sqrt{3})$
    f $\frac{1}{6}(3\sqrt{3} - \pi)$

3   a $\frac{1}{24}(2\pi - 5\sqrt{3})$
    b $\frac{1}{24}(6 + 2\pi + 3\sqrt{3})$
    c $\frac{1}{8}(2 + 5\pi)$
    d $\frac{5\pi}{3} - \frac{9}{4} - \sqrt{3}$
    e $\frac{1}{24}(2\pi + 11\sqrt{3})$
    f $\frac{5\sqrt{3}}{4}$

4   a Proof   b Proof

5   a Proof   b Proof

6   a Proof   b Proof

7   a Proof   b Proof

8   a Proof   b Proof

9   $\pi\left(2 + \frac{5\pi}{4}\right)$

10 a Proof   c Proof
    b Proof

Exercise 5E

1. a 5.22   b 17.09   c 0.92
   d 5.61   e 0.40   f 7.68

2. 1.55

3. a 6.76
   b Over-estimate since the top edges of the strips all lie above the curve

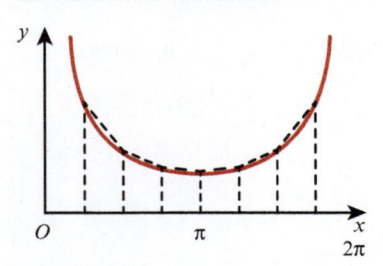

4. 1.77, under-estimate since top edges of the strips all lie below the curve

5. 4.07, over-estimate since top edges of the strips all lie above the curve

End-of-chapter review exercise 5

1. Proof

2. $12 + \ln\left(\dfrac{81}{25}\right)$

3. 1.81, under-estimate since top edges of the strips all lie below the curve

4. $\dfrac{1}{12}$

5. Proof

6. a $\dfrac{1}{6}(19 - 15e^{-2k} - 4e^{-3k})$
   b $\dfrac{19}{6}$

7. Proof

8. a $A = -20$ b Proof

9. i Proof   ii Proof

10. a $12e^x + \dfrac{4}{3}e^{3x} + c$ b Proof

11. i Proof   ii $\dfrac{5}{8}\pi + \dfrac{1}{4}$

12. i Proof   ii Proof
    iii $-\dfrac{1}{2}\cot x + c$

13. i $\dfrac{1}{2}\tan 2x + \dfrac{1}{2}x + \dfrac{1}{8}\sin 4x + c$
    ii $\ln(16e^{20})$

14. i Proof   ii $x\sin x$ iii 1

15. i Proof   ii Proof

6 Numerical solutions of equations

Prerequisite knowledge

1. a $\dfrac{35}{27}$ b $-48.7505\ldots$
   c 0.223130   d 5.405465

2. a $x = \sqrt{\dfrac{y-13}{4}}$ b $x = \dfrac{y+7}{3}$
   c $x = \sqrt[5]{\dfrac{1+y}{27}}$

3. Proof

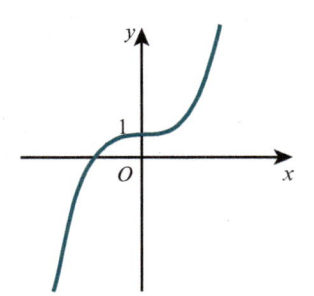

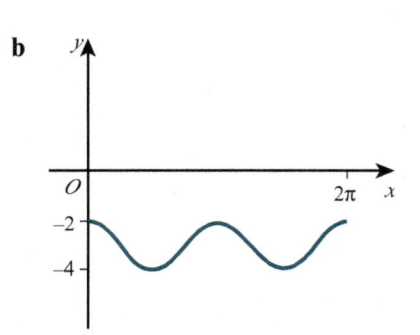

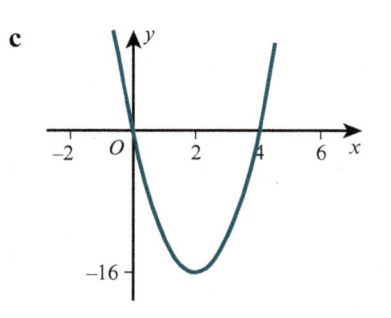

4. $x = 2$, $x = 5$

5. a, b, c

6 a

[Graph showing exponential-like curve passing through (0, 0.5)]

b

[Graph showing a curve with vertical asymptote near x=1, approaching horizontal asymptote]

c

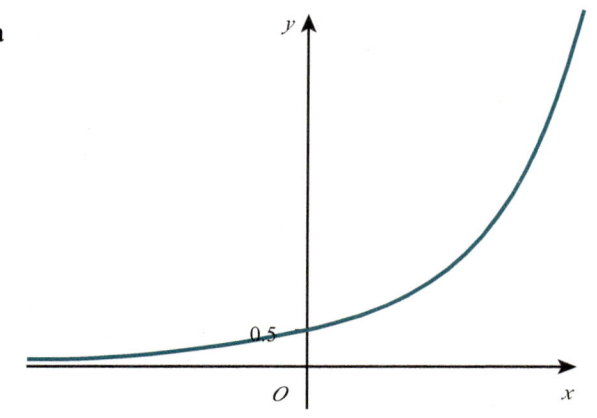

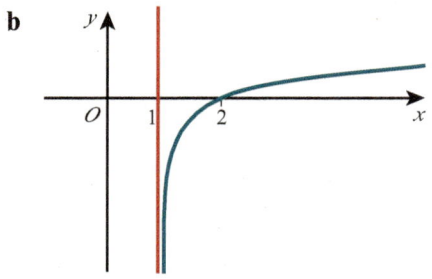

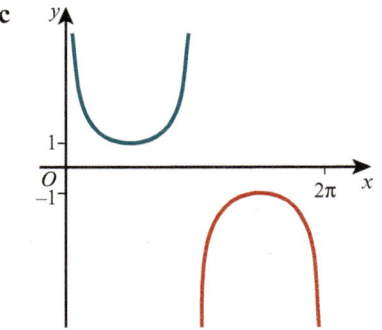

[Graph showing two curves: one U-shaped above x-axis starting near y=1, and another inverted curve below x-axis near 2π]

Exercise 6A

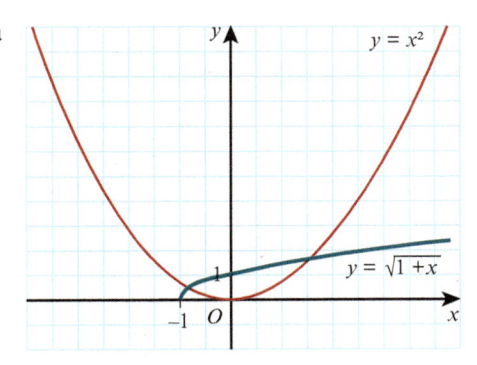

1 a

[Graph showing $y = x^2$ (parabola) and $y = \sqrt{1+x}$ starting at (−1, 0)]

b 2 points of intersection so 2 roots.

c Let $f(x) = x^2 - \sqrt{1+x} = 0$ then
$f(-1) = (-1)^2 - \sqrt{1+(-1)} = 1$ and
$f(0) = (0)^2 - \sqrt{1+0} = -1$. Change of sign indicates presence of root.

2 a

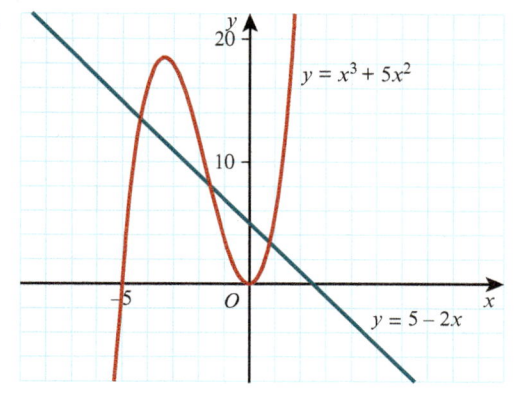

[Graph showing $y = x^3 + 5x^2$ and $y = 5 - 2x$ with 3 intersection points]

3 points of intersection, so 3 roots.

b Let $f(x) = x^3 + 5x^2 + 2x - 5 = 0$ then
$f(0) = 0^3 + 5(0)^2 + 2(0) - 5 = -5$ and
$f(2) = 2^3 + 5(2)^2 + 2(2) - 5 = 27$. Change of sign indicates presence of root.

3 a

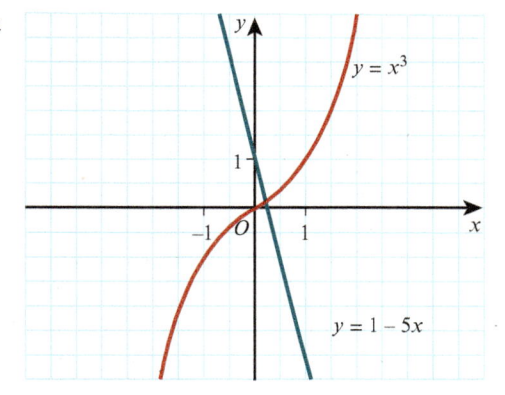

[Graph showing $y = x^3$ and $y = 1 - 5x$ with one intersection point]

One point of intersection so one solution only of $x^3 + 5x - 1 = 0$.

b Let $f(x) = x^3 + 5x - 1$ then
$f(0.1) = 0.1^3 + 5(0.1) - 1 = -0.499$ and
$f(0.5) = 0.5^3 + 5(0.5) - 1 = 1.625$. Change of sign indicates presence of root.

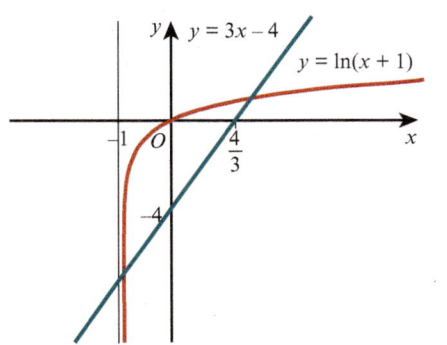

4 a

[Graph showing $y = 3x - 4$ and $y = \ln(x+1)$ with 2 intersection points]

b $\ln(x+1) = 3x - 4$ in 2 places and so $\ln(x+1) - 3x + 4 = 0$ has 2 roots.

5 a

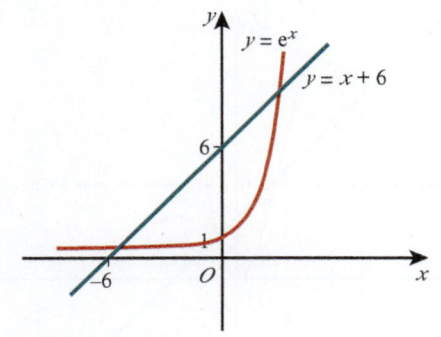

Graphs intersect at 2 points, so 2 roots.

b Let $f(x) = e^x - x - 6$ then
$f(2.0) = e^2 - 2 - 6 = -0.610...$ and
$f(2.1) = e^{2.1} - 2.1 - 6 = 0.0661...$ Change of sign indicates presence of root.

6 a Let $f(x) = (x+2)e^{5x} - 1 = 0$
then $f(0) = (0+2)e^0 - 1 = 1$ and
$f(-0.2) = (-0.2+2)e^{-1} - 1 = -0.337...$
Change of sign indicates presence of root.

b

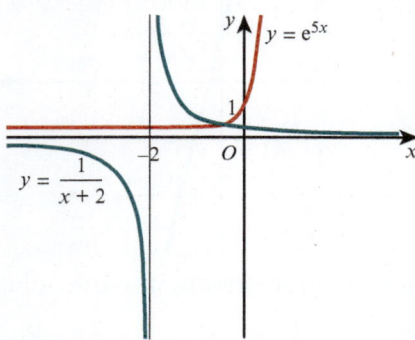

Graphs intersect at 1 point, so 1 root only.

7 a Let $f(x) = \cos^{-1} 2x - 1 + x = 0$ then
$f(0.4) = \cos^{-1}(0.8) - 1 + 0.4 = 0.0435...$ and
$f(0.5) = \cos^{-1} 1 - 1 + 0.5 = -0.5.$ Change of sign indicates presence of root.

b

One point of intersection for $-0.5 \leqslant x \leqslant 0.5$ and so only one root in this domain.

8 a

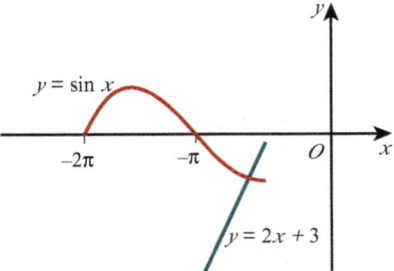

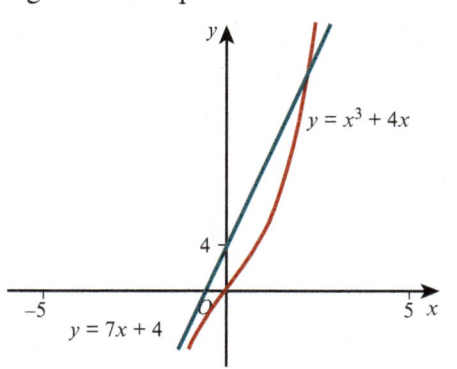

One point of intersection for $-2\pi < x < -\dfrac{\pi}{2}$ and so only one root of $1 = \dfrac{\sin x}{2x+3}$ on this domain. Also, should x be less than -2π or should x be greater than $-\dfrac{\pi}{2}$, the line and curve will not intersect again and so this is the only point of intersection of $y = \dfrac{\sin x}{2x+3}$ and $y = 1$.

b Let $f(x) = \dfrac{\sin x}{2x+3} - 1 = 0$ then
$f(-2) = \dfrac{\sin(-2)}{-4+3} - 1 = -0.09070...$ and
$f(-1.9) = \dfrac{\sin(-1.9)}{-3.8+3} - 1 = 0.1828...$ Change of sign indicates presence of root.

9 a

One point of intersection for $0 \leqslant x \leqslant 5$ and so only one root.

b Let $f(x) = x^3 - 3x - 4$ then
$f(2) = 2^3 - 3(2) - 4 = -2$ and
$f(3) = 3^3 - 3(3) - 4 = 14$

Change of sign indicates presence of root.

10 a

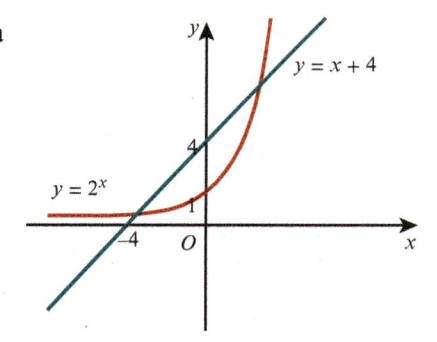

Two points of intersection, so two roots.

b Let $f(x) = 2^x - x - 4 = 0$ then
$f(2.7) = 2^{2.7} - 2.7 - 4 = -0.2019\ldots$ and
$f(2.8) = 2^{2.8} - 2.8 - 4 = 0.1644\ldots$ Change of sign indicates presence of root.

11 a

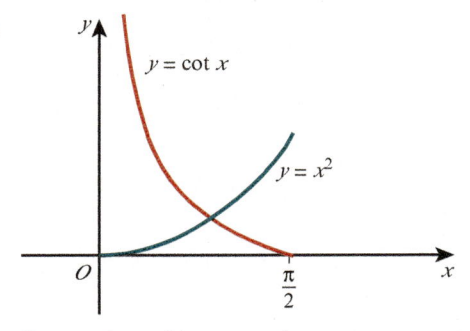

One point of intersection so one root.

b Let $f(x) = \cot x - x^2$ then
$f(0.8) = \cot 0.8 - 0.8^2 = 0.3312\ldots$ and
$f(1) = \cot 1 - 1^2 = -0.3579\ldots$ Change of sign indicates presence of root.

12 a

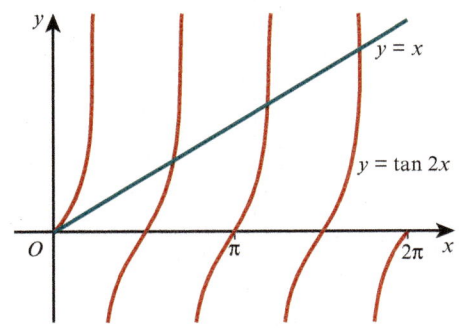

3 points of intersection for $0 < x < 2\pi$ and so 3 roots.

b Let $f(x) = x - \tan 2x = 0$ then
$f(2.1) = 2.1 - \tan 4.2 = 0.3222\ldots$ and
$f(2.2) = 2.2 - \tan 4.4 = -0.8963\ldots$ Change of sign indicates presence of root.

13 a

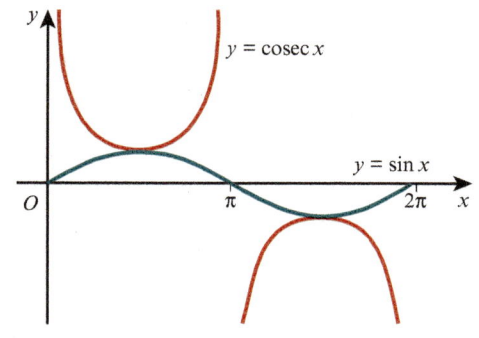

2 points of intersection for $0 < x < 2\pi$ and so 2 roots.

b Let $f(x) = \text{cosec } x - \sin x$
$f\left(\dfrac{3\pi}{2}\right) = \text{cosec } \dfrac{3\pi}{2} - \sin \dfrac{3\pi}{2} = 0$
$\dfrac{3\pi}{2} = 4.71$ correct to 3 significant figures.

14 a $f(x) = 20x^3 + 8x^2 - 7x - 3$ and so
$f(0.5) = 20(0.5)^3 + 8(0.5)^2 - 7(0.5) - 3 = -2$
and $f(1) = 20 + 8 - 7 - 3 = 18$. Change of sign indicates presence of root.

b

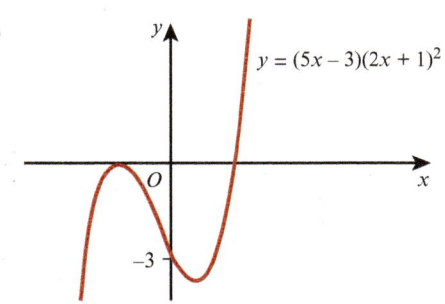

c Proof

15 a $800 = \dfrac{\frac{4}{3}\pi r^3}{2} + \pi r^2 (20)$

b $f(r) = \pi r^3 + 30\pi r^2 - 1200 = 0$ and
$f(3) = -266.946\ldots$ and $f(4) = 509.026\ldots$
Change of sign indicates presence of root.
Or a suitable pair of graphs drawn.

Exercise 6B

1 a 1.1338, 1.1085, 1.1276, 1.1133, 1.1240

b 1.12, $f(1.115) = (1.115)^3 + 5(1.115) - 7$
$= -0.0388\ldots$, $f(1.125) = (1.125)^3 + 5(1.125) - 7$
$= 0.0488\ldots$ Change of sign indicates presence of root.

2 a f(1) = ln(2) + 2 − 4 = −1.306 …
      f(2) = ln(3) + 6 − 4 = 3.0986 … Change of sign indicates presence of root.
   b 1.535

3 a Proof
   b 0.7231, 0.6142, 0.6584, 0.6387, 0.6472, 0.6435, 0.6451, 0.6444, 0.6447 so 0.64

4 a Proof   b 0.5382

5 a Proof   b Proof
   c $\frac{\pi}{15}$ → 0.23448, 0.24034, 0.24169, 0.24200, 0.24207 → 0.242
      or $\frac{\pi}{12}$ → 0.24656, 0.24311, 0.24233, 0.24215, 0.24210 → 0.242
      or 0.2 → 0.23224, 0.23982, 0.24157, 0.24197, 0.24206, 0.24209 → 0.242

6 a 0.6325, 0.8345, 0.7416, 0.7885, 0.7658, 0.7771, 0.7716, 0.7743, (0.7730, …) → 0.77
   b $x^3 + 5x^2 + 2x - 5 = 0$

7 a Proof   b $x_{n+1} = \sqrt[4]{1 + x_n}$
   c 1.2574, 1.2258, 1.2214, 1.2208, 1.2208… → 1.22

8 a $x_{n+1} = \sin^{-1}\left(\frac{1}{x_n^2}\right)$ or $x_{n+1} = \sqrt{\frac{1}{\sin x_n}}$
   b $x = \sin^{-1}\left(\frac{1}{x^2}\right)$ → 0.4605 …, which leads to inverse sine of a value > 1.
   c $x = \sqrt{\frac{1}{\sin x}}$ → 1.00125, 1.08970, 1.06210, 1.07004, 1.06769, 1.06838, (1.06818, 1.06824, 1.06822) → 1.068

9 a 2.0794, 2.1192, 2.1390, 2.1489, 2.1538, 2.1563, 2.1575, 2.1581 → 2.16
   b $x^2 = e^x - 4$

10 a Proof
   b 1 → 1.5, 1.2603, 1.3713, 1.3186, 1.3434, 1.3317, 1.3372, 1.3346, 1.3358, 1.3352, … → 1.34
      or 1.4 → 1.3053, 1.3497, 1.3287, 1.3386, 1.3339, 1.3361, 1.3351, 1.3356, … → 1.34
      or 1.2 → 1.4007, 1.3050, 1.3499, 1.3286, 1.3386, 1.3339, 1.3361, 1.3351, 1.3356, … → 1.34

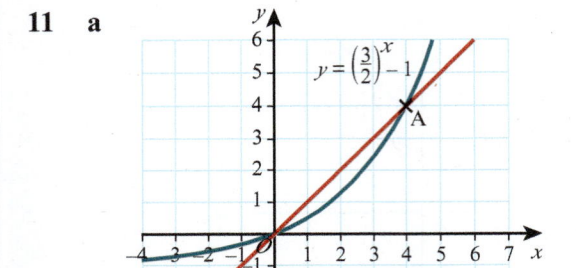

11 a [Graph showing $y = \left(\frac{3}{2}\right)^x - 1$ and $y = x$ intersecting at point A]
   b e.g. $x_n = \frac{\ln(x_n + 1)}{\ln(1.5)}$
   c $A(3.94, 3.94)$ → $OA = 5.6$

12 a $33 - 3r$ b Proof
   c Proof
   d 8 → 8.05869, 8.03200, 8.04419, 8.03863, 8.04117, 8.04001, 8.04054, 8.04030, 8.04041, (8.04036, 8.04038), … → 8.040
   e The radius of the cone that would give a container of the required volume

13 e.g $x_{n+1} = \sqrt{\frac{x_n^3 + 1}{7}}$ → $\alpha = 0.39$,
        $x_{n+1} = \sqrt[3]{7x_n^2 - 1}$ → $\beta = 6.98$

14 a Proof   b Proof
   c 0.9082 …, 0.9015 …, 0.9069 …, 0.9026 …, 0.9061 …, 0.9032 …, 0.9055 …, 0.9037 …, 0.9052 …, 0.9040 …, 0.9049 …, → 0.90
   d e.g. $x_{n+1} = \tan^{-1}\left(\frac{1}{\sin x_n}\right)$ → 0.9063 …, 0.9038 …, 0.9048 …, 0.9044 …, 0.9045 …, 0.9045 …, → 0.90

Exercise 6C

1 a Proof
   b e.g. $x_{n+1} = \frac{3}{2} + \frac{x_n}{e^{2x_n}}$ with $x_1 = 1.5$ → 1.574681, 1.567522, 1.568184, 1.568122, 1.568128 → 1.5681

2 **a** Proof
   b e.g. $1.5 \to 1.432164, 1.407497, 1.398602,$
       $1.395404, 1.394256, 1.393844, 1.393696,$
       $1.393643 \to 1.394$
   c $(8, -6)$

3 **a** Proof
   b e.g. $x_1 = \dfrac{\pi}{4} \to 1, 0.9093, 0.9695, 0.9330, 0.9567,$
       $0.9419, 0.9514, 0.9454, 0.9493 \to 0.95$

4 **a** Proof   **b** Proof
   c $0.3 \to 0.257594, 0.273768, 0.267462,$
       $(0.269900, 0.268954, 0.269321,$
       $0.269179) \to 0.27$
   d 0.00730 correct to 3 significant figures

End-of-chapter review exercise 6

1 **a** $1.5 \to 1.5397, 1.5546, 1.5606, 1.5632, 1.5643,$
       $1.5647, 1.5649, 1.5650, 1.5650, \ldots \to 1.57$
   b $x = \dfrac{6}{7}\left(x + \dfrac{1}{x^3}\right) \to 7x = 6\left(\dfrac{x^4 + 1}{x^3}\right) \to$
       $7x^4 = 6x^4 + 6 \to x^4 = 6 \to \alpha = \sqrt[4]{6}$

2 **a** $1, 2$ **b** Proof
   c e.g. $1.5 \to 2.1985, 1.7717, 2.0039, 1.8688,$
       $1.9445, (1.9011, 1.9256, 1.9117, 1.9196) \to 1.9$

3 **a** Proof
   b $f(x) = e^{2x+1} - 14 + x^3 = 0 \to f(0.5) =$
       $e^{2(0.5)+1} - 14 + (0.5)^3 = -6.4859 \ldots \to f(1) =$
       $e^{2(1)+1} - 14 + (1)^3 = 7.08553 \ldots$ change of sign
       indicates presence of root.
   c Proof
   d e.g. $x_{n+1} = \dfrac{\ln(14 - x_n^3) - 1}{2}$ with
       $x_1 = 0.75 \to 0.804230, 0.800597, 0.800858,$
       $0.800839, 0.800840, \ldots \to 0.8008$

4 **a** $1 \to 1.1825, 1.1692, 1.1662, 1.1657,$
       $1.1656, \ldots \to 1.17$
   b Proof

5 **a** Proof   **b** 3 and 4
   c $3 \to 3.09861, 3.13095, 3.14134, 3.14465,$
       $3.14570, 3.14604, 3.14614, 3.14618, \ldots \to 3.146$
   d $x = 5^y = 3.146, y = \log_5 3.146 = 0.71$

6 **a** Proof   **b** Proof
   c e.g. $1.5 \to 1.3083, 1.3689, 1.3495, 1.3557,$
       $1.3537, 1.3543, (1.3541), \ldots \to 1.35$
   d -1.35

7 **a** $\left[\dfrac{\ln(3x+5)}{3} + \dfrac{e^{6x}}{3}\right]_0^a = 0.6$
   b $0.2 \to 0.164717 \ldots, 0.165896 \ldots,$
       $0.165857 \ldots, (0.165858 \ldots), \to 0.166$

8 **a** Proof
   b $1 \to 1.06366, 1.07202, 1.07311, 1.07324,$
       $1.07327 \to 1.073$

9 **a** Proof
   b $f(\theta) = \theta - \cos^{-1}\sqrt{\dfrac{3\pi}{32\theta}} = 0 \to f(0.8) =$
       $0.8 - \cos^{-1}\sqrt{\dfrac{3\pi}{32(0.8)}} = -0.11882 \ldots, f(1.2) =$
       $1.2 - \cos^{-1}\sqrt{\dfrac{3\pi}{32(1.2)}} = 0.1475 \ldots$ Change of
       sign indicates presence of root.
   c $1 \to 0.99715, 0.99622, 0.99592, 0.99582,$
       $0.99579, 0.99578, 0.99578, \ldots \to 0.996$

10 **a** e.g. $f(x) = e^{x-2} - \sin x = 0 \to f(0.155) =$
       $e^{0.155-2} - \sin 0.155 = 0.00364 \ldots$
       $f(0.165) = e^{0.165-2} - \sin 0.165 = -0.00463 \ldots$
       Change of sign indicates presence of a root.
   b e.g. $e^{x-2} = \sin x$ (take logs to base e) $\to$
       $x - 2 = \ln(\sin x) \to x = 2 + \ln(\sin x) \to$
       $q = 2 + \ln(\sin q)$ when $x = q$.
   c e.g. $2 \to 1.9049, 1.9431, 1.9290, 1.9344,$
       $(1.9324, 1.9332, 1.9329, 1.9330, 1.9329,$
       $1.9329) \to 1.93$

11 **a** Proof
   b e.g. $1.2 \to 1.29439 \ldots, 1.23493 \ldots, 1.27370 \ldots,$
       $1.24893 \ldots, 1.26498 \ldots, 1.25467 \ldots,$
       $1.26133 \ldots, 1.25704 \ldots, (1.25981 \ldots,$
       $1.25802 \ldots, 1.25917 \ldots, 1.25843 \ldots,$
       $1.25891 \ldots, \ldots) \to 1.26$

12 **a** Proof
   b Proof

c e.g. $f(x) = \sec x - \left(\dfrac{\pi}{2} - x\right)\left(\dfrac{\pi}{4} + x\right)$

$f(-0.215) = \sec(-0.215) - \left(\dfrac{\pi}{2} + 0.215\right)\left(\dfrac{\pi}{4} - 0.215\right)$
$= 0.00495...$

$f(-0.205) = \sec(-0.205) - \left(\dfrac{\pi}{2} + 0.205\right)\left(\dfrac{\pi}{4} - 0.205\right)$
$= -0.00928...$

Change of sign indicates presence of root.

d $1 \to 0.4102, 0.6822, 0.6936, 0.6913, (0.6918, 0.6917, 0.6918, ...) \to 0.69$

13 a Proof
    b $0.2 \to 0.30889, 0.31470, 0.31509$
       $0.31511 \to 0.315$

14 a $[x \ln x - x]_1^a = 5$
    b $5 \to 5.59201, 5.57241, 5.57239,$
       $5.57239 \to 5.572$

Cross-topic review exercise 2

1 Proof

2 $\dfrac{\pi}{6}, \dfrac{\pi}{3}$

3 i Proof   ii $(\ln 3, -2)$

4 i Proof   ii $2x - 5y + 8 = 0$

5 a 0.11
   b $x + 4e^{-x} - 2e^{-2x} + c$

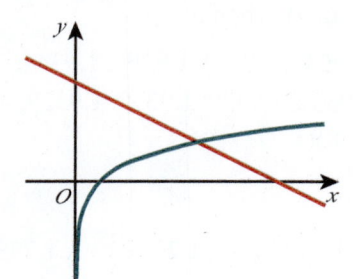

6 i

   ii Proof   iii 4.84

7 i Proof   ii Proof

8 i Proof

   ii 1.854

9 i $\dfrac{dy}{dx} = e^{-2x}(\sec^2 x - 2\tan x) = e^{-2x}(1 - \tan x)^2$
   ii $e^{-2x} > 0$ and $(1 - \tan x)^2 \geqslant 0$
   iii $x = \dfrac{1}{4}\pi$

10 i Proof
    ii a $2\sqrt{2}$ b 3

11 $(-3a, -a)$

12 i Proof   ii Proof   iii 1.54

13 i Proof   ii $y = -\dfrac{1}{2}x + \dfrac{1}{2}$

14 i $-\dfrac{11}{2}\sqrt{3}$ ii $-\dfrac{5}{6}$

15 i Proof   ii Proof
    iii Proof   iv Proof

16 i Proof   ii $(2, 3)$ iii $-\dfrac{3}{8}$

17 i $a = 9$ ii $8e - 14$

18 i Proof   ii $(-2, -1), (0, 1.44)$

19 i Proof   ii 0.678

20 i a $x - 3e^{-2x} + c$
       b $\dfrac{3 \sin 2x}{4} + \dfrac{3x}{2} + c$
    ii 4.84

21 i

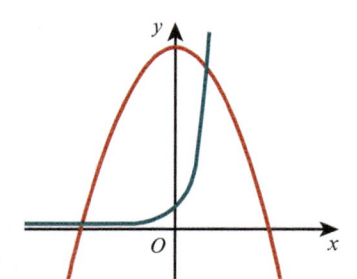

    ii Proof   iii Proof   iv 1.26

22 i

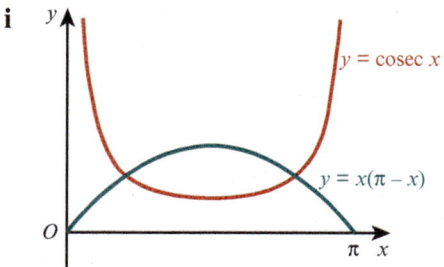

    ii Proof
    iii a 0.66
         b 2.48

23 i $\dfrac{5}{2}$

    ii a Proof b $(-5.15, -7.97)$

24 i Proof

    ii a $1.11, 2.03$

        b $\dfrac{1}{2}\sqrt{3}$

25 i Proof

    ii a $-0.572, 0.572$ b $\dfrac{3}{32}\pi + \dfrac{1}{4}$

26 i Proof ii Proof

27 i 0.362 or $20.7°$, 1.147 or $65.7°$

    ii 33

28 i Proof ii Proof iii 2.728

29 i Proof ii $a = 2, b = -1$

    iii $4 - 3\sqrt{2}$

7 Further algebra

Prerequisite knowledge

1 a $A = 6, B = -3, C = -9$

   b $A = -1, B = -\dfrac{5}{3}, C = 4$

2 a $1 + 14x + 84x^2$

   b $243 - 810x + 1080x^2$

3 Quotient $= x - 5$, remainder $= -11$

Exercise 7A

1 a $4 + \dfrac{20}{2x - 5}$ b $2 - \dfrac{3}{3x + 2}$

   c $2x^2 - x + \dfrac{1}{2} + \dfrac{7}{2(2x + 1)}$

   d $x + 2 - \dfrac{4x + 7}{x^2 + 2x + 3}$

   e $7x + 2 + \dfrac{30x + 11}{x^2 - 5}$ f $x^2 + 1 - \dfrac{6}{x^2 + 1}$

2 $A = 1, B = 4, C = 12, D = 29$

3 $A = 1, B = -1, C = 6, D = -6, E = 5$

4 $A = 2, B = 3, C = -1, D = 6$

Exercise 7B

1 a $\dfrac{4}{x + 3} + \dfrac{2}{x - 2}$ b $\dfrac{5}{x - 4} - \dfrac{3}{2x}$

   c $\dfrac{7}{x - 1} - \dfrac{6}{3x + 1}$

   d $\dfrac{1}{2(x - 3)} - \dfrac{1}{2(3x - 5)}$

   e $\dfrac{2}{x} + \dfrac{3}{x - 1} - \dfrac{4}{2x + 1}$

   f $\dfrac{2}{2x + 3} - \dfrac{2}{x + 2} + \dfrac{1}{x - 3}$

2 a $\dfrac{2}{x + 2} - \dfrac{4}{(x + 2)^2}$

   b $\dfrac{3}{2x + 1} + \dfrac{4}{x + 1} - \dfrac{2}{(x + 1)^2}$

   c $-\dfrac{2}{x} + \dfrac{3}{x - 1} - \dfrac{1}{(x - 1)^2}$

   d $\dfrac{5}{2x - 3} + \dfrac{4}{2x + 1} - \dfrac{1}{(2x + 1)^2}$

   e $\dfrac{3}{16(x + 2)} - \dfrac{3}{16(x - 2)} + \dfrac{3}{4(x - 2)^2}$

   f $-\dfrac{2}{9(x + 2)} + \dfrac{2}{9(x - 1)} + \dfrac{7}{3(x - 1)^2}$

3 a $\dfrac{2}{x} - \dfrac{3}{x^2 + 1}$

   b $\dfrac{3}{2x + 1} + \dfrac{2}{x^2 + 5}$

   c $\dfrac{1}{3x + 5} - \dfrac{2}{2x^2 + 1}$

   d $\dfrac{7}{2x^2 + 5} - \dfrac{3}{3x - 5}$

4 a $2 + \dfrac{3}{x - 1} - \dfrac{2}{x + 2}$

   b $1 + \dfrac{7}{4(x - 2)} - \dfrac{7}{4(x + 2)}$

   c $-4 + \dfrac{2}{x - 4} + \dfrac{3x - 1}{x^2 + 1}$

   d $2 + \dfrac{7}{2x} + \dfrac{1}{x - 1} - \dfrac{2}{(x - 1)^2}$

5 $A = 2, B = -3, C = 4, D = -1$

6 a $(2x - 1)(x - 2)(x + 1)$

   b $\dfrac{5}{2x - 1} - \dfrac{3}{x - 2} + \dfrac{1}{x + 1}$

7 a $(2x + 1)(x - 3)^2$

   b $\dfrac{2}{2x + 1} - \dfrac{1}{x - 3} + \dfrac{3}{(x - 3)^2}$

8   **a** $a = -4, b = -3$

      b $(2x+1)(x+3)(x-1)$

      c $\dfrac{6}{x+3} - \dfrac{32}{2x+1} + \dfrac{10}{x-1}$

9   **a** $\dfrac{1}{x} - \dfrac{1}{x+2}$

      b $\dfrac{3}{2} - \dfrac{1}{(n+1)} - \dfrac{1}{n+2}$ or $\dfrac{3n^2+5}{2(n+1)(n+2)}$

      c $\dfrac{3}{2}$

10   Telescoping series, $S_n = \dfrac{1}{2}\left[\dfrac{1}{2} - \dfrac{1}{n+1} + \dfrac{1}{n+2}\right]$,

        $S_\infty = \dfrac{1}{4}$

Exercise 7C

1   **a** $1 - 2x + 3x^2 - 4x^3, |x| < 1$

      b $1 - 3x + 9x^2 - 27x^3, |x| < \dfrac{1}{3}$

      c $1 + 8x + 40x^2 + 160x^3, |x| < \dfrac{1}{2}$

      d $1 - \dfrac{3}{2}x + \dfrac{3}{2}x^2 - \dfrac{5}{4}x^3, |x| < 2$

      e $1 + x - \dfrac{1}{2}x^2 + \dfrac{1}{2}x^3, |x| < \dfrac{1}{2}$

      f $1 - x - x^2 - \dfrac{5}{3}x^3, |x| < \dfrac{1}{3}$

      g $2 + 16x + 96x^2 + 512x^3, |x| < \dfrac{1}{4}$

      h $1 - 9x + 42x^2 - 152x^3, |x| < \dfrac{1}{2}$

      i $1 + 2x + \dfrac{5}{2}x^2 + 4x^3, |x| < \dfrac{1}{2}$

2   **a** $1 - 3x^2 + 6x^4, |x| < 1$

      b $1 - \dfrac{2}{3}x^2 - \dfrac{4}{9}x^4, |x| < \dfrac{1}{\sqrt{2}}$

      c $1 - 6x^2 + 6x^4, |x| < \dfrac{1}{2}$

3    $2 + 3x + 5x^2 + \dfrac{15}{2}x^3 + \dfrac{75}{4}x^4$

4   **a**   Yes, $(3x-1)^{-2} = (-1)^{-2}(1-3x)^{-2} = (1-3x)^{-2}$

      b   No, $\sqrt{2x-1} = \sqrt{-1}\sqrt{1-2x}$ and $\sqrt{-1}$ is not a real number.

5    $-1 - 6x - 24x^2 - 80x^3$

6   Proof

7    $k = \dfrac{177}{2}$

8    $a = 2, b = -5$

9   **a** $a = -\dfrac{1}{2}$

      b $1 + \dfrac{3}{2}x + \dfrac{3}{2}x^2 + \dfrac{5}{4}x^3 + \dfrac{15}{16}x^4$

10   $a = -9, a = 5$

11   **a** $a = 8, n = -3$ **b** $-5120x^3$

Exercise 7D

1   **a** $\dfrac{1}{4} - \dfrac{1}{4}x + \dfrac{3}{16}x^2 - \dfrac{1}{8}x^3, |x| < 2$

      b $\dfrac{1}{5} + \dfrac{2}{25}x + \dfrac{4}{125}x^2 + \dfrac{8}{625}x^3, |x| < \dfrac{5}{2}$

      c $3 - \dfrac{1}{6}x - \dfrac{1}{216}x^2 - \dfrac{1}{3888}x^3, |x| < 9$

      d $2 + \dfrac{1}{4}x - \dfrac{1}{32}x^2 + \dfrac{5}{768}x^3, |x| < \dfrac{8}{3}$

      e $\dfrac{4}{27} + \dfrac{4}{27}x + \dfrac{8}{81}x^2 + \dfrac{40}{729}x^3, |x| < 3$

      f $-\dfrac{1}{125} - \dfrac{16}{625}x - \dfrac{84}{3125}x^2 - \dfrac{64}{3125}x^3, |x| < \dfrac{5}{2}$

2   **a** $\dfrac{1}{4} - \dfrac{1}{4}x^2 + \dfrac{3}{16}x^4, |x| < \sqrt{2}$

      b $2 - \dfrac{1}{4}x^2 - \dfrac{1}{32}x^4, |x| < \sqrt{\dfrac{8}{3}}$

      c $9\sqrt{3} - \dfrac{15\sqrt{3}}{2}x^2 + \dfrac{15\sqrt{3}}{8}x^4, |x| < \sqrt{3}$

3    $2 - \dfrac{25}{6}x + \dfrac{23}{72}x^2 + \dfrac{67}{2592}x^3$

4   **a** $\dfrac{1}{2} + \dfrac{1}{4}x + \dfrac{1}{8}x^2, 1 - 6x + 27x^2$

      b $\dfrac{1}{2} - \dfrac{11}{4}x + \dfrac{97}{8}x^2, |x| < \dfrac{1}{3}$

5    $a = 2, \ b = \dfrac{75}{16}, \ \dfrac{125}{8}x^3$

6    $a = -2$

7   **a** $1 - \dfrac{2}{x} + \dfrac{4}{x^2} - \dfrac{8}{x^3}$ **b**   Proof

      c $\dfrac{x}{2} - \dfrac{x^2}{4} + \dfrac{x^3}{8} - \dfrac{x^4}{16}$

      d $\left|\dfrac{2}{x}\right| < 1$ gives $x < -2$ or $x > 2$, $\left|\dfrac{x}{2}\right| < 1$ gives $-2 < x < 2$. The two ranges do not overlap.

Exercise 7E

1. **a** $\dfrac{2}{1-x} - \dfrac{3}{1+2x}$
   b $-1 + 8x - 10x^2 + 26x^3$

2. **a** $\dfrac{2}{1-3x} + \dfrac{1}{1-x} - \dfrac{3}{(1-x)^2}$
   b $x + 10x^2 + 43x^3$

3. **a** $\dfrac{5}{1-x} + \dfrac{3x-1}{1+2x^2}$
   b $4 + 8x + 7x^2 - x^3$

4. **a** $1 + \dfrac{6}{1+2x} + \dfrac{5}{2-3x}$
   b $\dfrac{237}{8}$

5. **a** $\dfrac{3}{x-4} - \dfrac{3}{x+3}$
   b $-\dfrac{7}{4} + \dfrac{7}{48}x - \dfrac{91}{576}x^2$

6. **a** $\dfrac{5}{x+1} + \dfrac{1}{x-2} - \dfrac{3}{(x-2)^2}$
   b $\dfrac{15}{4} - 6x + \dfrac{69}{16}x^2$

End-of-chapter review exercise 7

1. $1 + 8x + 40x^2 + 160x^3$
2. $1 - 2x - 4x^2 - \dfrac{40}{3}x^3$
3. $2 - 7x + 18x^2$
4. $1 + 2x - \dfrac{3}{2}x^2$
5. $1 - \dfrac{3}{2}x + \dfrac{27}{8}x^2 - \dfrac{135}{16}x^3$
6. $\dfrac{5}{2} + \dfrac{5}{2}x + \dfrac{15}{8}x^2$
7. $\dfrac{1}{2} + \dfrac{5}{16}x + \dfrac{75}{256}x^2$
8. $A = -\dfrac{8}{3}$, $B = -4$, $C = \dfrac{16}{3}$
9. $A = 5$, $B = 3$, $C = -2$
10. $\dfrac{2}{x} + \dfrac{5x-3}{x^2+1}$
11. $A = 3$, $B = -1$, $C = 2$, $D = -2$
12. $2 - \dfrac{9}{5(2x-1)} - \dfrac{23}{5(x+2)}$
13. $k = 16$
14. **i** $\dfrac{1}{x+1} + \dfrac{3}{x-3} + \dfrac{12}{(x-3)^2}$
    ii $\dfrac{4}{3} - \dfrac{4}{9}x + \dfrac{4}{3}x^2$
15. **i** $-\dfrac{1}{x-2} + \dfrac{3x-1}{x^2+3}$
    ii $\dfrac{1}{6} + \dfrac{5}{4}x + \dfrac{17}{72}x^2$
16. **i** $-\dfrac{1}{1+x} + \dfrac{2x+1}{1+2x^2}$
    ii $3x - 3x^2 - 3x^3$

8 Further calculus

Prerequisite knowledge

1. **a** $3\cos 3x$ **b** $2xe^{x^2+1}$
   c $\dfrac{5}{5x-3}$ **d** $2\sec^2 2x + 5\sin x$

2. **a** $\dfrac{1}{5}e^{5x+1} + c$ **b** $\dfrac{x}{2} - \dfrac{1}{8}\sin 4x + c$
   c $\dfrac{5}{3}\ln(3x-2) + c$ **d** $\dfrac{1}{3}\tan 3x + c$

3. **a** $\dfrac{3}{2(x+3)} + \dfrac{1}{2(x-1)}$ **b** $\dfrac{7x}{x^2+2} - \dfrac{3}{x}$
   c $\dfrac{5}{x} - \dfrac{5}{x-1} + \dfrac{5}{(x-1)^2}$ **d** $3 - \dfrac{1}{x+1} + \dfrac{1}{x-2}$

Exercise 8A

1. **a** $\dfrac{2}{4x^2+1}$ **b** $\dfrac{5}{25x^2+1}$
   c $\dfrac{3}{x^2+9}$ **d** $\dfrac{1}{x^2-2x+2}$
   e $\dfrac{2x}{x^4+1}$ **f** $\dfrac{2}{5x^2+2x+1}$

2. **a** $\dfrac{x}{x^2+1} + \tan^{-1}x$
   b $\dfrac{2x - (4x^2+1)\tan^{-1}2x}{x^2(4x^2+1)}$
   c $e^x\left(\dfrac{1}{x^2+1} + \tan^{-1}x\right)$

3. $x - 4y = 2 - \pi$

4. $\dfrac{3+\pi}{5}$

Exercise 8B

1. **a** $\frac{1}{3}\tan^{-1}\left(\frac{x}{3}\right)$ **b** $\frac{1}{4}\tan^{-1}\left(\frac{x}{4}\right)$
   c $\frac{1}{2}\tan^{-1}(2x)$ **d** $\frac{1}{12}\tan^{-1}\left(\frac{3x}{4}\right)$
   e $\frac{\sqrt{3}}{6}\tan^{-1}\left(\frac{2\sqrt{3}x}{3}\right)$ **f** $\frac{\sqrt{6}}{6}\tan^{-1}\left(\frac{\sqrt{6}}{2}x\right)$

2. **a** $\frac{\pi}{12}$ **b** $\frac{\pi}{4}$ **c** $\frac{\sqrt{6}}{9}\pi$

3. $2\pi^2$

   e $\frac{20}{9}$ **f** $\frac{5}{32}(\pi+2)$
   g $\frac{1}{6}(2\pi - 3\sqrt{3})$ **h** $\frac{\pi}{2}$

5. π

6. $\frac{4}{15}$

7. $\frac{\pi}{3}$

8. $6e - \frac{6}{e}$

Exercise 8C

1. **a** $2\ln(x^3 - 1) + c$ **b** $\ln(1 + \sin x) + c$
   c $2\ln(x^2 - 5x + 1) + c$ **d** $\ln(\sin x) + c$
   e $-\frac{1}{2}\ln(2 - x^2) + c$ **f** $\ln(1 + \tan x) + c$

2. **a** $\frac{1}{2}\ln\left(\frac{3}{2}\right)$ **b** $\ln 5$
   c $\frac{1}{2}\ln\left(\frac{5}{2}\right)$ **d** $\frac{1}{2}\ln\left(\frac{7}{2}\right)$
   e $\frac{1}{2}\ln 2$ **f** $\ln 3$

3. Proof
4. Proof
5. $p = \sqrt{e^2 - 1}$

Exercise 8D

1. $\sqrt{x^2 - 3} + c$

2. **a** $-\frac{3x + 2}{6(x+2)^3} + c$ **b** $-\frac{1}{6}(1 - 2x^2)^{\frac{3}{2}} + c$
   c $\frac{\sin^6 x}{6} + c$ **d** $\frac{2}{3}(e^x + 2)^{\frac{3}{2}} + c$
   e $x + \frac{1}{5} - \frac{1}{5}\ln(5x + 1) + c$
   f $\frac{2}{45}(3x - 1)^{\frac{5}{2}} + \frac{2}{27}(3x - 1)^{\frac{3}{2}} + c$

3. $\frac{\pi}{4}$

4. **a** $\frac{2}{3}$ **b** $18\ln 3 - 16$
   c $\frac{3}{2}\ln(3 + 2\sqrt{2})$ **d** $2\sqrt{2}$

Exercise 8E

1. **a** $\frac{1}{2}\ln\left(\frac{x}{2-x}\right) + c$
   b $3\ln(x + 2) - \ln(1 - x) + c$
   c $2\ln(2x^2 - 9x - 5)$
   d $\ln(x - 3) + 2\tan^{-1}x + c$
   e $5\ln(x + 1) + \ln(2 - x) + \frac{2}{2-x} + c$
   f $2x - 3\ln(x - 1) + \ln(x + 4) + c$

2. **a** $\ln\frac{100}{27}$ **b** $\ln 10$
   c $\ln\frac{9\sqrt{3}}{16}$ **d** $\frac{7}{10}(\pi - 4\ln 3)$
   e $\frac{\pi}{4} + \ln 2$ **f** $\frac{1}{4}(\pi + \ln 4)$
   g $\frac{1}{6} + \ln 4$ **h** $2\ln\frac{9}{4} - \frac{3}{2}$
   i $\ln\frac{9}{2} - \frac{1}{3}$

3. **a** $1 + \ln\frac{16}{5}$ **b** $4 - \ln\frac{12}{5}$ **c** $2 - \ln 2$

4. Proof
5. Proof
6. Proof

Exercise 8F

1. **a** $3xe^x - 3e^x + c$
   b $x\sin x + \cos x + c$
   c $\frac{1}{2}x^2\ln 2x - \frac{1}{4}x^2 + c$
   d $\frac{1}{4}\sin 2x - \frac{1}{2}x\cos 2x + c$

  e $\frac{1}{4}x^4 \ln x - \frac{1}{16}x^4 + c$

  f $2\sqrt{x} \ln x - 4\sqrt{x} + c$

2   **a** $\frac{1}{18}(\pi - 2)$

  b $\ln 16 - \frac{1}{2}\ln 2 - \frac{3}{4}$

  c $\frac{1}{4}(e^2 + 1)$ **d** $\ln 27 - 2$

  e $\frac{5}{9} - \frac{2}{9e^3}$ **f** $\frac{8}{25}$

3   **a** $2(\ln 2 - 1)^2$ **b** $\frac{1}{32}(\pi^2 - 8)$

  c $\pi^2 - 4$ **d** $\frac{1}{4}e^2$

  e 2 **f** $\frac{1}{2}(1 + e^\pi)$

4   **a** $\frac{1}{9}(1 + 2e^3)$ **b** π

  c $\frac{1}{2}$ **d** $2 - \frac{26}{e^4}$

5 $\frac{1}{4}\pi(e^4 - 5)$

End-of-chapter review exercise 8

1 $\frac{1}{4} - \frac{1}{2}e^{-1}$

2 $4(\ln 4 - 1)$

3 $\frac{14}{9}$

4   **i**   Proof   **ii**   Proof

5   **i**   Proof   **ii** $\frac{1}{3}\pi - \frac{\sqrt{3}}{2}$

6   **i**   Proof   **ii** $15\ln 5 - 4$

7   **i** $\frac{1}{x+1} - \frac{1}{x+3}$ **ii**   Proof

  iii   Proof

8   **a** $3x + \frac{1}{2}\tan 2x + c$

  b $\frac{1}{8}\pi\sqrt{3} - \frac{1}{2}\ln\left(\frac{1}{\sqrt{2}}\right)$

9   **i**   Proof   **ii**   Proof

10   **i** $\sqrt{5} - 1$ **ii** $24 - 8e$

11   Proof

12   **i** $\left(e^{-\frac{1}{3}}, -\frac{1}{3e}\right)$ **ii** $4\ln 2 - \frac{15}{16}$

13   **i** $x = \sqrt[3]{2}$ **ii** $p = 3.40$

Cross-topic review exercise 3

1 $1 + 15x + 135x^2 + 945x^3$

2 $\frac{\sqrt{3}}{9}\pi$

3 $4 - 6x + 6x^2 - 5x^3$

4 $\frac{1}{2} + \frac{1}{8}x + \frac{3}{64}x^2$

5 $\pi - 2$

6 $1 + \frac{3}{2}x + \frac{3}{8}x^2 + \frac{15}{16}x^3$

7   Proof

8   **i** $1 + 2x + 6x^2$ **ii** 5

9   **i** $a = \frac{\sqrt{2}}{2}$ **ii** $1 - \sqrt{2}x + \frac{3}{2}x^2$

10   **i**   Proof   **ii** $8 + 2\ln\frac{1}{2}$

11   Proof

12   **a**   Proof   **b**   Proof

13   **a**   Proof   **b** 6.56

14   **i**   Proof   **ii** $2e^2 - 10$

15   **i** $\frac{3}{2-x} + \frac{4x}{4+x^2}$ **ii**   Proof

16   **i**   Proof   **ii** 1.94

17   **i** $\frac{1}{3-x} + \frac{3}{2(1+2x)} - \frac{1}{2(1+2x)^2}$

  ii $\frac{4}{3} - \frac{8}{9}x + \frac{1}{27}x^2$

18   **i** $\frac{2}{1-x} - \frac{1}{2-x} + \frac{3}{(2-x)^2}$

  ii $\frac{9}{4} + \frac{5}{2}x + \frac{39}{16}x^2$

19   **a** $\frac{3}{x-1} - \frac{6x+1}{2x^2 - 1}$

  b $-2 + 3x - x^2 + 9x^3$

20   **a** $A = \frac{1}{2}, B = 4, C = 2, D = -\frac{15}{2}$

  b   Proof

21   **i** $\left(\frac{1}{\sqrt{e}}, -\frac{1}{2e}\right)$ **ii** $\frac{1}{9}(2e^3 + 1)$

22 i Proof   ii $\dfrac{11}{96}$

23 i $5\cos(\theta - 0.6435)$

   ii a $1.80, 5.77$

   b $2\tan(\theta - 0.6435) + c$

9 Vectors

Prerequisite knowledge

1 $36.7°$ (correct to 1 decimal place)

2 a $y = \dfrac{3}{2}x - 1$ b $y = -\dfrac{2}{3}x - 6$

3 a $4\,\text{cm}$ b $\dfrac{\sqrt{61}}{2}$

4 $(2, 12)$

Exercise 9A

1 a $\vec{AB} = \begin{pmatrix} 5 \\ -3 \end{pmatrix}$ $\vec{BC} = \begin{pmatrix} 5 \\ 2 \end{pmatrix}$

   b $\vec{AC} = \begin{pmatrix} 10 \\ -1 \end{pmatrix}$

2 a $\vec{EF} = \begin{pmatrix} -7 \\ 3 \end{pmatrix}$

   b $\vec{DF} - \vec{DE} = \begin{pmatrix} 3 \\ 5 \end{pmatrix} - \begin{pmatrix} 10 \\ 2 \end{pmatrix} = \begin{pmatrix} -7 \\ 3 \end{pmatrix}$

3 $\vec{QR} = \vec{PR} - \vec{PQ} = \vec{PR} + \vec{QP} = \vec{QP} + \vec{PR}$ QED

4 a $\vec{XY} = \mathbf{b} - \mathbf{a}$ and $\vec{BC} = 2\mathbf{b} - 2\mathbf{a} = 2(\mathbf{b} - \mathbf{a})$
   BC is a scalar multiple of XY, therefore BC is parallel to XY.

   b $k = \dfrac{1}{2}$

5 a $\begin{pmatrix} 12 \\ 2 \\ 0 \end{pmatrix}$ b $\begin{pmatrix} 3 \\ 4 \\ 2 \end{pmatrix}$

6 a i $\mathbf{q} + \mathbf{s} - \mathbf{p}$ ii $\mathbf{p} - 2\mathbf{q} - \mathbf{r} - \mathbf{s}$

   b For example, angle $AHC = 45°$ (Interior angle of a regular octagon $= 135°$ and angle $GHC = 90°$) and the exterior angle (at A) is $45°$ and so the line segments AB and HC are parallel.
   $k = 1 + \sqrt{2}$

7 Proof

Exercise 9B

1 a $\dfrac{1}{\sqrt{58}}(-3\mathbf{j} + 7\mathbf{k})$ b $\lambda = 3$

2 a $d = 2$ b $\vec{ON} = 6\mathbf{i} + 4\mathbf{j} + 2\mathbf{k}$

   c $\dfrac{2}{3}\mathbf{i} + \dfrac{2}{3}\mathbf{j} + \dfrac{1}{3}\mathbf{k}$

3 a i $|\vec{AD}| = |\mathbf{d} - \mathbf{a}| = \left|\begin{pmatrix} -6 \\ -6 \\ 0 \end{pmatrix} - \begin{pmatrix} 2 \\ 2 \\ 0 \end{pmatrix}\right| = \left|\begin{pmatrix} -8 \\ -8 \\ 0 \end{pmatrix}\right|$
   $= \sqrt{128} = 8\sqrt{2}$,

   $|\vec{AB}| = |\mathbf{b} - \mathbf{a}| = \left|\begin{pmatrix} 13 \\ 5 \\ 4 \end{pmatrix} - \begin{pmatrix} 2 \\ 2 \\ 0 \end{pmatrix}\right| = \left|\begin{pmatrix} 11 \\ 3 \\ 4 \end{pmatrix}\right|$
   $= \sqrt{11^2 + 3^2 + 4^2} = \sqrt{146}$,

   $|\vec{BC}| = |\mathbf{c} - \mathbf{b}| = \left|\begin{pmatrix} 5 \\ -3 \\ 4 \end{pmatrix} - \begin{pmatrix} 13 \\ 5 \\ 4 \end{pmatrix}\right| = \left|\begin{pmatrix} -8 \\ -8 \\ 0 \end{pmatrix}\right|$
   $= 8\sqrt{2}$,

   $|\vec{DC}| = |\mathbf{c} - \mathbf{d}| = \left|\begin{pmatrix} 5 \\ -3 \\ 4 \end{pmatrix} - \begin{pmatrix} -6 \\ -6 \\ 0 \end{pmatrix}\right| = \left|\begin{pmatrix} 11 \\ 3 \\ 4 \end{pmatrix}\right|$
   $= \sqrt{11^2 + 3^2 + 4^2} = \sqrt{146}$

   ii Opposite sides are parallel and equal in length.

   b i $\vec{OM} = \vec{OA} + \dfrac{1}{2}\vec{AB} = \begin{pmatrix} 2 \\ 2 \\ 0 \end{pmatrix} + \dfrac{1}{2}\begin{pmatrix} 11 \\ 3 \\ 4 \end{pmatrix} = \begin{pmatrix} 7.5 \\ 3.5 \\ 2 \end{pmatrix}$

   so $M(7.5, 3.5, 2)$

   ii $\vec{OP} = \vec{OB} + \dfrac{1}{3}\vec{BD}$

   $= \begin{pmatrix} 13 \\ 5 \\ 4 \end{pmatrix} + \dfrac{1}{3}\left[\begin{pmatrix} -6 \\ -6 \\ 0 \end{pmatrix} - \begin{pmatrix} 13 \\ 5 \\ 4 \end{pmatrix}\right]$

   $= \begin{pmatrix} 13 \\ 5 \\ 4 \end{pmatrix} + \dfrac{1}{3}\begin{pmatrix} -19 \\ -11 \\ -4 \end{pmatrix} = \dfrac{1}{3}\begin{pmatrix} 20 \\ 4 \\ 8 \end{pmatrix}$

   so $P\left(\dfrac{20}{3}, \dfrac{4}{3}, \dfrac{8}{3}\right)$

Answers

4 **a** $|2\mathbf{i} + 6\mathbf{j} - 4\mathbf{k}| = \sqrt{4 + 36 + 16} = \sqrt{56} = 2\sqrt{14}$
   b $|\overrightarrow{OA}| = \sqrt{1 + 4 + 25} = \sqrt{30}$,
   $|\overrightarrow{OB}| = \sqrt{9 + 16 + 1} = \sqrt{26}$ and $30 + 26 = 56$
   c $\frac{1}{2} \times \sqrt{30} \times \sqrt{26} = \sqrt{195}$

5 $16 + (q - 2)^2 = 22$ and so $q = 2 \pm \sqrt{6}$.

6 **a** 25 cm
   b $\overrightarrow{ON} = 9.6\mathbf{i} + 20\mathbf{j} + 4.2\mathbf{k}$

7 $\sqrt{4 + 25 + a^2} = \sqrt{1 + (1+a)^2 + (-3)^2}$ so $a = 9$.

8 **a** $\overrightarrow{OP} = \lambda \overrightarrow{OQ}$ and using the y-component, $\lambda = \frac{1}{4}$.
   Hence, $-6k = \frac{1}{4}(2k + 13)$, $k = -\frac{1}{2}$ and checking $8(1 + k) = \frac{1}{4}(-32k)$ gives $k = -\frac{1}{2}$.
   b $\overrightarrow{OP} = \begin{pmatrix} 3 \\ -2 \\ 4 \end{pmatrix} = 3\mathbf{i} - 2\mathbf{j} + 4\mathbf{k}$ and
   $\overrightarrow{OQ} = \begin{pmatrix} 12 \\ -8 \\ 16 \end{pmatrix} = 12\mathbf{i} - 8\mathbf{j} + 16$.
   c $\overrightarrow{PQ} = 9\mathbf{i} - 6\mathbf{j} + 12\mathbf{k}$ and
   $|\overrightarrow{PQ}| = \sqrt{9^2 + (-6)^2 + 12^2} = 3\sqrt{29}$

9 Home is the null displacement $\begin{pmatrix} 0 \\ 0 \\ 0 \end{pmatrix}$. Total vector sum is $\begin{pmatrix} 13 \\ 8 \\ 0 \end{pmatrix}$ so to get home the displacement is $\begin{pmatrix} -13 \\ -8 \\ 0 \end{pmatrix}$. The distance home is $10\sqrt{169 + 64} = 153$ cm, correct to the nearest cm.

Exercise 9C

1 **a** $\mathbf{a} \cdot \mathbf{b} = 0$, $\mathbf{a} \perp \mathbf{b}$
   b $\theta = \cos^{-1}\left(\frac{12}{13\sqrt{2}\sqrt{19}}\right) = 81.4°$
   c $\mathbf{e} \cdot \mathbf{f} = 0$, $\mathbf{e} \perp \mathbf{f}$

2 $\overrightarrow{OB} \cdot \overrightarrow{OA} = (-5)(1) + (0)(7) + (3)(2) = 1$
   $|\overrightarrow{OA}| = \sqrt{34}$, $|\overrightarrow{OB}| = \sqrt{54}$
   $BOA = \cos^{-1}\left(\frac{1}{\sqrt{34}\sqrt{54}}\right) = 88.7°$ correct to 1 decimal place.

3 $6a + (-2)(4) + (5)(-2) = 0$ $a = 3$

4 **a** $5k^2 - 3(k + 2) - (7k + 9) = 0$
   $5k^2 - 10k - 15 = 0$ $k^2 - 2k - 3 = 0$
   $(k + 1)(k - 3) = 0 \rightarrow k = -1$ or $k = 3$
   b $\overrightarrow{OP} = \begin{pmatrix} 10 \\ -3 \\ 23 \end{pmatrix}$, $\overrightarrow{OQ} = \begin{pmatrix} 2 \\ 4 \\ -1 \end{pmatrix}$
   $\overrightarrow{OP} \cdot \overrightarrow{OQ} = 10(2) + (-3)(4) + (23)(-1) = -15$
   $|\overrightarrow{OP}| = \sqrt{10^2 + (-3)^2 + 23^2} = \sqrt{638}$,
   $|\overrightarrow{OQ}| = \sqrt{2^2 + 4^2 + (-1)^2} = \sqrt{21}$
   $\rightarrow \theta = \cos^{-1}\left(\frac{-15}{\sqrt{638}\sqrt{21}}\right) = 97.4°$

5 $\overrightarrow{NP} = 2\mathbf{j} + 3\mathbf{k}$ and $\overrightarrow{MP} = -3\mathbf{i} - 2\mathbf{j} + \mathbf{k}$ and so
   $\overrightarrow{NP} \cdot \overrightarrow{MP} = 2(-2) + 3(1) = -1$, $|\overrightarrow{NP}| = \sqrt{13}$,
   $|\overrightarrow{MP}| = \sqrt{14}$
   $NPM = \cos^{-1}\left(\frac{-1}{\sqrt{13}\sqrt{14}}\right) = 94.2509\ldots = 94.3°$

6 $\mathbf{a} \cdot \mathbf{j} = (4)(0) + (-8)(1) + (1)(0) = -8$
   $|\mathbf{a}| = \sqrt{4^2 + (-8)^2 + 1^2} = \sqrt{81} = 9$, $|\mathbf{j}| = 1$
   $\theta = \cos^{-1}\left(\frac{-8}{9}\right) = 152.733\ldots = 152.7°$ correct to 1 decimal place.

7 $\mathbf{a} \cdot \mathbf{b}$ is a scalar and the dot product is a product of two vectors.

8 **a** $\overrightarrow{OM} = 2\mathbf{i} + 4\mathbf{j} + 4\mathbf{k}$, $\overrightarrow{NG} = -4\mathbf{i} + 3\mathbf{j} + 4\mathbf{k}$
   b $\overrightarrow{OM} \cdot \overrightarrow{NG} = 2(-4) + 4(3) + 4(4) = 20$,
   $|\overrightarrow{OM}| = 6$, $|\overrightarrow{NG}| = \sqrt{41}$, $\cos^{-1}\left(\frac{20}{6\sqrt{41}}\right) = 58.6°$
   correct to 1 decimal place.

9 $\overrightarrow{AM} = -77\mathbf{i} + 30\mathbf{j} + 36\mathbf{k}$ and
   $|\overrightarrow{DB}| = \sqrt{77^2 + 36^2} = 85$ so
   $\overrightarrow{AN} = 60\mathbf{j} + \frac{1}{5}\overrightarrow{BD} = 60\mathbf{j} + \frac{1}{5}(-77\mathbf{i} + 36\mathbf{k})$
   $= -\frac{77}{5}\mathbf{i} + 60\mathbf{j} + \frac{36}{5}\mathbf{k}$
   $\overrightarrow{AM} \cdot \overrightarrow{AN} = -77\left(-\frac{77}{5}\right) + 30(60) + 36\left(\frac{36}{5}\right) = 3245$
   $|\overrightarrow{AM}| = 25\sqrt{13}$

$|\overrightarrow{AN}| = \sqrt{3889}$

$MAN = \cos^{-1}\left(\dfrac{3245}{25\sqrt{13}\sqrt{3889}}\right) = 54.7°$ correct to 1 decimal place.

10 a $\overrightarrow{AN} = \begin{pmatrix} -3 \\ 1.5 \\ 4.5 \end{pmatrix}, |\overrightarrow{AN}| = \dfrac{3\sqrt{14}}{2}$

$\cos^{-1}\left(\dfrac{4.5}{3 \times 1.5\sqrt{14}}\right) = 74.5°$ correct to 1 decimal place.

b $\overrightarrow{MN} = \begin{pmatrix} -3 \\ 0 \\ 4.5 \end{pmatrix}$

c $\overrightarrow{PN} \cdot \overrightarrow{MN} = 0, \overrightarrow{PN} = \begin{pmatrix} 3 \\ 1.5 \\ 4.5 - p \end{pmatrix}$,

$3(-3) + 4.5(4.5 - p) = 0, p = 2.5, \overrightarrow{OP} = \begin{pmatrix} 0 \\ 0 \\ 2.5 \end{pmatrix}$

Exercise 9D

1 a $\mathbf{r} = -\mathbf{j} + 5\mathbf{k} + \lambda(2\mathbf{i} + 6\mathbf{j} - \mathbf{k})$
  b $\mathbf{r} = \lambda(7\mathbf{i} - \mathbf{j} - \mathbf{k})$
  c $\mathbf{r} = 7\mathbf{i} + 2\mathbf{j} - 3\mathbf{k} + \lambda(3\mathbf{i} - 4\mathbf{k})$

2 a $x = 2\lambda$ b $x = 7\lambda$
    $y = -1 + 6\lambda$ $y = -\lambda$
    $z = 5 - \lambda$ $z = -\lambda$
  c $x = 7 + 3\lambda$
    $y = 2$
    $z = -3 - 4\lambda$

3 Direction of line through $9\mathbf{i} + 2\mathbf{j} - 5\mathbf{k}$ and $\mathbf{i} + 7\mathbf{j} + \mathbf{k}$ is, for example, $8\mathbf{i} - 5\mathbf{j} - 6\mathbf{k}$. The direction of this line is a scalar multiple of $16\mathbf{i} - 10\mathbf{j} - 12\mathbf{k}$ and so the lines are parallel.

4 a $x = 2 + t$ b $x = 2t$
    $y = 13 + t$ $y = 10 + 5t$
    $z = 1 - t$ $z = 0$
  c $x = 1 + 2t$
    $y = -3 + 3t$
    $z = 4t$

5 a $\overrightarrow{AB} = \begin{pmatrix} 1 \\ -3 \\ 8 \end{pmatrix}$ so

$\overrightarrow{OA} + t\overrightarrow{AB} = \begin{pmatrix} 0 \\ 4 \\ -2 \end{pmatrix} + t\begin{pmatrix} 1 \\ -3 \\ 8 \end{pmatrix} = \begin{pmatrix} t \\ 4 - 3t \\ -2 + 8t \end{pmatrix}$ and so

$x = t$
$y = 4 - 3t$
$z = -2 + 8t$

b Oxy plane $\to z = 0, -2 + 8t = 0, t = \dfrac{1}{4}$

$\to x = \dfrac{1}{4}, y = 4 - \dfrac{3}{4} = \dfrac{13}{4} \to \left(\dfrac{1}{4}, \dfrac{13}{4}, 0\right)$

6 a $\mathbf{r} = (\mu + 4)\mathbf{i} + (\mu - 7)\mathbf{j} + (3\mu)\mathbf{k}$
    $\to \mathbf{r} = 4\mathbf{i} - 7\mathbf{j} + \mu(\mathbf{i} + \mathbf{j} + 3\mathbf{k}) \to$ direction is $\mathbf{i} + \mathbf{j} + 3\mathbf{k}$ which is not a scalar multiple of $6\mathbf{i} + \mathbf{j} + 3\mathbf{k}$, so the lines are not parallel.

b $\cos^{-1}\left(\dfrac{16}{\sqrt{11}\sqrt{46}}\right) = 44.7°$ correct to 1 decimal place.

7 a $\begin{pmatrix} x \\ y \\ z \end{pmatrix} = \begin{pmatrix} 5 \\ -3 \\ 2 \end{pmatrix} + t\begin{pmatrix} 4 \\ -1 \\ -3 \end{pmatrix}$

b $\overrightarrow{BA} \cdot \mathbf{d}_{L_2} = \begin{pmatrix} 1 \\ 4 \\ 0 \end{pmatrix} \cdot \begin{pmatrix} 4 \\ -1 \\ -3 \end{pmatrix}$

$= 1(4) + (4)(-1) + (0)(-3) = 0$ ✓

8 a $\overrightarrow{AB} = \begin{pmatrix} -3 \\ -2 \\ -3 \end{pmatrix}$ so $\overrightarrow{OB} + t\overrightarrow{AB} = \begin{pmatrix} 0 \\ -1 \\ 2 \end{pmatrix} + t\begin{pmatrix} -3 \\ -2 \\ -3 \end{pmatrix}$

b $\overrightarrow{ON} = \begin{pmatrix} -3t + 3 \\ -2t + 1 \\ -3t + 5 \end{pmatrix}$, and then $\overrightarrow{CN} = \begin{pmatrix} -3t + 2 \\ -2t - 1 \\ -3t + 2 \end{pmatrix}$

Since $\overrightarrow{CN}$ is perpendicular to L,

$(-3t + 2)(-3) + (-2t - 1)(-2) + (-3t + 2)(-3) = 0$

$22t - 10 = 0$, so $t = \dfrac{5}{11}$,

$\overrightarrow{ON} = \begin{pmatrix} 3 \\ 1 \\ 5 \end{pmatrix} + \dfrac{5}{11}\begin{pmatrix} -3 \\ -2 \\ -3 \end{pmatrix} = \dfrac{1}{11}\begin{pmatrix} 18 \\ 1 \\ 30 \end{pmatrix}$

Answers

Exercise 9E

1. a Skew
   b Parallel
   c Intersecting (8, 5, 15)
   d Intersecting (5, −1, 3)

2. $p = -80$, $P(5, -3, 16)$

3. a $4\mathbf{i} - 3\mathbf{k}$, $-8\mathbf{i} + 4\mathbf{j}$, $12\mathbf{j} + 5\mathbf{k}$
   b $55.8°$, $72.3°$, $51.9°$
   c 13, $4\sqrt{14}$, $3\sqrt{17}$

4. a $\mathbf{r} = 3\mathbf{i} + 7\mathbf{j} + 9\mathbf{k} + \lambda(4\mathbf{i} + 4\mathbf{j} + 5\mathbf{k})$
   b Proof

5. a $\overrightarrow{AB} = -2\mathbf{i} + 2\mathbf{j} + 4\mathbf{k}$
   b e.g. $\mathbf{r} = \mathbf{i} + 5\mathbf{k} + \lambda(-2\mathbf{i} + 2\mathbf{j} + 4\mathbf{k})$
   c $\cos^{-1}\left(\dfrac{10}{2\sqrt{6}\sqrt{14}}\right) = 56.938\ldots° = 56.9°$ correct to 1 decimal place
   d $(0, 1, 7)$

End-of-chapter review exercise 9

1. a $\cos^{-1}\left(\dfrac{40}{2\sqrt{14}\sqrt{62}}\right) = 47.2466\ldots° = 47.2°$ correct to 1 decimal place
   b $\mathbf{r} = 2\mathbf{i} + 3\mathbf{j} + 7\mathbf{k} + \lambda(2\mathbf{i} - 5\mathbf{j} - 13\mathbf{k})$

2. a Not perpendicular as
   $\overrightarrow{OA} \cdot \overrightarrow{OB} = (-2)(1) + (0)(-1) + (6)(4) = 22 \neq 0$
   b i $\overrightarrow{AB} = \begin{pmatrix} 3 \\ -1 \\ -2 \end{pmatrix}$
      ii $\mathbf{r} = -2\mathbf{i} + 6\mathbf{k} + \lambda(3\mathbf{i} - \mathbf{j} - 2\mathbf{k})$
   c $(4, -2, 2)$

3. a $\overrightarrow{AH} = -9\mathbf{i} + 15\mathbf{j} + 12\mathbf{k}$, $\overrightarrow{NH} = 2.5\mathbf{i} + 15\mathbf{j} + 6\mathbf{k}$
   b $\cos^{-1}\left(\dfrac{274.5}{\frac{\sqrt{1069}}{2} \times 15\sqrt{2}}\right) = 37.6695\ldots° = 37.7°$ correct to 1 decimal place
   c e.g. $\mathbf{r} = 9\mathbf{i} + \lambda(-9\mathbf{i} + 15\mathbf{j} + 12\mathbf{k})$

4. a $n = 7$
   b $\cos^{-1}\left(\dfrac{19}{\sqrt{41}\sqrt{41}}\right) = 62.3923\ldots° = 62.4°$ correct to 1 decimal place

5. a $8 - 4 + 5p = 0 \rightarrow p = -\dfrac{4}{5}$
   b i $\mathbf{r} = -3\mathbf{i} + \mathbf{j} + 5\mathbf{k} + \lambda(7\mathbf{i} - \mathbf{j} - \mathbf{k})$
      ii Proof

6. a 6
   b $\mathbf{r} = -4\mathbf{i} + 6\mathbf{j} - 6\mathbf{k} + \lambda(-2\mathbf{i} - 14\mathbf{j} + 2\mathbf{k})$
   c $-5\mathbf{i} - \mathbf{j} - 5\mathbf{k}$
   d $50.6°$

7. a $\overrightarrow{AB} \cdot \overrightarrow{CB} = \begin{pmatrix} -2 \\ -4 \\ 6 \end{pmatrix} \cdot \begin{pmatrix} 2 \\ -4 \\ -2 \end{pmatrix}$
   $= (-2)(2) + (-4)(-4) + (6)(-2) = 0$ ✓
   b $\overrightarrow{AD} = \begin{pmatrix} -10 \\ 20 \\ 10 \end{pmatrix}$ and $\overrightarrow{BC} = \begin{pmatrix} -2 \\ 4 \\ 2 \end{pmatrix}$, $\overrightarrow{AD} = 5\overrightarrow{BC}$.
   The lines AD and BC are parallel.
   c $\overrightarrow{OE} = \begin{pmatrix} -1 \\ 12 \\ 4 \end{pmatrix}$ $\mathbf{r} = -\mathbf{i} + 12\mathbf{j} + 4\mathbf{k} + \lambda(\mathbf{i} - 10\mathbf{j} + 3\mathbf{k})$

8. a $36.3°$ correct to 1 decimal place.
   b Point of intersection is $(4, 0, 1)$
   c Foot of perpendicular is $N(3, 2, 4)$ and
   $|\overrightarrow{EN}| = |-2\mathbf{i} - \mathbf{j}| = \sqrt{2^2 + 1^2} = \sqrt{5}$

9. a $\overrightarrow{PQ} = \begin{pmatrix} -9.5 \\ 4 \\ 2.5 \end{pmatrix}$ $\overrightarrow{PS} = \begin{pmatrix} -4.5 \\ -6 \\ -7.5 \end{pmatrix}$
   b $R(-5, 0, -1)$
   c Proof and side length $= 7.5\sqrt{2}$
   d $T(2, 1, 1.5)$
   e i e.g. $\mathbf{r} = \mathbf{v} + \lambda(\mathbf{t} - \mathbf{v})$
   $\mathbf{r} = \begin{pmatrix} 5 \\ 17.5 \\ -13.5 \end{pmatrix} + \lambda \begin{pmatrix} -3 \\ -16.5 \\ 15 \end{pmatrix}$
      ii Proof
      iii Right, squared-based pyramid

10   a   When $\lambda = -1$ the position vector given is **P**.
     b $\sqrt{185}$
     c $143.0°$ correct to 1 decimal place
     d   Foot of perpendicular is $(-3, 29, -4)$ so perpendicular distance is $\sqrt{67}$.

11   a $\mathbf{r} = 7\mathbf{i} + \mathbf{j} + 6\mathbf{k} + \lambda(3\mathbf{i} + 4\mathbf{j} - 5\mathbf{k})$
     b $P(4, -3, 11)$ and $\mathbf{r} = -5\mathbf{j} + 7\mathbf{k} + \mu(4\mathbf{i} + 2\mathbf{j} + 4\mathbf{k})$
     c $|\overrightarrow{PQ}| = 6$

12   a $p = 2, q = -1$
     b $7\mathbf{i} + 4\mathbf{j} + 8\mathbf{k}$
     c $\cos^{-1}\left(\dfrac{51}{\sqrt{29}\sqrt{99}}\right) = 17.8584\ldots° = 17.9°$ correct to 1 decimal place.

10 Differential equations

Prerequisite knowledge

1   $w = 0.02r^3$

2   $y = \dfrac{60e^{\frac{x}{4}}}{1 + 3e^{\frac{x}{4}}}$

3   a $\dfrac{1}{3}\ln|3x - 1| + c$
    b $-\ln|\cos x| + c$
    c $-xe^{-x} - e^{-x} + c$

4   $\dfrac{2}{7}(\ln|x - 2| - \ln|3x + 1|) + c$

5   a

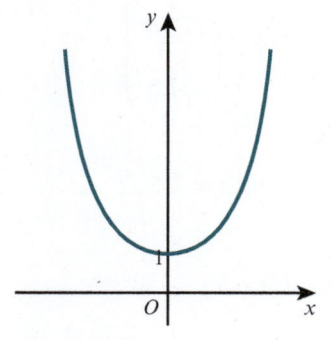

    b $k = 12$

Exercise 10A

1   a $y = \dfrac{x^4}{4} - 6x + C$
    b $A = \pi r^2 + C$

    c $\dfrac{s^3}{3} = \sin(t + 5) + C$
    d $\ln|V| = 2t + c$ or $V = 0$
    e $\ln|y + 1| = \ln|x| + C$ or $y = -1$
    f $-\cos y = (1 - x)\sin x - \cos x + C$

2   a $y = 5e^{\tan x}$
    b $\dfrac{y^2}{2} - \dfrac{2y^{\frac{5}{2}}}{5} = \dfrac{x^3}{3} - x - \dfrac{54}{5}$

3   $y = \ln(e^x + 1)$

4   a $\dfrac{dy}{dx} = 5y^3 x$
    b $y^2 = \dfrac{1}{6 - 5x^2}$

5   a $\dfrac{1}{3(2 - x)} + \dfrac{1}{3(x + 1)}$
    b   i $x = \dfrac{2(e^{3t} - 1)}{e^{3t} + 2}$ ii $x \to 2$

6   Proof

7   $x = \dfrac{2}{6 - e^{\frac{1}{4}\sin t}}$

8   $ye^{y+1} - e^{y+1} = \dfrac{x^2}{2} - \ln x + C$

9   $y = \tan\left(\dfrac{1}{2}\tan^{-1}\left(\dfrac{x}{2}\right) + \dfrac{\pi}{4}\right)$

10 $\ln|1 - 2y| = \dfrac{1}{x^2} + B$

11 $x = e^{t + \frac{1}{2}\sin 2t}$

12 a $\dfrac{1}{2}e^{t^2} + C$
    b $\dfrac{x^3}{3} = \dfrac{1}{2}e^{t^2} + \dfrac{5t^2}{2} - \dfrac{1}{6}$

13 $y = \dfrac{3}{2 - x^2}$

14 $k = \ln 7, \dfrac{1}{2}\ln(3x^2 + x^4) = t\ln 7 + \ln 2$

15 a $\ln|\ln x| + C$
    b $|y| = B|\ln x|$ where $B = e^C$

16 48 hours (to the nearest hour)

Exercise 10B

1   a $\dfrac{dh}{dt} = -kh^2$ where $k > 0$

b $\dfrac{dn}{dt} = kn$ where $k > 0$

c $\dfrac{dv}{dt} = -kv(v+1)$ where $k > 0$

d $\dfrac{dV}{dt} = -kV$ where $k > 0$

e $\dfrac{dC}{dt} = kC^3$ where $k > 0$

f $\dfrac{dv}{dt} = \dfrac{k}{e^{\frac{t}{2}}}$ where $k > 0$

2 a $\dfrac{dx}{dt} = -k\sqrt{x}$ where $k > 0$

b $t = 600 - 240\sqrt{x}$

c 176 seconds correct to 3 significant figures

3 a $\dfrac{dA}{dt} = kA$ where k is a positive constant

b Proof

c \$334 correct to 3 significant figures

4 a $\dfrac{dx}{dt} = k(100 - x)$ where k is a positive constant

b $x = 100 - 75e^{-kt}$

c 86.9 °C

d 100 °C

5 a Proof

b $\dfrac{dh}{dt} = \dfrac{dV}{dt} \times \dfrac{dh}{dV} = \dfrac{-36}{\pi h^2}$

c $t = \dfrac{\pi}{36}\left(1125 - \dfrac{h^3}{3}\right)$

d 85.9 seconds correct to 3 significant figures

6 a $\dfrac{dP}{dt} = kP$ where k is a positive constant

b Proof

c 5.42 minutes correct to 3 significant figures

7 a $\dfrac{dr}{dt} = \dfrac{3.92}{\sqrt{r}}$

b $r = \sqrt[3]{(5.88t - 1.568)^2}$

c $V = 10.3\,\text{cm}^3$ correct to 3 significant figures

8 At about 6.34 pm

9 Approximately $\dfrac{37}{50}$ or 74%

10 $\dfrac{dL}{dt} = k$ where k is a positive constant; $L = 0.3t + 20$

11 a 7.47 minutes correct to 3 significant figures

b 24 °C

12 a $\dfrac{dn}{dt} = k\sqrt{n}$ where k is a positive constant; $2\sqrt{n} = kt + C$

b 9

c $t = 11 \Rightarrow n = 12\,100$ and $t = 12 \Rightarrow n = 14\,400$

13 a 0.458 hours correct to 3 significant figures

b Unlimited growth, unrealistic

End-of-chapter review exercise 10

1 $y = -\ln\left(\dfrac{5}{4} - \dfrac{1}{4}e^{2x} + \dfrac{x}{2}e^{2x} - \dfrac{1}{2}x^2e^{2x}\right)$

2 a $x = \dfrac{2500e^{\frac{t}{2}}}{4 + e^{\frac{t}{2}}}$

b As $t \to \infty$, $x \to 2500$

3 $y^4 = 17e^{4x} - 1$

4 a $\dfrac{dy}{dx} = kx\sqrt{y}$; $24\sqrt{y} = 93 - 5x^2$

b 5

5 a $10(\ln|10-x| - \ln|5-x|) + C$

b i Proof

ii $x = \dfrac{10(e^{\frac{t}{10}} - 1)}{2e^{\frac{t}{10}} - 1}$

iii 5 grams

6 $y^5 = xe^x - e^x + 1025$; $y = 4.06$ correct to 3 significant figures

7 a $\dfrac{1}{2}\sqrt{y^4 - 1} + C$

b $\dfrac{1}{2}\sqrt{y^4 - 1} = 2x + \ln|x| + \sqrt{2} - 2$

8 a $\dfrac{dh}{dt} = k(8-h)$ where $k > 0$;

   $h = 8 - 7.5e^{\left(\frac{1}{5}\ln 0.8\right)t}$

   b As $t \to \infty$, $h \to 8$

9 $y = -\dfrac{1}{3}\ln(2 - e^{3x})$

10 a $\dfrac{dx}{dt} = -kxt$

   b $\ln x = -\dfrac{kt^2}{2} + C$

   c 47.4 seconds correct to 3 significant figures

11 a $\dfrac{1}{5}\left(\ln|P| - \ln|5 - P|\right) + C$

   b $P = \dfrac{15e^{5t}}{2 + 3e^{5t}}$

   c As $t \to \infty$, $P \to 5$

12 73.4 million correct to 3 significant figures

13 a $-\dfrac{1}{2}\cos x^2 + c$

   b $\dfrac{y^2}{2} = -\dfrac{1}{2}\cos x^2 + 1$

14 $y = \dfrac{\ln(1 - 3\ln|\cos x|)}{3}$

15 a $\dfrac{dx}{dt} = k(2000 - x)x$, $t = \dfrac{15}{2}\ln\left(\dfrac{3x}{2000 - x}\right)$

   b 30.3 hours correct to 3 significant figures

16 $y = \dfrac{x^2 - 4}{x^2 + 4}$

17 $\ln|x + 2| = \dfrac{\theta}{2} - \dfrac{1}{8}\sin 4\theta + \ln 2$, $x = 1.09$ correct to 3 significant figures

11 Complex numbers

Prerequisite knowledge

1 a $3a - 2bx$ b $2a^2 - abx - 3b^2x^2$

2 a $-2\sqrt{2}$ b -1

   c $2 - \sqrt{3}$

3

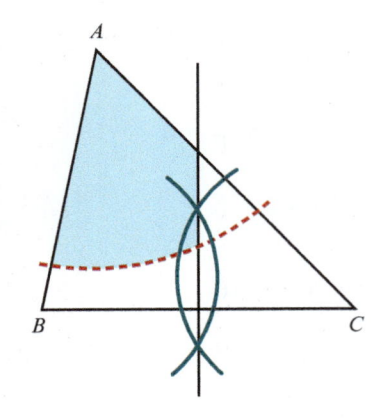

4 a $-\dfrac{\pi}{4}, \dfrac{3\pi}{4}$

   b 2.29 radians correct to 3 significant figures

5 a 5

   b 0.927 radians correct to 3 significant figures or 53.1° correct to 1 decimal place

   c $-2\mathbf{i} + 11\mathbf{j}$

Exercise 11A

1 a $12\mathbf{i}$ b $\dfrac{2}{3}\mathbf{i}$

   c $(3\sqrt{10})\mathbf{i}$ d $13\mathbf{i}$

2 a 8   b $(9 + 2\sqrt{2})\mathbf{i}$

   c $\sqrt{29}$ d $\dfrac{5}{6}$

3 a $\dfrac{8}{5}\mathbf{i}$ b $\mathbf{i}\dfrac{\sqrt{7}}{2}$

   c $\dfrac{1}{2}\mathbf{i}$

Exercise 11B

1 a $6 - 5\mathbf{i}$ b $4 + \mathbf{i}$

   c $11 + 7\mathbf{i}$ d $-\dfrac{1}{5} - \dfrac{13}{5}\mathbf{i}$

2 a $-1 \pm (2\sqrt{3})\mathbf{i}$ b $-2 \pm \mathbf{i}$

   c $\dfrac{1}{2} \pm \dfrac{3}{2}\mathbf{i}$ d $3 \pm \mathbf{i}\sqrt{6}$

   e $-\dfrac{4}{3} \pm \dfrac{\sqrt{14}}{3}\mathbf{i}$ f $-\dfrac{5}{4} \pm \dfrac{\sqrt{7}}{4}\mathbf{i}$

3 a $x = 3$, $y = -1$ b $x = 1$, $y = 3$

   c $x = 1$, $y = 2$

4 a $5 + 5\mathbf{i}$ b 41

- **c** $40 - 42i$
- **d** $28 - 96i$
- **e** $5 - 2i$
- **f** $-1 + i$
- **g** $5 - i$
- **h** $\frac{1}{2} - \frac{5}{2}i$

5
- **a** Proof
- **b** $z = \frac{1}{5} - \frac{2\sqrt{6}}{5}i$

6
- **a** $z^2 + 49 = 0$
- **b** $z^2 - 2z + 26 = 0$
- **c** $z^2 - 4z + 13 = 0$
- **d** $z^2 + 5z + 14 = 0$

7 $x = \frac{1}{2}, y = \frac{3}{2}$

8 $1 + 2i$

9 $z^2 - 10z + 28 = 0$

10 $3.2 - 2.4i$ amps

Exercise 11C

All angles are given in radians correct to 3 significant figures where rounded.

1 a

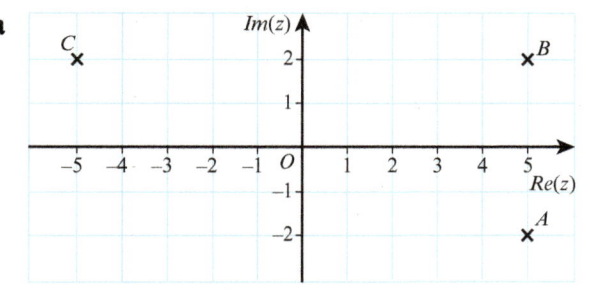

b $(-u)^* = -5 - 2i$

2 a

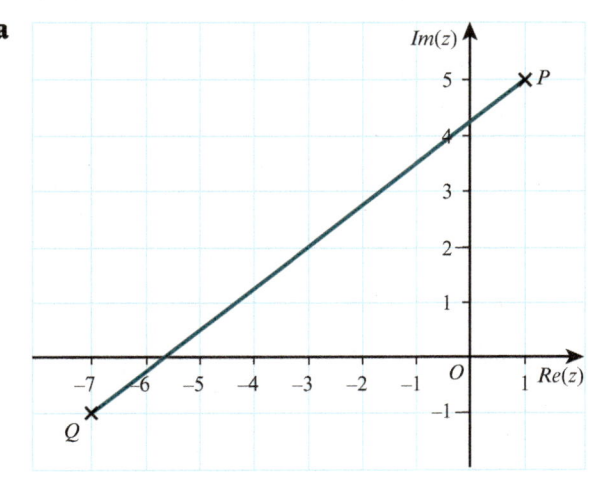

b $-3 + 2i$

3
- **a** $(13, 2.75)$
- **b** $\left(5, \frac{\pi}{2}\right)$
- **c** $(17, 1.08)$
- **d** $(61, -0.181)$
- **e** $(41, -1.79)$
- **f** $\left(2, -\frac{2\pi}{3}\right)$
- **g** $(3, 0.730)$
- **h** $(25, -2.86)$
- **i** $\left(\sqrt{2}k, -\frac{\pi}{4}\right)$

4
- **a** $\sqrt{10}(\cos(1.89) + i\sin(1.89))$
  $\sqrt{10}(\cos(0.322) + i\sin(0.322))$
  $\sqrt{10}(\cos(-1.25) + i\sin(-1.25))$
- **b** AC is a straight line, midpoint O as $|z_1| = |z_3|$.
  Angle $AOB = \pi - \tan^{-1}(3) - \tan^{-1}\left(\frac{1}{3}\right) = \frac{\pi}{2}$.
  Triangles AOB and AOC are isosceles, since $|z_1| = |z_2|$ and $|z_2| = |z_3|$. QED.

5
- **a** $\frac{3}{2} + \frac{3\sqrt{3}}{2}i$
- **b** $1.91 + 4.62i$
- **c** $-\frac{1}{2}$
- **d** $\frac{3\sqrt{2}}{2} - \frac{3\sqrt{2}}{2}i$

6 $|z^2| = \frac{73}{97}$, $\arg z^2 = 3.02$

7
- **a** $5e^{\frac{i\pi}{6}}$
- **b** $\sqrt{2}e^{-\frac{i\pi}{4}}$
- **c** $\frac{\sqrt{2}e^{-\frac{5\pi i}{12}}}{5}$

8 $a = \frac{5\sqrt{3}}{2}, b = \frac{5}{2}$

9
- **a** r^2
- **b** $\cos 2\theta + i\sin 2\theta$

10
- **a** Proof
- **b** Proof

Exercise 11D

All angles are given in radians correct to 3 significant figures where rounded.

1
- **a** 1
- **b** $z_2 = i, z_3 = -1$ and real

2
- **a** $z = 7, z = 4 + i\sqrt{3}, z = 4 - i\sqrt{3}$
- **b** $z = -\frac{11}{8}, z = \frac{-25 + i\sqrt{3}}{16}, z = \frac{-25 - i\sqrt{3}}{16}$

3 $\frac{\sqrt{6}}{2}(\cos 1.78 + i\sin 1.78)$,
$\frac{\sqrt{6}}{2}(\cos(-1.78) + i\sin(-1.78))$

4 $z = 3, z = 5i, z = -5i$

5   $x = 8$, $y = 3$

6   $z = -0.5$, $z = 3 + i$, $z = 3 - i$

7   $z = 3i$, $z = -3i$, $z = 1 + 2i$, $z = 1 - 2i$

8   a $-5 + i$, $5 - i$

    b $-3 - i\sqrt{2}$, $3 + i\sqrt{2}$

    c $\sqrt{2} - i\dfrac{\sqrt{3}}{2}$, $-\sqrt{2} + i\dfrac{\sqrt{3}}{2}$

    d $4 - 3i$, $-4 + 3i$

    e $1 + i\sqrt{5}$, $-1 - i\sqrt{5}$

    f $\dfrac{\sqrt{2}}{2} e^{\frac{\pi}{4}i}$, $\dfrac{\sqrt{2}}{2} e^{\frac{-3\pi}{4}i}$

9   a $z = 2 - i$ b $p = 37$, $q = -40$

    c

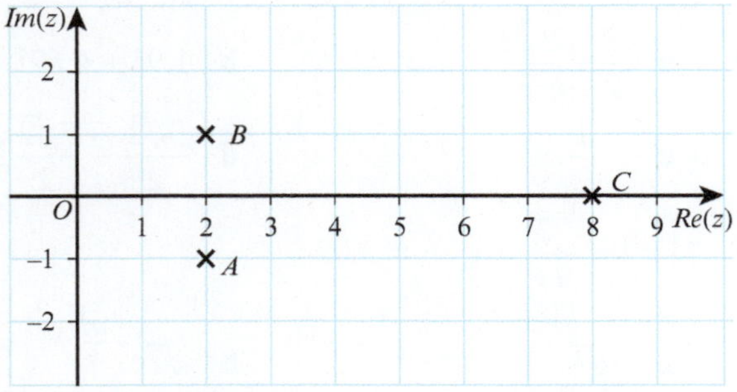

10 $a = 16$, $b = -1$, $z = 4i$, $z = -4i$, $z = \dfrac{1}{2} + \dfrac{\sqrt{15}}{2} i$, $z = \dfrac{1}{2} - \dfrac{\sqrt{15}}{2} i$

Exercise 11E

1   a   A half-line from $(2, -3)$ at an angle of $\dfrac{\pi}{12}$ radians.

    b   The region to the right of the perpendicular bisector of the points $(0, 6)$ and $(10, 0)$.

    c   A circle, centre $(-6, 1)$, radius 7.

    d   A half-line from $(0, 0)$ at an angle of $\dfrac{5\pi}{12}$ radians.

2   a

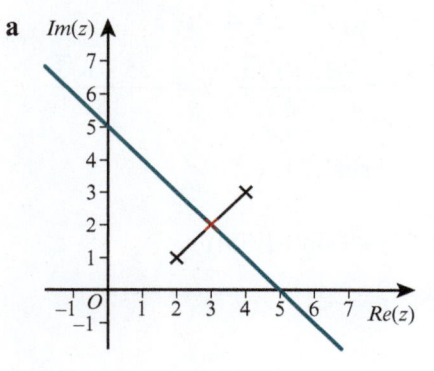

    b

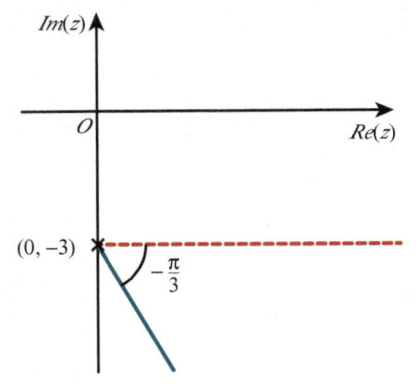

    c

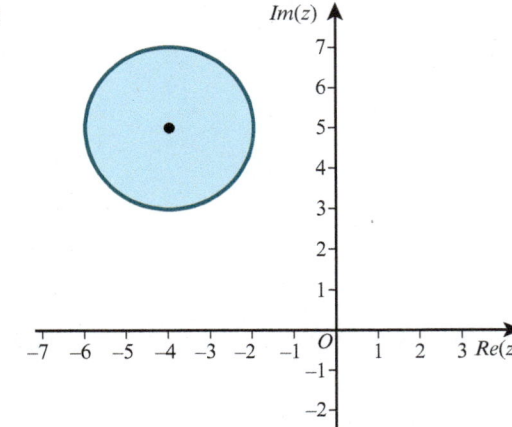

    d

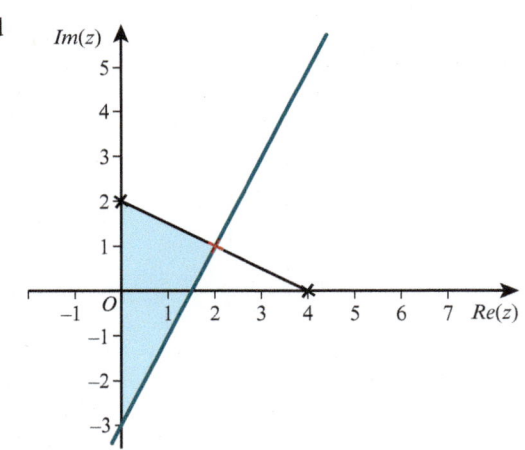

3     $1 \pm i\sqrt{15}$

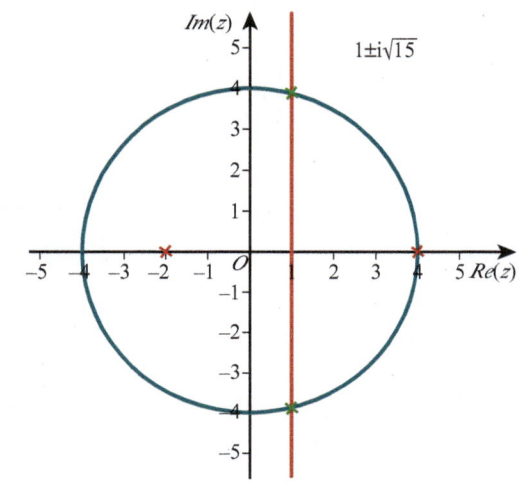

4

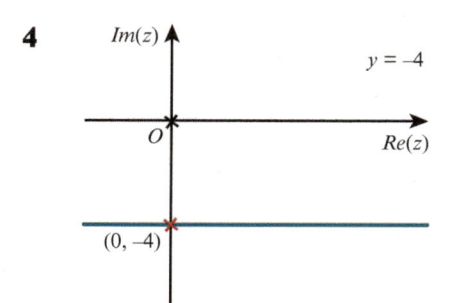

[Diagram: Im(z)-Re(z) plane showing line $y = -4$ with points $(0, -4)$ and $(0, -8)$ marked]

5 $(x-3)^2 + (y+6)^2 = 9$

[Diagram: circle centered at $(3, -6)$ with radius 3]

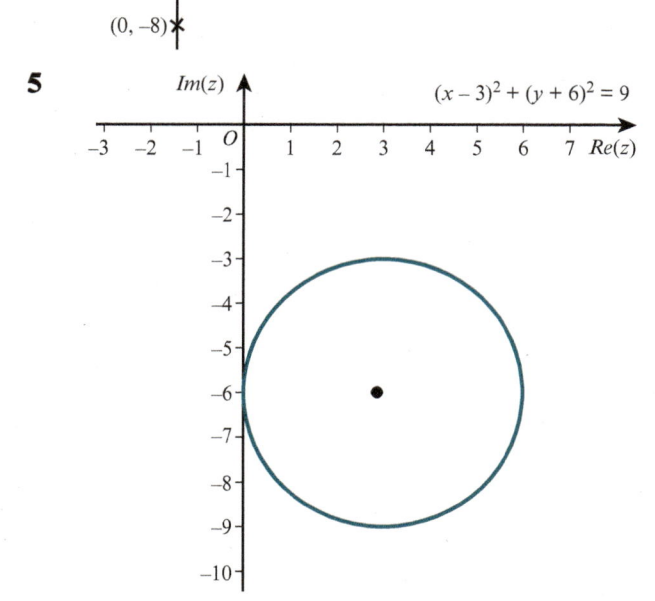

6 No

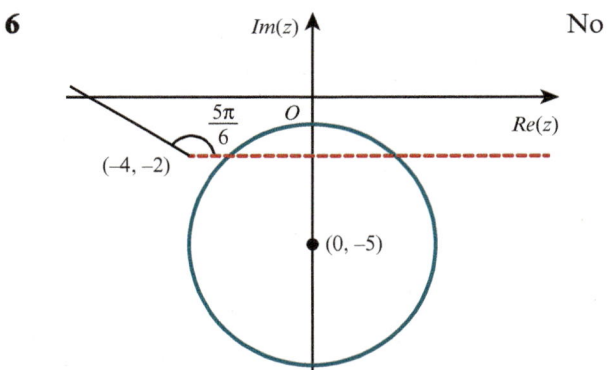

[Diagram: Im(z)-Re(z) plane showing a ray at angle $\frac{5\pi}{6}$ from point $(-4, -2)$, and a circle centered at $(0, -5)$]

7 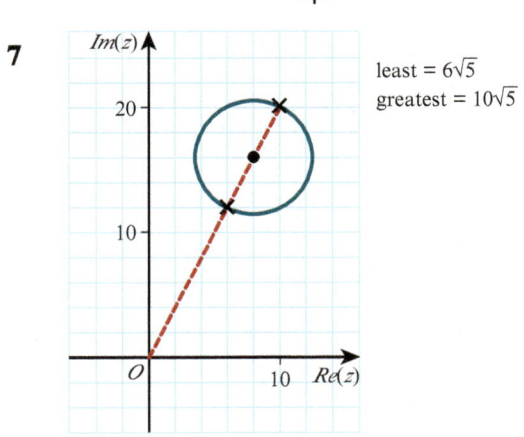 least $= 6\sqrt{5}$, greatest $= 10\sqrt{5}$

[Diagram: circle with line from origin through it]

8

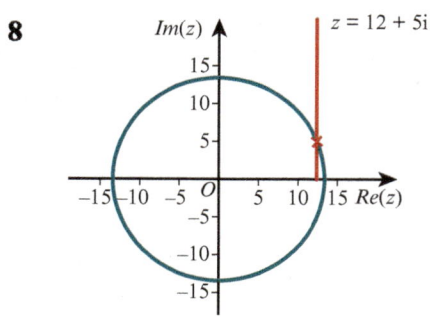

[Diagram: circle of radius 13 centered at origin with $z = 12 + 5i$ marked]

9

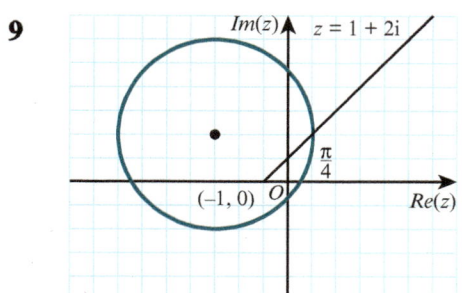

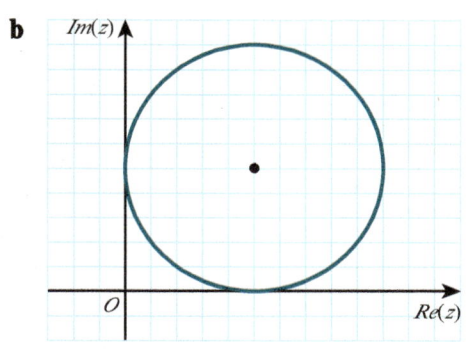

[Diagram: circle with $z = 1 + 2i$, point $(-1, 0)$, angle $\frac{\pi}{4}$]

10 a $(x-5)^2 + (y-5)^2 = 25$

b [Diagram: circle centered at $(5, 5)$ with radius 5, tangent to both axes]

c Least arg $z = 0$, greatest arg $z = \dfrac{\pi}{2}$

End-of-chapter review exercise 11

All angles are given in radians correct to 3 significant figures where rounded.

1 a $-0.1 - 1.7i$

b $w = 1 - 5i$, $w = 1 + 5i$

c [Diagram: shaded disc centered near $(1, 5)$]

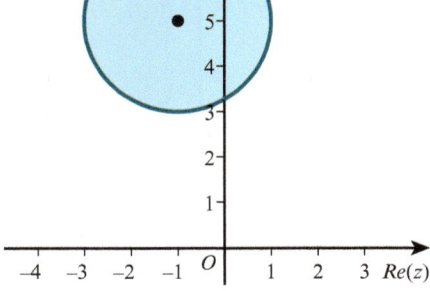

2 a $zz^* = k^2 + 36$, $\dfrac{z}{z^*} = \dfrac{(k^2 - 36) - 12ki}{k^2 + 36}$

   b $uw = 8e^{\left(-\frac{7\pi}{12}\right)i}$, $\dfrac{u}{w} = 2e^{\left(-\frac{7\pi}{12}\right)i}$

3 a Isosceles triangle

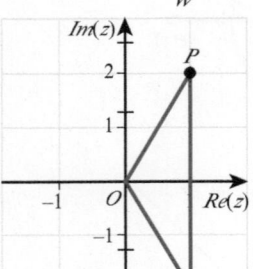

   b $\dfrac{u}{w} = \sqrt{2}\left(\cos\left(\dfrac{3}{4}\pi\right) + i\sin\left(\dfrac{3}{4}\pi\right)\right)$

4 a $x = \dfrac{1}{2}$, $y = 3$

   b Right-angled

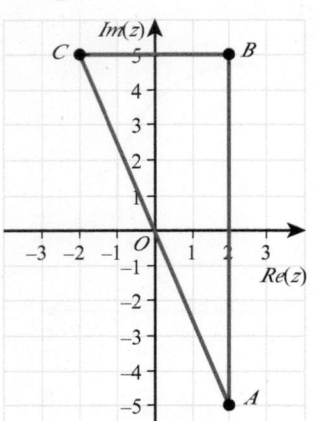

   c i $\dfrac{21}{29} - \dfrac{20}{29}i$

      ii $\cos(-0.761) + i\sin(-0.761)$

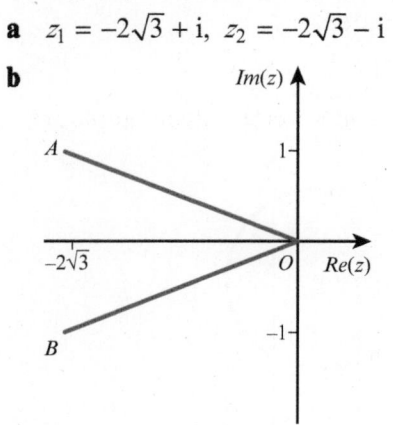

5 a $z_1 = -2\sqrt{3} + i$, $z_2 = -2\sqrt{3} - i$

   b Reflection in the real axis

   c $|z_1| = |z_2| = \sqrt{13}$, $\arg z_1 = 2.86$, $\arg z_2 = -2.86$

6 a $\left(8, -\dfrac{\pi}{6}\right)$

   b $2\sqrt{2}e^{\left(-\frac{\pi}{4}\right)i}$

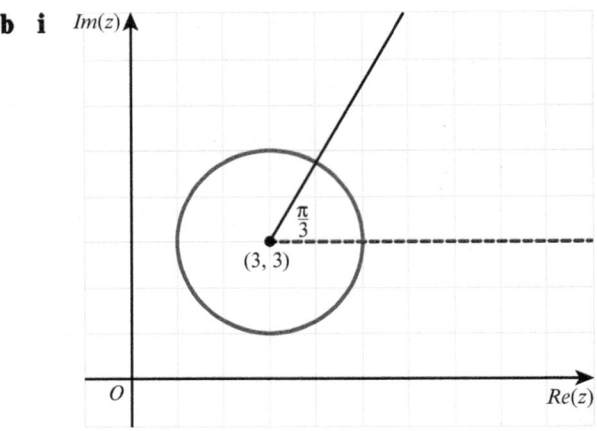

7 a $w = -1 + i$

   b ii $z = 4 + (3 + \sqrt{3})i$

8 a $x = -3$, $y = \sqrt{2}$ or $x = 3$, $y = -\sqrt{2}$

   b i Proof

      ii $z_1 = 3$, $z_2 = -\dfrac{1}{2} + \dfrac{i}{2}$, $z_3 = -\dfrac{1}{2} - \dfrac{i}{2}$

9 a $z_2 = 3 + 2i$

   b $z = -\dfrac{4}{3}$, $z = \dfrac{1}{6} - \dfrac{\sqrt{3}}{2}i$, $z = \dfrac{1}{6} + \dfrac{\sqrt{3}}{2}i$

10 a i $f(-3) = 0$

       ii $z = 1$, $z = -3$, $z = -\dfrac{1}{4} + \dfrac{\sqrt{15}}{4}i$, $z = -\dfrac{1}{4} - \dfrac{\sqrt{15}}{4}i$

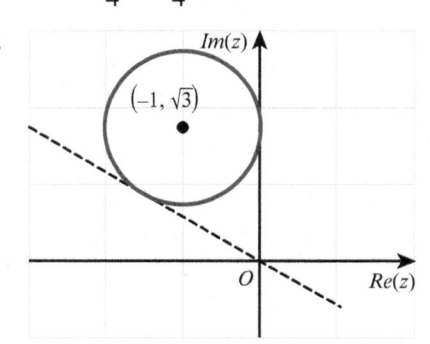

    b ii Min. $\arg z = \dfrac{\pi}{2}$

       iii Max. $\arg z = \dfrac{\pi}{2} + \dfrac{\pi}{6} + \dfrac{\pi}{6} = \dfrac{5\pi}{6}$

11 a i $z^2 + 3z + 4 = 0$

       ii $|z_1| = 2$

b i $z = -1$, $z = \dfrac{1 + i\sqrt{3}}{2}$, $z = \dfrac{1 - i\sqrt{3}}{2}$

ii Equilateral

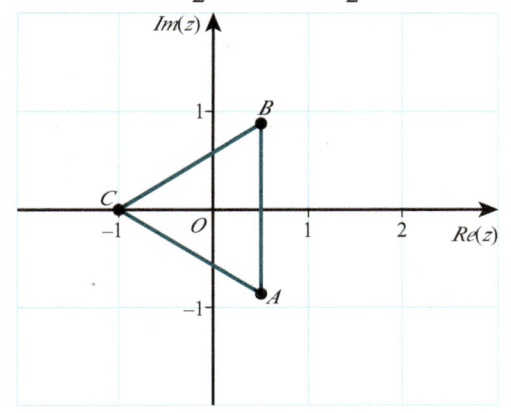

12 a Proof

b $\left|\dfrac{z}{z^*}\right| = 1$, $\arg\left(\dfrac{z}{z^*}\right) = -0.841$

c $3z^2 - 4z + 3 = 0$

13 a $k = -1$ **b** $\arg z = 0.862$

14 i $7 - 2i$

ii $6.69 e^{\frac{1}{4}\pi i}$

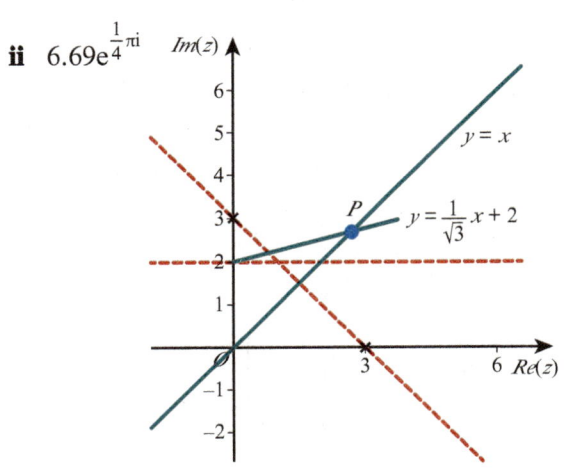

15 i $u = -2 - 2i$, $v = 1 + 2i$

ii Least $|z - w| = 3\sin\left(\dfrac{\pi}{4}\right) - 1 = \dfrac{3\sqrt{2}}{2} - 1$

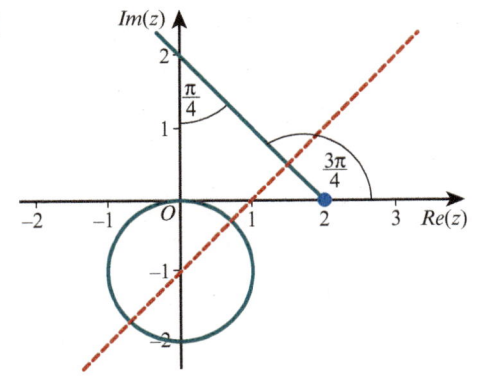

16 i $\dfrac{3}{2} + \dfrac{1}{2}i$ **ii** $-\dfrac{\sqrt{3}}{2} + \dfrac{1}{2}i$

Cross-topic review exercise 4

1 a $\mathbf{r} = \begin{pmatrix} -5 \\ 0 \\ 3 \end{pmatrix} + \lambda \begin{pmatrix} 6 \\ 7 \\ -1 \end{pmatrix}$

b i $m = -2$

ii Proof

2 $y^2 = 4(x^2 - 1)$

3 a $p = -2$, $q = -25$

b i $\dfrac{\sqrt{14}}{42}(-10\mathbf{i} + \mathbf{j} - 5\mathbf{k})$

ii Angle $POQ = 90°$; $63\sqrt{5}$

4 i $|z| = 2$, $\arg z = \dfrac{\pi}{6}$ or $30°$ or 0.524 radians

ii a $3\sqrt{3} + i$

b $\dfrac{\sqrt{3}}{2} + \dfrac{i}{2}$

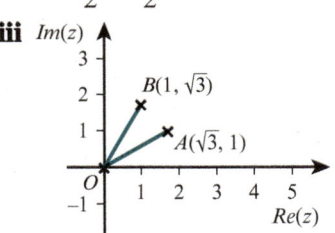

5 $x = \sqrt{2\sin 2\theta} - 1$

6 a $\overrightarrow{OS} = \begin{pmatrix} 4 \\ -3 \\ 0 \end{pmatrix}$

b $\sqrt{33}$, $\sqrt{26}$

c $65.8°$, $114.2°$ correct to 1 decimal place

7 i $-\dfrac{2}{5} + \dfrac{11}{5}i$

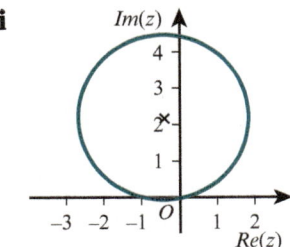

8 i $|u| = \sqrt{8}$ or $2\sqrt{2}$, $\arg u = \dfrac{\pi}{4}$ or 45°

   ii

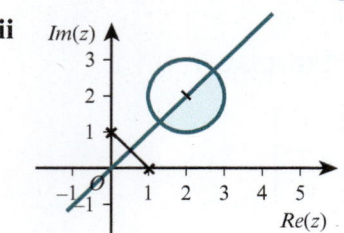

   iii $|z| = \sqrt{8-1} = \sqrt{7}$

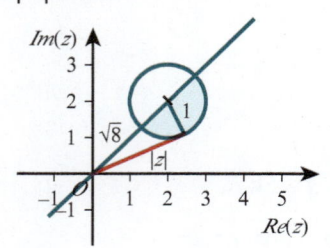

9 i Proof

   ii $-\ln(20 - x) = 0.05t - \ln 20$

   iii 7.9

   iv t becomes very large, x approaches 20

10 a $\mathbf{r} = -2\mathbf{i} + \mathbf{j} - \mathbf{k} + \mu(\mathbf{i} + 3\mathbf{k})$ b $\dfrac{\sqrt{265}}{5}$

11 i $\ln R = \ln x - 0.57x + 3.80$ $R = xe^{-0.57x+3.80}$

   ii $R = \dfrac{1}{0.57}e^{-1+3.80} = 28.850...$

12 i $9e^{\frac{\pi}{3}i}$

   ii $re^{i\theta} = 3e^{\frac{\pi}{6}i}$ or $re^{i\theta} = 3e^{\frac{\pi}{6}i - \pi i} = 3e^{-\frac{5}{6}\pi i}$

13 $y = 4(2 + e^{3x})^2$

14 a $\begin{pmatrix} -7 \\ 3 \\ -13 \end{pmatrix}$ b 50.1°

15 $\ln y = \ln x - x^2 + 1 + \ln 2$ and $y = 2xe^{-x^2+1}$.

16 i $|u| = 3$

   ii Half-line from $(0, -3)$

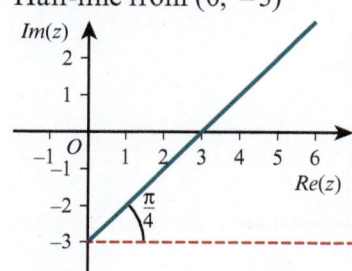

Least $|z| = 3\sin\left(\dfrac{\pi}{4}\right) = \dfrac{3\sqrt{2}}{2}$

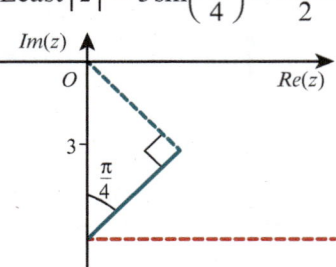

iii Circle, centre $3 - 3i$, radius 1.

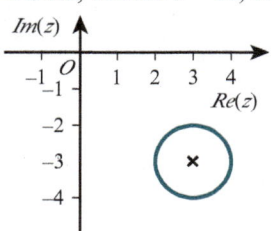

Greatest $|z| = \sqrt{3^2 + (-3)^2} + 1 = 3\sqrt{2} +$

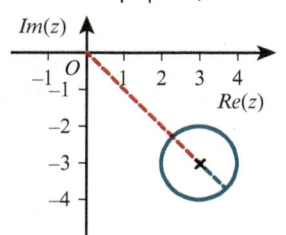

17 a $m = -1$, $m = -4$

    b $m = -3$

    c AB and CD do not intersect.

18 i Proof

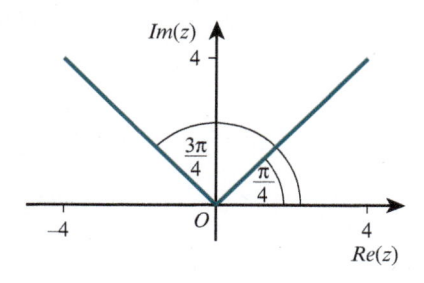

    ii $-6 \leqslant p \leqslant 2$

iii $w^* = 2 - 4i$

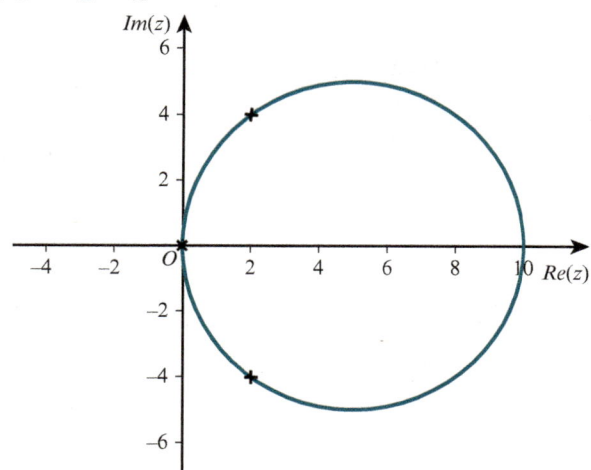

$|z - 5| = 5$

19 $y = \dfrac{3e^{16x^2} - 1}{3 - e^{16x^2}}$

20 i Proof

ii $\dfrac{2}{5} h^{\frac{5}{2}} = -\dfrac{1}{150} H^{\frac{5}{2}} t + \dfrac{2}{5} H^{\frac{5}{2}}$ $t = \dfrac{60 H^{\frac{5}{2}} - 60 h^{\frac{5}{2}}}{H^{\frac{5}{2}}}$

iii $t = 49.3933... = 49.4$

21 a $\mathbf{r} = 2\mathbf{i} + 5\mathbf{j} + 7\mathbf{k} + \lambda(7\mathbf{i} - 6\mathbf{j} - 9\mathbf{k})$

b $127.0°$

22 i

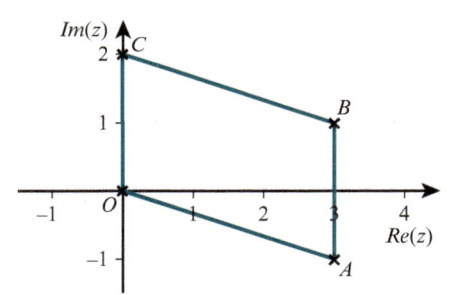

$OABC$ is a parallelogram.

ii $\dfrac{u^*}{u} = 0.8 + 0.6i$ iii Proof

23 i Second root is $1 - (\sqrt{2})i$.

ii Other two roots are $-1 \pm i$.

24 a $h = 3$

b $\dfrac{\sqrt{10}}{10} (\mathbf{j} + 3\mathbf{k})$

c $94.9°$

d $\mathbf{r} = 5\mathbf{i} + 4\mathbf{j} + \lambda(5\mathbf{i} - 3\mathbf{k})$

25 $\dfrac{y^3}{3} = 2xe^{3x} - \dfrac{2}{3} e^{3x} + \dfrac{10}{3}$, $y = 2.437... = 2.44$ correct to 3 significant figures

26 i The roots are: $-3 + i\sqrt{2}$ and $3 - i\sqrt{2}$.

ii $|w - (1 + 2i)| = 1$: circle, centre $(1, 2)$ radius 1

$\arg(z - 1) = \dfrac{3}{4}\pi$: half-line, above and to the left of the point $(1, 0)$

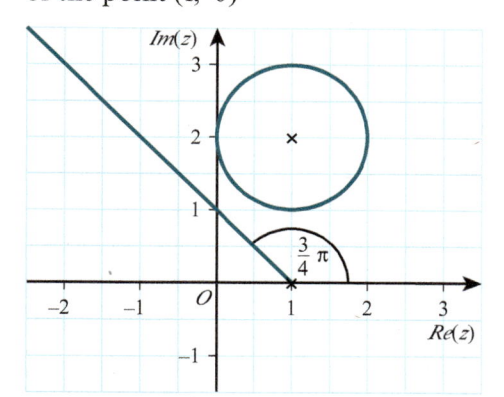

b Least $|w - z| = \sqrt{2} - 1$

27 a Proof

b $m = -4$

c $73.2°$

28 i $N = \dfrac{1800 e^{\frac{1}{2}t}}{5 + e^{\frac{1}{2}t}}$

ii 1800

29 i The roots are: $-\sqrt{3} + i\sqrt{2}$ and $-\sqrt{3} + i\sqrt{2}$.

ii Circle, centre $(0, 3)$, radius 2. Greatest value of $\arg z = 2.3005... = 2.30$ correct to 3 significant figures.

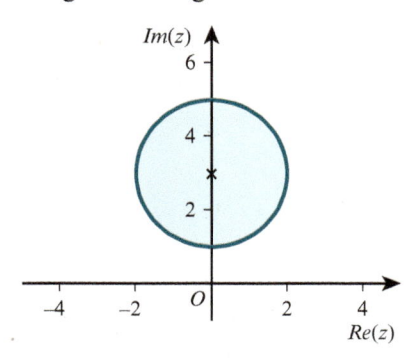

30 i $\dfrac{dV}{dt} = 80 - kV$

ii $k_{n+1} = \dfrac{4 - 4e^{-15k_n}}{25}$; 0.14

iii $V = 540 \text{ cm}^3$ correct to 2 significant figures, V approaches the value given by $\dfrac{80}{k} = 570 \text{ cm}^3$ correct to 2 significant figures.

Pure Mathematics 2
Practice exam-style paper

1. Proof
2. $x \leqslant \frac{1}{5}$ or $x \geqslant 1$
3. 1.46
4. **a** $p = -17$ **b** 71
5. **a** $1 \to 1.2003, 1.1794, 1.1895, 1.1849, 1.1871, 1.1861, (1.1865, 1.863), \ldots \to 1.19$
   b Answer $x = \sqrt{\dfrac{8x^2}{3 \sec x}}$ obtains $\sec x = \dfrac{8}{3}$
   $\cos x = \dfrac{3}{8} \to 1.1863996$
6. **a** $\dfrac{1}{2}(2t+1)(1-2e^{2t})$ **b** $y = 2x - 1$
7. **a** $10\sin(\theta + 36.87)°$ **b** $7.56°, 98.7°$
   c 13
8. **a** Proof   **b** Proof

Pure Mathematics 3
Practice exam-style paper

1. 2.652
2. Proof
3. $a = 7, b = -12$
4. $-101.9°, 78.1°$
5. $2\sqrt{11}$
6. **a** $\dfrac{5}{x-2} + \dfrac{2x-1}{x^2+5}$
   b $-\dfrac{27}{10} - \dfrac{17x}{20} - \dfrac{117x^2}{200}$
7. **a** $\dfrac{dy}{dx} = \sin 2x + 2x\cos 2x$
   b $\dfrac{\pi}{4}$
8. **a** $u = 2 + i, w = 3 + 9i$
   b $A(1, -1), B(1, 1), C(0, -2); BC = \sqrt{10}$
9. **a** $p = \pm 3$
   b **i** $\mathbf{r} = 5\mathbf{i} + 2\mathbf{k} + \lambda(\mathbf{i} - 6\mathbf{j} + \mathbf{k})$
       ii Proof
10. **a** Proof
    b Proof
    c $0.1 \to 0.07719, 0.07146, 0.06998, 0.06959, 0.06949, 0.06946 \to 0.069$
    d $90.5°C$

Glossary

A

Absolute value: the non-negative value of a number without regard to its sign

Algebraic improper fraction: the algebraic fraction $\frac{P(x)}{Q(x)}$, where $P(x)$ and $Q(x)$ are polynomials in x, is said to be an improper fraction if the degree of $P(x) \geq$ the degree of $Q(x)$

Arbitrary constant: a constant that may be assumed to be any value

Argument of a complex number: the argument of a complex number $x + iy$ is the direction of the position vector $\begin{pmatrix} x \\ y \end{pmatrix}$

C

Complex number: a number that can be written in the form $x + iy$ where x and y are real (this form of a complex number is called the Cartesian form)

Compound angle formulae:
$\sin(A + B) \equiv \sin A \cos B + \cos A \sin B$
$\sin(A - B) \equiv \sin A \cos B - \cos A \sin B$
$\cos(A + B) \equiv \cos A \cos B - \sin A \sin B$
$\cos(A - B) \equiv \cos A \cos B + \sin A \sin B$
$\tan(A + B) \equiv \dfrac{\tan A + \tan B}{1 - \tan A \tan B}$
$\tan(A - B) \equiv \dfrac{\tan A - \tan B}{1 + \tan A \tan B}$

Converging: the sequence of values generated by an iterative formula is convergent or said to be converging to a root if the values in the sequence approach the actual value of the root

Cosecant: $\operatorname{cosec} \theta = \dfrac{1}{\sin \theta}$

Cotangent: $\cot \theta = \dfrac{1}{\tan \theta} \left(= \dfrac{\cos \theta}{\sin \theta} \right)$

D

Decompose: split a single algebraic fraction into two or more **partial fractions**

Degree: the highest power of x in the polynomial is called the **degree** of the polynomial.
For example, $5x^3 - 6x^2 + 2x - 6$ is a polynomial of degree 3.

Direction of a vector: this is represented by the arrow on the vector and can be, for example, the angle of the path that takes you from one point to another along the length of the vector

Dividend: the quantity being divided by another quantity

Division algorithm for polynomials:
dividend = divisor × quotient + remainder

Divisor: the quantity by which another quantity is to be divided

Double angle formulae: $\sin 2A \equiv 2 \sin A \cos A$
$\cos 2A \equiv \cos^2 A - \sin^2 A$
$\tan 2A \equiv \dfrac{2 \tan A}{1 - \tan^2 A}$

E

Explicit functions: functions of the form $y = f(x)$ are called explicit functions as y is given explicitly in terms of x

F

Factor: a portion of a quantity that, when multiplied by other factors, gives the entire quantity

Factor theorem: if for a polynomial $P(x)$, $P(c) = 0$ then $x - c$ is a factor of $P(x)$

G

General binomial theorem:
$(1 + x)^n = 1 + nx + \dfrac{n(n-1)}{2!} x^2 + \dfrac{n(n-1)(n-2)}{3!} x^3 + \ldots$
where n is rational and $|x| < 1$

General solution: a solution for a differential equation that works for any value of the constant C

I

Imaginary number: any multiple of the unit imaginary number i where i^2 is defined to be -1

Implicit function: when a function is given as an equation connecting x and y, where y is not the subject

Integration by parts: $\int u \dfrac{dv}{dx} dx = uv - \int v \dfrac{du}{dx} dx$

Integration by substitution: can be considered as the reverse process of differentiation by the chain rule

Iteration: an iteration is one trial in a process that is repeated such as when using an iterative formula (see iterative process)

Iterative formula: this type of formula is one that is used repeatedly. The output from each stage is used as the input for the next stage. It is commonly used in numerical methods.

Iterative process: a process for finding a particular result by repeating trials of operations. Each repeat trial is called an iteration and each iteration should produce a value closer to the result being found.

L

Laws of logarithms:

Multiplication law	Division law	Power law
$\log_a(xy) =$ $\log_a x + \log_a y$	$\log_a\left(\dfrac{x}{y}\right) =$ $\log_a x - \log_a y$	$\log_a(x)^m =$ $m \log_a x$

Locus: a locus is a path traced out by a point as it moves following a particular rule. The rule is expressed as an inequality or an equation.

Logarithm: the power to which a base needs to be raised to produce a given value

M

Magnitude of a vector: the length or size of the vector

Modulus: the magnitude of the number without a sign attached. The modulus of a number is also called the absolute value.

Modulus of a complex number: the modulus of the complex number $x + iy$ is the magnitude of the position vector $\begin{pmatrix} x \\ y \end{pmatrix}$

N

Natural exponential function: the function $y = e^x$

Natural logarithms: logarithms to the base of e. $\ln x$ is used to represent $\log_e x$

P

Parallel vectors: two vectors are parallel when one is a scalar multiple of the other

Parameter: sometimes variables x and y are given as a function of a third variable t. The variable t is called a parameter and the two equations are called the parametric equations of the curve.

Partial fraction: each of two or more fractions into which a more complex fraction can be decomposed as a sum

Particular solution: a solution for a differential equation that works for a specific value of C

Polynomial: an expression of the form $a_n x^n + a_{n-1} x^{n-1} + a_{n-2} x^{n-2} + \ldots + a_2 x^2 + a_1 x^1 + a_0$

Position vector: a means of locating a point in space relative to an origin

Principal argument: the principal argument, θ, of a complex number is an angle such that $-\pi < \theta \leq \pi$. Sometimes it might be more convenient to give θ as an angle such that $0 \leq \theta < 2\pi$. (Usually the argument is given in radians.)

Product rule: $\dfrac{d}{dx}(uv) = u\dfrac{dv}{dx} + v\dfrac{du}{dx}$

Q

Quotient: a result obtained by dividing one quantity by another

Quotient rule: $\dfrac{d}{dx}\left(\dfrac{u}{v}\right) = \dfrac{v\dfrac{du}{dx} - u\dfrac{dv}{dx}}{v^2}$

R

Real number: the set of all rational and irrational numbers

Remainder: the amount left over after a division

Remainder theorem: if a polynomial $P(x)$ is divided by $x - c$, the remainder is $P(c)$

Resultant vector: the combination of two or more single vectors

S

Scalar: a quantity (often a number) that is used to scale a vector

Secant: $\sec\theta = \dfrac{1}{\cos\theta}$

T

Trapezium rule: this numerical method involves splitting the area under the curve $y = f(x)$ between $x = a$ and $x = b$ into equal width strips

U

Unit vector: a vector with magnitude or length one

V

Vector addition: the process of finding the sum of two or more vectors

Vector subtraction: the addition of the negative of a vector

Index

absolute value *see* modulus function
area under a curve
   trapezium rule 126–8
   see also integration
Argand diagrams 273
argument of a complex number 274

binomial expansion 181
   of $(a + x)^n$ where n is not a positive integer 177–8
   of $(1 + x)^n$ where n is not a positive integer 174–6
   and partial fractions 179–80

calculus *see* differential equations; differentiation; integration
chain rule 88
change of sign method, numerical solutions 136
circles, loci 289–92
cobweb pattern, iterative processes 142
complex conjugate pairs 284
complex conjugates 270
complex numbers (z) 267, 269–70
   argument of 274
   calculating with 271, 280–1
   Cartesian form 273
   cube roots of one 287
   equal numbers 270
   loci 288–94
   modulus of 273
   polar forms 277
   roots of polynomials 283–5
   square roots of 286
complex plane
   Argand diagrams 273
   exponential form 277–81
   modulus-argument form 273–7
components of a displacement 212
compound angle formulae 58–60
cosecant (cosec) 53–6
   graph of 54
cosine (cos)
   compound angle formulae 58–60
   derivative of $\cos x$ 96
   derivative of $\cos(ax + b)$ 97
   double angle formula 61, 62–4
   graph of 55
   integration 118–19, 122
cotangent (cot) 53–6
   graph of 55
cube roots of one (unity) 287
cubic equations 284–5

degree of a polynomial 11
differential equations 244–5
   forming an equation from a problem 252–5
   general solution 246
   modelling growth 259–60
   Newton's law of heating or cooling 251–2
   particular solution 246
   separating the variables 245–9
differentiation
   chain rule 88
   derivative of $e^{f(x)}$ 88–90
   derivative of e^x 87–8
   derivative of $\ln f(x)$ 92–3
   derivative of $\ln x$ 91–2
   derivative of $\tan^{-1} x$ 186–7
   derivatives of trigonometric functions 95–8
   implicit 99–101
   parametric 103–5
   product rule 81–3
   quotient rule 84–6
displacement (translation) vectors 211–12
   addition and subtraction 213–15
   components of 212
   magnitude of 215–16
   notation 212, 213
   unit vectors 217–18
dividend 12
division law, logarithms 33
divisor 12
dot product *see* scalar product

e (Euler's number) 42
   derivative of $e^{f(x)}$ 88–90
   derivative of e^x 87–8
   integration of exponential functions 112–13
equations
   exponential 38–9
   of the form $a\sin\theta + b\cos\theta = c$ 68–70
   involving the modulus function 3–6
   logarithmic 35–6
   parametric 103, 232–4
   quadratic 283
   see also differential equations; numerical solutions of equations
Euler, Leonhard 42, 277
explicit functions 99
exponential equations 38–9
exponential form of the complex plane 277–81
exponential growth 259–60
exponential inequalities 40–1

factor theorem 14–17
factors 12
first order differential equations 244
Fourier, Joseph 70
fractals 267

general binomial theorem 174
general solution of a differential equation 244, 246
golden ratio (1:ϕ) 153
golden spiral 153
graph sketching 135–7
graphs
    of the modulus function 7
    of reciprocal trigonometric functions 54–6
    transforming a relationship to linear form 44–6
growth models 259–60

i 267–9
imaginary numbers 266–9
implicit differentiation 99–101
implicit functions 99
improper algebraic fractions 166–7
    partial fractions 171–2
indices, relationship to logarithms
    in base 10 26–9
    in other bases 30–2
inequalities
    exponential 40–1
    involving the modulus function 8–10
integration
    of $\frac{1}{ax+b}$ 115–17
    of $\frac{1}{x}$ 115–17
    of $\frac{1}{x^2+a^2}$ 187–8
    of exponential functions 112–13
    of $\frac{k\,\mathrm{f}'(x)}{\mathrm{f}(x)}$ 189–90
    solving differential equations 245–9
    trapezium rule approximation 126–8
    of trigonometric functions 118–19, 121–4
    use of partial fractions 194–6
    using different techniques 201–2
integration by parts 197–200
integration by substitution 191–3
intersecting lines, vector equations 236–7
iterations 141
iterative formulae 142
iterative processes 140–1
    cobweb pattern 142
    staircase pattern 145

subscript notation 142–3
trigonometric equations 149–52
using a calculator 143
worked examples 143–5

loci 288
    circles with centre (0, 0) 289–90
    circles with centre (a, b) 290–2
    half-lines and part-lines 292–3
    perpendicular bisectors 293–4
logarithmic equations 35–6
logarithms (logs) 26
    to base 10 26–9
    to base a 30–2
    laws of 33–4
    natural 42
        derivative of ln f(x) 92–3
        derivative of ln x 91–2
    solving exponential equations 38–9
    solving exponential inequalities 40–1
    transforming a relationship to linear form 44–6
logistic growth models 259–60

magnitude (modulus) of a vector 215–16
modulus function (absolute value) 2–3
    graphs of $y = |\mathrm{f}(x)|$ 7
    solving equations 3–6
    solving inequalities 8–10
modulus-argument form of the complex plane 273–7
multiplication law, logarithms 33

Napier, John 42
natural exponential function 42
natural logarithms 42
    derivative of ln f(x) 92–3
    derivative of ln x 91–2
Newton, Isaac 175
    law of heating or cooling 251–2
numerical solutions of equations 135
    iterative processes 140–5
    starting points 135–7

parallel lines, vector equations 236
parallel vectors 213
    scalar product 227
parametric differentiation 103–5
parametric equations 103
    of a line 232–4
partial fractions 181
    application of 172–3
    and binomial expansions 179–80
    Heaviside's cover-up method 174

Index

improper 171–2
   with a quadratic factor in the denominator that cannot be factorised 171
   with repeated linear factor in the denominator 169–70
   use in integration 194–6
   where the denominator has distinct linear factors 168–9
particular solution of a differential equation 244, 246
perpendicular bisectors 293–4
perpendicular vectors, scalar product 227
polar coordinates 277
polar forms of a complex number 277
polynomials 11
   complex roots 283–5
   division algorithm 13
   division of 11–13
   factor theorem 14–17
   remainder theorem 18–20
population growth models 259–60
position vectors 220
   Cartesian components 220–3
power law 33, 46
principal argument of a complex number 274
problem solving, forming differential equations 252–5
product rule 81–3

quadratic equations, complex roots 283
quartic equations 285
quotient 12
quotient rule 84–6

reciprocal trigonometric ratios 53–6
remainder theorem 18–20
resultant vectors 213–15
roots of polynomials 283–5

scalar product (dot product) 225–6
   for component form 227–9
   parallel vectors 227
   perpendicular vectors 227
secant (sec) 53–6
   graph of 55
   integration 118–19, 123–4
separating the variables, differential equations 245–9
sine (sin)
   compound angle formulae 58–60
   derivative of $\sin x$ 95–6
   derivative of $\sin(ax + b)$ 97
   double angle formula 61, 62–4
   graph of 54
   integration 118–19, 122, 124

skew lines, vector equations 237–8
square roots of a complex number 286
staircase pattern, iterative processes 145
subscript notation, iterative processes 142–3
substitution, integration by 191–3

tangent (tan)
   compound angle formulae 58–60
   derivative of $\tan x$ 96
   derivative of $\tan^{-1} x$ 186–7
   derivative of $\tan(ax + b)$ 97
   double angle formula 62–3
   graph of 55
   integration 118–19, 123
telescoping series 172–3
translation vectors *see* displacement vectors
trapezium rule 126–8
trigonometry
   compound angle formulae 58–60
   derivatives of trigonometric functions 95–8
   double angle formulae 61–4
   equations of the form $a\sin\theta + b\cos\theta = c$ 68–70
   integration of trigonometric functions 118–19, 121–4
   proving identities 66
   reciprocal functions 53–6
   using iterative processes 149–52

unit vectors 217–18

vector equation of a line 231–2
   intersection of two lines 236–7
   parametric form 232–4
   skew lines 237–8
vector spaces 225
vectors
   addition and subtraction 213–15
   angle between 225
   displacement (translation) vectors 211–18
   magnitude of 215–16
   multiplication by a scalar 213
   notation 212, 213
   parallel 213, 227
   perpendicular 227
   position vectors 220–3
   scalar product 225–9
   unit vectors 217–18
   uses of 211